More Praise for *This Is How They Tell Me the World Ends*

"Possibly the most important book of the year . . . Perlroth's precise, lucid, and compelling presentation of mind-blowing disclosures about the underground arms race is a must-read exposé." —*Booklist* (starred review)

"[A] wonderfully readable new book. Underlying everything Perlroth writes is the question of ethics: What is the right thing to do? Too many of the people she describes never seemed to think about that; their goals were short-term or selfish or both. A rip-roaring story of hackers and bug-sellers and spies that also looks at the deeper questions."
—Steven M. Bellovin, professor of computer science, Columbia University

"The murky world of zero-day sales has remained in the shadows for decades, with few in the trade willing to talk about this critical topic. Nicole Perlroth has done a great job tracing the origin stories, coaxing practitioners into telling their fascinating tales, and explaining why it all matters."
—Kim Zetter, author of *Countdown to Zero Day*

"Nicole Perlroth does what few other authors on the cyber beat can: She tells a highly technical, gripping story as if over a beer at your favorite local dive bar. A page-turner."
—Nina Jankowicz, author of *How to Lose the Information War*

"From one of the literati, a compelling tale of the digerati: Nicole Perlroth puts arresting faces on the clandestine government-sponsored elites using 1s and 0s to protect us or menace us—and profit."
—Glenn Kramon, former *New York Times* senior editor

"Reads like a thriller. A masterful inside look at a highly profitable industry that was supposed to make us safer but has ended up bringing us to the brink of the next world war."
—John Markoff, former *New York Times* cybersecurity reporter

THIS IS
HOW THEY
TELL ME
THE WORLD
ENDS

THIS IS HOW THEY TELL ME THE WORLD ENDS

The Cyber-Weapons Arms Race

Nicole Perlroth

BLOOMSBURY PUBLISHING

NEW YORK • LONDON • OXFORD • NEW DELHI • SYDNEY

BLOOMSBURY PUBLISHING
Bloomsbury Publishing Inc.
1385 Broadway, New York, NY 10018, USA

BLOOMSBURY, BLOOMSBURY PUBLISHING, and the Diana logo are trademarks
of Bloomsbury Publishing Plc

First published in the United States 2021

ISBN: HB: 978-1-63557-605-4; eBook: 978-1-63557-606-1

LIBRARY OF CONGRESS CATALOGING-IN-PUBLICATION DATA IS AVAILABLE

2 4 6 8 10 9 7 5 3

Typeset by Westchester Publishing Services
Printed and bound in the U.S.A. by Berryville Graphics Inc., Berryville, Virginia

To find out more about our authors and books visit www.bloomsbury.com and sign up
for our newsletters.

Bloomsbury books may be purchased for business or promotional use. For information on
bulk purchases please contact Macmillan Corporate and Premium Sales Department at
specialmarkets@macmillan.com.

For Tristan, who always pulled me out of my secret hiding spots.
For Heath, who married me even though I couldn't
tell him where I was hiding.
For Holmes, who hid in my belly.

There's something happening here.
What it is ain't exactly clear.
There's a man with a gun over there
Telling me I got to beware
I think it's time we stop, children, what's that sound,
Everybody look what's going down

—BUFFALO SPRINGFIELD

CONTENTS

AUTHOR'S NOTE

THIS BOOK IS the product of more than seven years of interviews with more than three hundred individuals who have participated in, tracked, or been directly affected by the underground cyberarms industry. These individuals include hackers, activists, dissidents, academics, computer scientists, American and foreign government officials, forensic investigators, and mercenaries.

Many generously spent hours, in some cases days, recalling the details of various events and conversations relayed in these pages. Sources were asked to present documentation, whenever possible, in the form of contracts, emails, messages, and other digital crumbs that were considered classified or, in many cases, privileged through nondisclosure agreements. Audio recordings, calendars, and notes were used whenever possible to corroborate my own and sources' recollection of events.

Because of the sensitivities of the subject matter, many of those interviewed for this book agreed to speak only on the condition that they not be identified. Two people only spoke with me on the condition that their names be changed. Their accounts were fact-checked with others whenever possible. Many agreed to participate only to fact-check the accounts provided to me by others.

The reader should not assume that any individual named in these pages was a source for the events or dialogue described. In several cases accounts came from the person directly, but in others they came from eyewitnesses, third parties, and, as much as possible, written documentation.

And even then, when it comes to the cyberarms trade, I have learned that hackers, buyers, sellers, and governments will go to great lengths to avoid any written documentation at all. Many accounts and anecdotes were omitted from the following pages simply because there was no way to back up their version of events. I hope readers will forgive those omissions.

I have done my best, but to this day, so much about the cyberarms trade remains impenetrable that it would be folly to claim that I have gotten everything right. Any errors are, of course, my own.

My hope is that my work will help shine even a glimmer of light on the highly secretive and largely invisible cyberweapons industry so that we, a society on the cusp of this digital tsunami called the Internet of Things, may have some of the necessary conversations now, before it is too late.

—Nicole Perlroth
November 2020

PROLOGUE

Kyiv, Ukraine

B Y THE TIME my plane touched down in Kyiv—in the dead of winter 2019—nobody could be sure the attack was over, or if it was just a glimpse of what was to come.

A note of attenuated panic, of watchful paranoia, had gripped our plane from the moment we entered Ukrainian airspace. Turbulence had knocked us upward so suddenly I could hear bursts of nausea in the back of the plane. Beside me, a wisp of a Ukrainian model gripped my arm, shut her eyes, and began to pray.

Three hundred feet below, Ukraine had gone into orange alert. An abrupt windstorm was ripping roofs off apartment buildings and smashing their dislodged fragments into traffic. Villages on the outskirts of the capital and in western Ukraine were losing power—again. By the time we jerked onto the runway and started to make our way through Boryspil International Airport, even the young, gangly Ukrainian border guards seemed to be nervously asking one another: Freak windstorm? Or another Russian cyberattack? These days, no one could be sure.

One day earlier, I had bid my baby adieu and traveled to Kyiv as a kind of dark pilgrimage. I came to survey the rubble at ground zero for the most devastating cyberattack the world had ever seen. The world was still reeling from the fallout of a Russian cyberattack on Ukraine that less than two years earlier had shut down government agencies, railways, ATMs, gas stations, the postal service, even the radiation monitors at the old Chernobyl nuclear site, before the code seeped out of Ukraine and haphazardly zigzagged its way around the globe. Having escaped, it paralyzed factories in the far reaches of

Tasmania, destroyed vaccines at one of the world's largest pharmaceutical companies, infiltrated computers at FedEx, and brought the world's biggest shipping conglomerate to a halt, all in a matter of minutes.

The Kremlin had cleverly timed the attack to Ukraine's Constitution Day in 2017—the equivalent of our Fourth of July—to send an ominous reminder to Ukrainians. They could celebrate their independence all they wished, but Mother Russia would never let them out of its grip.

The attack was the culmination of a series of escalating, insidious Russian cyberattacks, revenge for Ukraine's 2014 revolution, when hundreds of thousands of Ukrainians took to Kyiv's Independence Square to revolt against the Kremlin's shadow government in Ukraine and ultimately oust its president, and Putin's puppet, Viktor Yanukovych.

Within days of Mr. Yanukovych's fall, Putin had pulled Yanukovych back to Moscow and sent his forces to invade the Crimean Peninsula. Before 2014, the Crimean Peninsula was a Black Sea paradise, a diamond suspended off the south coast of Ukraine. Churchill once coined it "the Riviera of Hades." Now it belonged to Russia, the infernal epicenter of Vladimir Putin's standoff with Ukraine.

Putin's digital army had been messing with Ukraine ever since. Russian hackers made a blood sport of hacking anyone and anything in Ukraine with a digital pulse. For five long years, they shelled Ukrainians with thousands of cyberattacks a day and scanned the country's networks incessantly for signs of weakness—a weak password, a misplaced zero, pirated and unpatched software, a hastily erected firewall—anything that could be exploited for digital mayhem. Anything to sow discord and undermine Ukraine's pro-Western leadership.

Putin laid down only two rules for Russia's hackers. First, no hacking inside the motherland. And second, when the Kremlin calls in a favor, you do whatever it asks. Otherwise, hackers had full autonomy. And oh, how Putin loved them.

Russian hackers are "like artists who wake up in the morning in a good mood and start painting," Putin told a gaggle of reporters in June 2017, just three weeks before his hackers laid waste to Ukraine's systems. "If they have patriotic leanings, they may try to add their contribution to the fight against those who speak badly about Russia."

Ukraine had become their digital test kitchen, a smoldering hellscape where they could test out every hacking trick and tool in Russia's digital arsenal without fear of reprisal. In the first year, 2014, alone, Russian state media and trolls barraged Ukraine's presidential election with a disinformation campaign that alternately blamed the country's mass pro-Western uprisings on an illegal coup, a military "junta," or "deep states" in America and Europe. Hackers stole campaign emails, prowled for voter data, infiltrated Ukraine's election authority, deleted files, and implanted malware in the country's election reporting system that would have claimed victory for a far-right fringe candidate. Ukrainians discovered the plot just before the results were reported to Ukraine's media. Election security experts called it the most brazen attempt to manipulate a national election in history.

In retrospect, this should have all set off louder alarm bells in the United States. But in 2014, Americans' gaze was elsewhere: the violence in Ferguson, Missouri; the horrors of ISIS and its seeming emergence out of nowhere; and, on my beat, the North Korean hack of Sony Pictures that December, when Kim Jong-un's hackers exacted revenge on the movie studio for a Seth Rogen–James Franco comedy depicting the assassination of their Dear Leader. North Korean hackers torched Sony's servers with code, then selectively released emails to humiliate Sony executives in an attack that offered Putin the perfect playbook for 2016.

For most Americans, Ukraine still felt a world way. We caught passing glimpses of Ukrainians protesting in Independence Square, and later celebrating as a new pro-Western leadership replaced Putin's puppet. Some kept an eye on the battles in eastern Ukraine. Most can recall the Malaysian airplane—filled with Dutch passengers—that Russian separatists shot out of the sky.

But had we all been paying closer attention, we might have seen the blaring red warning lights, the compromised servers in Singapore and Holland, the blackouts, the code spiking out in all directions.

We might have seen that the end game wasn't Ukraine. It was us.

RUSSIA'S INTERFERENCE IN Ukraine's 2014 elections was just the opening salvo for what would follow—a campaign of cyberaggression and destruction the world had never seen.

They were stealing a page from their old Cold War playbooks, and as my taxi made its way from Boryspil to Kyiv's center, Independence Square, the bleeding heart of Ukraine's revolution, I wondered which page they might read from next, and if we'd ever get to a place where we might anticipate it.

The crux of Putin's foreign policy was to undercut the West's grip on global affairs. With every hack and disinformation campaign, Putin's digital army sought to tie Russia's opponents up in their own politics and distract them from Putin's real agenda: fracturing support for Western democracy and, ultimately, NATO—the North Atlantic Treaty Organization—the only thing holding Putin in check.

The more disillusioned Ukrainians became—where were their Western protectors, after all?—the better the chance they might turn away from the West and return to the cold embrace of Mother Russia.

And what better way to aggravate Ukrainians and make them question their new government than to turn off Ukraine's heat and power in the dead of winter? On December 23, 2015, just ahead of Christmas Eve, Russia crossed a digital Rubicon. The very same Russian hackers that had been laying trapdoors and virtual explosives in Ukrainian media outlets and government agencies for months had also silently embedded themselves in the nation's power stations. That December they made their way into the computers that controlled Ukraine's power grid, meticulously shutting off one circuit breaker after another until hundreds of thousands of Ukrainians were without power. For good measure, they shut down emergency phone lines. And for added pain, they shut off the backup power to Ukraine's distribution centers, forcing operators to fumble around in the dark.

The power wasn't out long in Ukraine—less than six hours—but what happened in western Ukraine that day is without precedent in history. The digital Cassandras and the tinfoil-hat crowd had long warned that a cyberattack would hit the grid, but until December 23, 2015, no nation-state with the means had the balls to actually pull it off.

Ukraine's attackers had gone to great lengths to hide their true whereabouts, routing their attack through compromised servers in Singapore, the Netherlands, and Romania, employing levels of obfuscation that forensics investigators had never seen. They'd downloaded their weapon onto Ukraine's networks in benign-looking bits and pieces to throw off intrusion detectors and carefully randomized their code to evade antivirus software. And yet Ukraine officials immediately knew who was behind the attack. The time and resources required to pull off a grid attack with that level of sophistication were simply beyond that of any four-hundred-pound hacker working from his bed.

There was no financial profit to be gleaned from turning off the power. It was a political hit job. In the months that followed, security researchers confirmed as much. They traced the attack back to a well-known Russian intelligence unit and made their motives known. The attack was designed to remind Ukrainians that their government was weak, that Russia was strong, that Putin's digital forces were so deep into Ukraine's every digital nook and cranny that Russia could turn the lights off at will.

And just in case that message wasn't clear, the same Russian hackers followed up one year later, turning off Ukraine's power again in December 2016. Only this time they shut off heat and power to the nation's heart—Kyiv—in a display of nerve and skill that made even Russia's counterparts at the National Security Agency headquarters in Fort Meade, Maryland, wince.

FOR YEARS CLASSIFIED national intelligence estimates considered Russia and China to be America's most formidable adversaries in the cyber realm. China sucked up most of the oxygen, not so much for its sophistication but simply because its hackers were so prolific at stealing American trade secrets. The former director of the NSA, Keith Alexander, famously called Chinese cyberespionage the "greatest transfer of wealth in history." The Chinese were stealing every bit of American intellectual property worth stealing and handing it to their state-owned enterprises to imitate.

Iran and North Korea were high up on the list of cyber threats, too. Both demonstrated the will to do the United States harm. Iran had brought down U.S. banking websites and obliterated computers at the Las Vegas Sands casino after Sands CEO Sheldon Adelson publicly goaded Washington into bombing Iran, and—in a wave of ransomware attacks—Iranian cybercriminals had held American hospitals, companies, entire towns hostage with code. North Korea had torched American servers simply because Hollywood had offended Kim Jong-un's film tastes, and later, Jong-un's digital minions managed to steal $81 million from a bank in Bangladesh.

But there was no question that in terms of sophistication, Russia was always at the top of the heap. Russian hackers had infiltrated the Pentagon, the White House, the Joint Chiefs of Staff, the State Department, and Russia's Nashi youth group—either on direct orders from the Kremlin or simply because they were feeling patriotic—knocked the entire nation of Estonia offline after Estonians dared to move a Soviet-era statue. In one cyberattack Russian hackers, posing as Islamic fundamentalists, took a dozen French televisions channels off the air. They were caught dismantling the safety controls at a Saudi petrochemical company—bringing Russian hackers one step closer to triggering a cyber-induced explosion. They bombarded the Brexit referendum, hacked the American grid, meddled with the 2016 U.S. elections, the French elections, the World Anti-Doping Agency, and the holy goddamn Olympics.

But for the most part, by 2016 the U.S. intelligence community still assumed that America's capabilities far exceeded those of the opposition. The Kremlin was testing out the best of its cyber arsenal in Ukraine, and as far as American counterintelligence specialists could tell, Russia was still nowhere close to the cyber skills of the USA.

And it might have stayed that way for some time. For how long exactly, no one could predict; but between 2016 and 2017 the gap between the United States' cyber capabilities and those of every single other nation and bad-faith actor on earth closed substantially. Starting in 2016, the U.S. National Security Agency's own cyber arsenal—the sole reason the United States maintained its offensive advantage in cyberspace—was dribbled out online by a mysterious group whose identity remains unknown to this day. Over a period of nine months a cryptic hacker—or hackers; we still don't know who the NSA's torturers are—calling itself the Shadow Brokers started trickling out NSA hacking tools and code for any nation-state, cybercriminal, or terrorist to pick up and use in their own cyber crusades.

The Shadow Brokers' leaks made headlines, but like most news between 2016 and 2017, the news did not stick in the American conscience for long. The public's understanding of what was transpiring was—to put it mildly—a mismatch to the gravity of the situation, and to the impact those leaks would soon have on the NSA, on our allies, and on some of America's biggest corporations and smallest towns and cities.

The Shadow Brokers leaks were the world's first glimpse of the most powerful and invisible cyber arsenal on earth. What these cryptic hackers had exposed was the largest government program you have never heard of, a cyber-weapons and espionage operation so highly classified that for decades it was kept off the books completely, hidden from the public via shell companies, mercenaries, black budgets, nondisclosure agreements, and, in the early days, giant duffel bags of cash.

By the time the Shadow Brokers started dribbling out the NSA's cyber-weapons, I had been closely tracking the agency's offensive program for four years—ever since I'd caught a privileged glimpse of it in documents leaked by former NSA contractor Edward J. Snowden. I had tracked the program's three-decade-long history. I had met its Godfather. I had met its hackers, its suppliers, its mercenaries. And I had become intimately acquainted with its copycats as they sprung up all over the world. Increasingly, I was coming across the very men and women whose lives had been ruined by their tools.

In fact, the only thing I had not seen—up close—is what happened when the NSA's most powerful cyberweapons got into our adversary's hands.

So in March 2019 I went to Ukraine to survey the ruins for myself.

RUSSIA'S ATTACKS ON Ukraine's power grid thrust the world into a new chapter of cyberwar. But even those 2015 attacks did not compare to what happened when Russia got ahold of the NSA's best-kept hacking tools two years later.

On June 27, 2017, Russia fired the NSA's cyberweapons into Ukraine in what became the most destructive and costly cyberattack in world history. That afternoon Ukrainians woke up to black screens everywhere. They could not take money from ATMs, pay for gas at stations, send or receive mail, pay for a train ticket, buy groceries, get paid, or—perhaps most terrifying of all—monitor radiation levels at Chernobyl. And that was just in Ukraine.

The attack hit any company that did any business in Ukraine. All it took was a single Ukrainian employee working remotely for the attack to shut down entire networks. Computers at Pfizer and Merck, the pharmaceutical compa-nies; at Maersk, the shipping conglomerate; at FedEx, and at a Cadbury choc-olate factory in Tasmania were all hijacked. The attack even boomeranged back on Russia, destroying data at Rosneft, Russia's state-owned oil giant, and Evraz, the steelmaker owned by two Russian oligarchs. The Russians had used the NSA's stolen code as a rocket to propel its malware around the globe. The hack that circled the world would cost Merck and FedEx, alone, $1 billion.

By the time I visited Kyiv in 2019, the tally of damages from that single Russian attack exceeded $10 billion, and estimates were still climbing. Shipping and railway systems had still not regained full capacity. All over Ukraine, people were still trying to find packages that had been lost when the shipment tracking systems went down. They were still owed pension checks that had been held up in the attack. The records of who was owed what had been obliterated.

Security researchers had given the attack an unfortunate name: NotPetya. They initially assumed it was ransomware called Petya, only to find out later that Russian hackers had specifically designed NotPetya to appear as run-of-the-mill ransomware, though it was not ransomware at all. Even if you paid the ransom, there was no chance of getting any data back. This was a nation-state weapon designed to exact mass destruction.

I spent the next two weeks in Ukraine ducking frigid air blasts from Siberia. I met with journalists. I walked Independence Square with protesters as they recounted the bloodiest days of the revolution. I trekked out to the industrial zone to meet with digital detectives who walked me through the digital wreckage of NotPetya. I met with Ukrainians whose family business—tax reporting software used by every major Ukrainian agency and company—had been Patient Zero for the attacks. The Russians had cleverly disguised their malware as a tax software update from the company's systems, and now all its mom-and-pop operators could do was half cry, half laugh at the role they had played in nation-state cyberwar. I spoke with the head of Ukraine's cyber police force and with any Ukrainian minister who would have me.

I visited with American diplomats at the U.S. embassy just before they became entangled in the impeachment of President Donald Trump. On the day I visited, they were overwhelmed by Russia's latest disinformation campaign: Russian trolls had been inundating Facebook pages frequented by young Ukrainian mothers with anti-vaccination propaganda. This, as the country reeled from the worst measles outbreak in modern history. Ukraine now had one of the lowest vaccination rates in the world and the Kremlin was capitalizing on the chaos. Ukraine's outbreak was already spreading back to the States, where Russian trolls were now pushing anti-vaxxer memes on Americans. American officials seemed at a loss for how to contain it. (And they were no better prepared when, one year later, Russians seized on the pandemic to push conspiracy theories that Covid-19 was an American-made bioweapon, or a sinister plot by Bill Gates to profit off vaccines.) There seemed no bottom to the lengths Russia was willing to go to divide and conquer.

But that winter of 2019, most agreed that NotPetya was the Kremlin's boldest work to date. There was not a single person I met in Kyiv over the course of those two weeks who did not remember the attack. Everyone remembered where they were and what they were doing the moment the screens went dark. It was their twenty-first-century Chernobyl. And at the old nuclear plant, some ninety miles north of Kyiv, the computers went "black, black, black," Sergei Goncharov, Chernobyl's gruff tech administrator, told me.

Goncharov was just returning from lunch when the clock struck 1:12 P.M., and twenty-five hundred computers went dark over the course of seven minutes. Calls started flooding in; everything was down. As Goncharov tried to get Chernobyl's networks back up, he got the call that the computers that monitored radiation levels—over the very same site of the blast more than three decades earlier—had gone dark. Nobody knew if the radiation levels were safe, or if they were being sabotaged in some sinister way.

"In the moment, we were so consumed with making our computers work again that we didn't think too much about where this was coming from," Goncharov told me. "But once we pulled up our heads and saw the sheer speed with which the virus had spread, we knew we were looking at something bigger, that we were under attack."

Goncharov got on a loudspeaker and told anyone who could still hear him to rip their computers out of the wall. He instructed others to go out and manually start monitoring the radiation levels atop the Nuclear Exclusion Zone.

Goncharov was a man of few words. Even as he described the worst day of his life, he spoke in monotone. He was not one prone to loud emotion. But the day of the NotPetya attack, he told me, "I went into psychological shock." Two years later, I could not tell if he'd come out of it.

"We are now living in a totally different era," he told me. "There is now only Life before NotPetya and Life after NotPetya."

Everywhere I went in Ukraine over the course of those two weeks, Ukrainians felt the same. At a bus stop, I met a man who was in the midst of buying a car when the dealership turned him away—a first in used car sales history?—after the registration systems shut down. At a coffee shop I met a woman whose livelihood, a small online knitting supply shop, went bankrupt after the postal service lost track of her packages. Many shared tales of running out of cash or gas. But for the most part, like Goncharov, they all just remembered the sheer speed at which everything went down.

Given the timing—on the eve of Ukraine's Independence Day—it didn't take long to connect the dots. It was that old, bitter scoundrel, Mother Russia, messing with them again. But the Ukrainians are a resilient bunch. Over twenty-seven years of tragedy and crises, they'd coped with a dark humor. Some joked that with everything down, Vova—their nickname for Putin—had tacked a few extra days onto their Independence Day holiday. Others said the attack was the only thing that had gotten them off Facebook in years.

For all the psychological shock and financial cost of the events that June 2017, Ukrainians seemed to recognize that things could have been a lot worse. Front office systems were badly damaged. Important records would never be retrieved. But the attack had stopped short of the kind of deadly calamity that could derail passenger jets and planes or ignite a deadly explosion of some kind. Beyond the radiation monitors at Chernobyl, Ukraine's nuclear stations were still fully operational.

Moscow had pulled some punches in the end. Like the earlier power-grid attacks before it—when the lights went off just long enough to send a message—the damage from NotPetya was immeasurably low compared to what Russia could have done, with the access it had and the American weapons at its disposal.

Some surmised that Russia had used the NSA's stolen arsenal to thumb its nose at the agency. But the Ukrainian security experts I spoke with had a disturbing alternate theory: the NotPetya attack, and the power-grid attacks before it, were just a dry run.

This is what Oleh Derevianko, the blond Ukrainian cybersecurity entrepreneur, told me one evening over *vareniki* dumplings in aspic, Ukrainian meat encased in some kind of fatty Jell-O. Derevianko's firm had been on the front lines of the attacks. Over and over again, the forensics showed that the Russians were just experimenting. They were employing a cruel version of the scientific method: testing one capability here, one method there, honing their skills in Ukraine, demonstrating to their Russian overlords what could be done, earning their stripes.

There was a reason that the NotPetya attack was so destructive, why it wiped 80 percent of Ukraine's computers clean, Derevianko told me. "They were just cleaning up after themselves. These are new weapons in a new war. Ukraine was just their testing ground. How do they plan to use their weapons in the future? We don't know."

But the country had not had a cyberattack of that magnitude in two years, and although there was some evidence that Russia was planning to interfere in Ukraine's 2019 elections in two short weeks, the wave of cyber destruction had slowed to a trickle.

"That means they've moved on," he said.

We poked at our meat Jell-O in silence, got the check, and ventured outside. For the first time, it seemed, the violent windstorms had subsided. Even so, the typically lively cobblestone streets of Old Kyiv were empty. We walked up Andrew's Descent, Kyiv's equivalent of Paris's Montmartre, a famous narrow, winding cobblestone slope, past art galleries, antique shops, and art studios toward St. Andrew's Church, a glimmering white, blue, and gold vision originally designed as a summer residence for the Russian empress Elizabeth in the 1700s.

As we reached the church, Derevianko stopped. He looked up at the yellow glow of the lamppost above us.

"You know," he began, "if they switch off the lights here, we might be without power for a few hours. But if they do the same to you . . ."

He didn't finish his sentence. But he didn't have to. I'd already heard the same, over and over again, from his countrymen and my sources back in the United States.

We all knew what came next.

WHAT HAD SAVED Ukraine is precisely what made the United States the most vulnerable nation on earth.

Ukraine wasn't fully automated. In the race to plug everything into the internet, the country was far behind. The tsunami known as the Internet of Things, which had consumed Americans for the better part of the past decade, had still not washed up in Ukraine. The nation's nuclear stations, hospitals, chemical plants, oil refineries, gas and oil pipelines, factories, farms, cities, cars, traffic lights, homes, thermostats, lightbulbs, refrigerators, stoves, baby monitors, pacemakers, and insulin pumps were not yet "web-enabled."

In the United States, though, convenience was everything; it still is. We were plugging anything we could into the internet, at a rate of 127 devices a second. We had bought into Silicon Valley's promise of a frictionless society. There wasn't a single area of our lives that wasn't touched by the web. We could now control our entire lives, economy, and grid via a remote web control. And we had never paused to think that, along the way, we were creating the world's largest attack surface.

At the NSA—whose dual mission is gathering intelligence around the world and defending U.S. secrets—offense had eclipsed defense long ago. For every hundred cyberwarriors working on offense, there was only one lonely analyst playing defense. The Shadow Brokers leak was by far the most damaging in U.S. intelligence history. If Snowden leaked the PowerPoint bullet points, the Shadow Brokers handed our enemies the actual bullets: the code.

The biggest secret in cyberwar—the one our adversaries now know all too well—is that the same nation that maintains the greatest offensive cyber advantage on earth is also among its most vulnerable.

Ukraine had another edge on the USA: its sense of urgency. After five years of getting beat up and blacked out by one of the world's greatest predators, Ukraine knew that its future depended on a vigilant cyber defense. NotPetya had in many ways been a chance to start over, to build new systems from the ground up, and to keep Ukraine's most critical systems from ever touching the web. In the weeks after I left the country, Ukrainians would cast their vote in the presidential election on paper. There would be no fancy ballot-marking machines; every vote would be marked by hand. Paper ballots would be counted manually. Of course, that wouldn't stop allegations of nationwide vote-buying. But the idea of migrating Ukraine's elections to computers struck everyone I met during my time there as pure insanity.

Over and over again, the United States failed to reach the same sobering conclusions. We failed to see that the world of potential war has moved from land to sea to air to the digital realm. A few months after I left Ukraine, it wasn't the Russian attacks in Ukraine that stuck in Americans' memory, but the country's role in Trump's looming impeachment. We seemed to have somehow forgotten that, in addition to Russia's disinformation campaign in 2016—the dumping of Democratic emails, the Russians who posed as Texan secessionists and Black Lives Matter activists to sow discord—they had also probed our back-end election systems and voter registration data in all fifty states. They may have stopped short of hacking the final vote tallies, but everything they did up to that point, American officials concluded, was a trial run for some future attack on our elections.

And yet, Trump was still blaming Russian interference in the 2016 election on, alternately, a four-hundred-pound hacker on his bed and China. With Putin grimacing merrily by his side at a press conference in Helsinki in 2018, Trump not only snubbed the findings of his own intelligence community—"I have President Putin; he just said it's not Russia. I will say this, I don't see any reason why it would be"—but welcomed Putin's offer to allow Russia to join the U.S. hunt for its 2016 meddlers. And with the next election steadily approaching, Putin and Trump met once more, this time in Osaka in June 2019, where they chuckled together like old college buddies. When a reporter asked Trump if he would warn Russia not to meddle in 2020, Trump sneered and waved his finger jovially at his friend: "Don't meddle in the election, President."

And now here we are. As of this writing, the 2020 election is still being litigated, foreign actors have seized on our domestic chaos, our cyberweapons have leaked, with Russian hackers inside our hospitals, the Kremlin's agents deep in the American grid, determined attackers probing our computer networks millions of times a day, fighting a pandemic that has virtualized our lives in ways previously unimaginable, and more vulnerable to the kind of "Cyber Pearl Harbor" attacks security experts warned me about for seven tumultuous years.

Back in Kyiv, the Ukrainians wouldn't let me forget it. They stopped just short of grabbing me by the ears and screaming "You're next!" The warning lights were blinking red once again. And we were no more enlightened than we were the last go-around.

If anything, we were more exposed. Worse, our own cyberweapons were coming for us. The Ukrainians knew it. Our enemies certainly knew it. The hackers had always known it.

This is how they tell me the world ends.

THIS IS
HOW THEY
TELL ME
THE WORLD
ENDS

PART I

Mission Impossible

Be careful. Journalism is more addictive than crack cocaine.
Your life can get out of balance.

—Dan Rather

CHAPTER I

Closet of Secrets

Times Square, Manhattan

I WAS STILL COVERED in dust when my editors told me to surrender my devices, take an oath of silence, and step into Arthur Sulzberger's storage closet in July of 2013.

Just days earlier I'd been driving across the Maasai Mara in an open jeep, wrapping up a three-week trek across Kenya. I had hoped a few weeks off the grid would help repair nerves frayed by two years covering cyberterrorism. My sources kept insisting that this was just the beginning—that things would only get worse.

I was only thirty then, but already felt the immense burden of my assigned subject. When I got the call to join the *New York Times* in 2010, I was writing magazine cover stories from Silicon Valley about venture capitalists who, by sheer luck or skill, had invested early in Facebook, Instagram, and Uber and were now all too aware of their celebrity status. The *Times* had caught notice and was interested in hiring me, only for a different beat. "You're the *Times*," I told them. "I'll cover anything you want me to cover. How bad could it be?" When they told me they were considering me for the cybersecurity beat, I was sure they were joking. Not only did I not know anything about cybersecurity, I had actively gone out of my way to not know anything about cybersecurity. Surely they could find cybersecurity reporters who were more qualified.

"We interviewed those people," they told me. "We didn't understand anything they were saying."

A few short months later, I found myself in a dozen half-hour interviews with senior editors at *Times* headquarters, trying not to let my panic show. When the interviews wrapped up that evening, I marched across the street to the nearest bodega, bought the cheapest twist-cap wine I could find, and chugged it straight from the bag. I told myself that one day I would at least be able to tell my grandchildren that the holy *New York Times* had once invited me into the building.

But to my surprise, I was hired. And three years later I was still trying not to let my panic show. In those three years, I'd covered Chinese hackers as they infiltrated thermostats, printers, and takeout menus. I'd covered an Iranian cyberattack that replaced data at the world's richest oil company with an image of a burning American flag. I'd watched as Chinese military hackers and contractors crawled through thousands of American systems in search of everything from plans for the latest stealth bomber to the formula for Coca-Cola. I'd covered an escalating series of Russian attacks on American energy companies and utilities. And I'd embedded with the *Times'* own IT security team as the Chinese hacker we came to refer to as "the summer intern" popped up on our networks each morning at 10:00 A.M. Beijing time and rolled out by 5:00 P.M. in search of our sources.

All the while I clung desperately to the idea that I could live a normal life. But the deeper I ventured into this world, the more I found myself adrift. Breaches happened around the clock. Weeks went by when I rarely slept; I must have looked ill. The unpredictable hours cost me more than one relationship. And it wasn't long before the paranoia began to seep in. One too many times I caught myself staring suspiciously at anything with a plug, worried it was a Chinese spy.

By mid-2013 I was determined to get as far away as possible from anything to do with computers. Africa seemed like the only logical place. After three weeks of sleeping in tents, running with giraffes, and finishing each day with a sundowner in hand, watching the sun dip behind a slow-moving parade of elephants and, later, cozied by a campfire as my safari guide, Nigel, narrated each lion's roar, I was just beginning to feel the salve of remoteness.

But when I arrived back in Nairobi, my phone resumed its incessant buzzing. Standing outside an elephant orphanage in Karen, Kenya, I took one last deep breath and scrolled through the thousands of unread messages in my inbox. One screamed louder than the others: "Urgent. Call me." It was my editor at the *Times*. Our connection was already spotty, but he insisted on whispering, burying his words in the din of the newsroom. *"How quickly can you get to New York? . . . I can't say over the phone . . . They need to tell you in person . . . Just get here."*

Two days later I stepped out of an elevator onto the upper, executive floor at *Times* headquarters in tribal sandals I'd bought off a Maasai warrior. This was July 2013, and Jill Abramson and Dean Baquet—the then and very-soon-to-be executive editors of the *Times*—were waiting. Rebecca Corbett, the *Times* investigative editor, and Scott Shane, our veteran national security reporter, had been summoned as well. There were also three faces I did not yet recognize but would come to know all too well: James Ball and Ewen MacAskill from the British newspaper the *Guardian* and Jeff Larson from ProPublica.

James and Ewan relayed how days earlier, British intelligence officers had stormed the *Guardian*'s headquarters back in London and forced the newspaper to take drills and whirring blades to Snowden's hard drives of classified secrets, but not before a copy had been smuggled to the *New York Times*. Together, Jill and Dean said Scott and I were to work alongside the *Guardian* and ProPublica to write two stories based off the leaks from Edward Snowden, the infamous NSA contractor who had removed thousands of classified documents from the agency's computers before fleeing to Hong Kong, and later taking up exile in Moscow. Snowden had given his trove of classified secrets to Glenn Greenwald, a *Guardian* columnist. But we were reminded that day, Britain lacked the same free speech protections as the United States. Collaborating with an American paper, especially one with top First Amendment lawyers, like the *New York Times*, gave the *Guardian* some cover.

But first, the *Guardian* had rules. We were not to speak a word of the project to anyone. "No fishing," which meant that we were forbidden from searching the documents for keywords not directly related to our assignments. There would be no phones, no internet. Oh, and no windows.

That last bit proved particularly problematic. Italian architect Renzo Piano had designed the *Times* headquarters as a model of full transparency. The entire building—every floor, every conference room, every office—is encased in floor-to-ceiling glass, with the exception of one space: Arthur Sulzberger's tiny storage closet.

This last demand struck me as absurdly paranoid, but the Brits were insistent. There was the possibility that the National Security Agency, its British counterpart, the Government Communications Headquarters (GCHQ), or some foreign something somewhere would shoot laser beams at our windows and intercept our conversations. The same GCHQ technicians who looked on as the *Guardian* smashed Snowden's hard drives had told them so.

And so began my first taste of post-Snowden reality.

Day after day, for the next six weeks, I bid adieu to my devices, crawled into this strange, undisclosed semisecure location, wedged myself between Scott, Jeff, and the Brits, pored over top-secret NSA documents, and told no one.

To be honest, my reaction to the leaked NSA documents was probably very different from that of most Americans, who were shocked to discover that our nation's spy agency was indeed spying. After three years of covering nonstop Chinese espionage, a big part of me was reassured to see that our own hacking capabilities far exceeded the misspelled phishing emails Chinese hackers were using to break into American networks.

Scott's assignment was to write a sweeping account of the NSA's capabilities. My assignment was more straightforward, but—given that I had no phone, no internet, and was forbidden from calling any sources—also insanely tedious: I was to find out how far the world's top intelligence agencies had come in cracking digital encryption.

As it turned out, not very far. After several weeks of sorting through documents, it was becoming clear that the world's digital encryption algorithms were—for the most part—holding up quite nicely. But it was also clear that the NSA didn't need to crack those encryption algorithms when it had acquired so many ways to hack around them.

In some cases, the NSA was back-channeling with the international agencies that set the cryptographic standards adopted by security companies and their clients. In at least one case, the NSA successfully convinced Canadian bureaucrats to advocate for a flawed formula for generating the random numbers in encryption schemes that NSA computers could easily crack. The agency was even paying major American security companies, like RSA, to make its flawed formula for generating random numbers the default encryption method for widely used security products. When paying companies off didn't do the trick, the NSA's partners at the CIA infiltrated the factory floors at the world's leading encryption chip makers and put backdoors into the chips that scrambled data. And in other cases still, the agency hacked its way into the internal servers at companies like Google and Yahoo to grab data before it was encrypted.

Snowden would later say he leaked the NSA data to draw the public's attention to what he viewed as unlimited surveillance. The most troubling of his revelations seemed to be the NSA's phone call metadata collection program—a log of who called whom, when, and how long they spoke—and the lawful interception programs that compelled companies like Microsoft and Google to secretly turn over their customer data. But for all the shock and outrage those programs churned up on cable television and on Capitol Hill, it was becoming apparent that Americans were missing something bigger.

The documents were littered with references to NSA backdoors in nearly every piece of commercial hardware and software on the market. The agency appeared to have acquired a vast library of invisible backdoors into almost every major app, social media platform, server, router, firewall, antivirus software, iPhone, Android phone, BlackBerry phone, laptop, desktop, and operating system.

In the hacking world, these invisible backdoors have sci-fi names: they call them zero-days (or 0 days), pronounced "oh-days." *Zero-day* is one of those cyber terms like *infosec* and *man-in-the-middle attack* that security professionals throw around to make it all too easy for the rest of us to tune them out.

For the unindoctrinated: zero-days offer digital superpowers. They are a cloak of invisibility, and for spies and cybercriminals, the more invisible you can make yourself, the more power you will have. At the most basic level a zero-day is a software or hardware flaw for which there is no existing patch. They got their name because, as with Patient Zero in an epidemic, when a zero-day flaw is discovered, software and hardware companies have had zero days to come up with a defense. Until the vendor learns of the flaw in their system, comes up with a fix, disseminates its patch to users around the globe, and users run their software updates—*Dear reader: run your software updates!*—or swaps out, or otherwise mitigates, the vulnerable hardware, everyone who relies on that affected system is vulnerable.

Zero-days are the most critical tool in a hacker's arsenal. Discovering one is like discovering the secret password to the world's data. A first-rate zero-day in Apple's mobile software allows spies and hackers with the requisite skills to exploit it, to remotely break into iPhones undetected, and glean access to every minutiae of our digital lives. A series of seven zero-day exploits in Microsoft Windows and Siemens' industrial software allowed American and Israeli spies to sabotage Iran's nuclear program. Chinese spies used a single Microsoft zero-day to steal some of Silicon Valley's most closely held source code.

Finding a zero-day is a little like entering God mode in a video game. Once hackers have figured out the commands or written the code to exploit it, they can scamper through the world's computer networks undetected until the day the underlying flaw is discovered. Zero-day exploitation is the most direct application of the cliché "Knowledge is power if you know how to use it."

Exploiting a zero-day, hackers can break into any system—any company, government agency, or bank—that relies on the affected software or hardware and drop a payload to achieve their goal, whether it be espionage, financial theft, or sabotage. It doesn't matter if that system is fully patched. There are no patches for zero-days, until they are uncovered. It's a little like having the spare key to a locked building. It doesn't matter if you're the most vigilant IT administrator on earth. If someone has a zero-day for a piece of software that runs on your computer system, and knows how to exploit it, they can use it to break in to your computers, unbeknownst to you, making zero-days one of the most coveted tools in a spy or cybercriminal's arsenal.

For decades, as Apple, Google, Facebook, Microsoft, and others introduced more encryption to their data centers and customers' communications, the only way to intercept unencrypted data was to break into someone's device before its contents had been scrambled. In the process, "zero-day exploits" became the blood diamonds of the security trade, pursued by nation-states, defense contractors, and cybercriminals on one side, and security defenders on the other. Depending where the vulnerability is discovered, a zero-day exploit can grant the ability to invisibly spy on iPhone users the world over, dismantle the safety controls at a chemical plant, or send a spacecraft hurtling into earth. In one of the more glaring examples, a programming mistake, a single missing hyphen, sent the *Mariner 1*—the first American spacecraft to attempt an exploration of Venus—off-course, forcing NASA to destroy its $150 million spacecraft 294 seconds after launch, or risk it crashing into a North Atlantic shipping lane or worse, a heavily populated city. In our virtual world, the equivalent of the missing-hyphen error was everywhere, and now I was seeing just how critical to our nation's premier spies they had become. According to the descriptions littered before me, the NSA's expansive catalog meant that they could break into and spy on devices when they were offline, or even turned off. The agency could skirt most anti-intrusion detection systems and turn antivirus products—the very software designed to keep spies and criminals out—into a powerful spy tool. The Snowden documents only alluded to these hacking tools. They did not contain the tools themselves, the actual code and algorithmic breakthroughs.

The tech companies weren't giving the agency unlawful backdoors into their systems. When the first Snowden documents dropped, my sources at the nation's top technology companies—Apple, Google, Microsoft, Facebook—swore up and down that, yes, they complied with legal requests for specific customer information, but no, they had never granted the NSA, or any other government agency for that matter, a backdoor to any of their apps, products, or software. (Some companies, like Yahoo, were later found to be going above and beyond to comply with lawful NSA requests.)

The NSA was searching for and honing its own zero-days inside the agency's Tailored Access Operations (TAO) unit. But as I pored through the Snowden documents, it also became apparent that many of these zero-days and exploits were sourced from outside the agency too. The documents hinted at a lively outsourcing trade with the NSA's "commercial partners" and "security partners," though it never named names or spelled those relationships out in detail. There had long been a black market for cybercriminals looking to procure hacking tools off the dark web. But over the past few years, there were growing reports of a murky but legal gray market between hackers and government agencies, their zero-day brokers and contractors. Reporters had only scratched the surface. The Snowden documents confirmed the NSA played in it too, but like so many of Snowden's leaks, the documents withheld critical context and detail.

I came back to the questions at the heart of this again and again. There were only two explanations that made any sense: either Snowden's access as a contractor didn't take him far enough into the government's systems for the intel required, or some of the government's sources and methods for acquiring zero-days were so confidential, or controversial, that the agency never dared put them in writing.

That closet would be my first real glimpse into the most secretive, highly classified, and invisible market on earth.

THE STORAGE CLOSET made it impossible to think. After a month of arid African winds and open savanna, I was having an especially hard time without windows.

It was also becoming painfully clear that documents that would prove critical to our encryption story were missing from our trove. Early on in the project, James and Ewan had referenced two memos that laid out, in clear detail, the steps the NSA had taken to crack, weaken, and hack around encryption. But after weeks of searching, it was obvious that those memos were missing from our stash. The Brits conceded as much, and promised to retrieve them from Glenn Greenwald, the *Guardian* writer, who was now living in the jungles of Brazil.

We only had a slice of the documents Snowden had stolen. Greenwald had the entire trove—including the two memos the Brits told us were critical to our encryption story—but apparently he was holding them hostage. Greenwald was no fan of the *New York Times*—to put it mildly—and Ewen and James told us he was furious the *Guardian* had brought the *Times* into the project.

Greenwald was still reeling from a *Times* decision a decade earlier to delay publication of a 2004 story detailing how the NSA was wiretapping American phone calls without the court-approved warrants ordinarily required for domestic spying. The *Times* had held the story for a year after the Bush administration argued that it could jeopardize investigations and tip off suspected terrorists. Like Greenwald, Snowden was also furious with the *Times* for holding the story. That was the reason, Snowden said, that he had not brought the stolen NSA documents to the *Times* in the first place. He falsely assumed we'd sit on the trove or sit idly by as the government thwarted publication. So when Snowden and Greenwald learned that we had been brought into the project, James and Ewan told us, they were apoplectic.

James and Ewan assured us that Greenwald was more reasonable than the screeching mess of a man we saw on Twitter every day. But despite their repeated promises to fetch the missing memos from Greenwald's Brazilian compound, it was clear that somebody was in no mood to share his toys.

*   *   *

IT WOULD BE weeks before we got our hands on the missing memos. In the meantime, this little game got old fast. The tight quarters, the lack of oxygen, and the buzz of the fluorescent lights began taking a toll.

It was becoming painfully obvious that we were being used. To the *Guardian*, the *Times* was an insurance policy against their troubles with the British intelligence officers back home. The *Times* gave them safe cover and free lunch every day, but they did not want us as actual partners. We were supposed to be working in tandem, but the Brits had started publishing their own stories without giving us a heads-up. At one point, someone leaked details of our secret joint venture to Buzzfeed. And once the cat was out of the bag, the fact we were all still hiding in a windowless storage closet struck us as particularly absurd.

My faith in the *Guardian*, in journalism, was being sorely tested.

I missed the elephants.

Each night I would return to my hotel room and stare suspiciously at my hotel key card and anyone roaming the halls. The paranoia had started to follow me home.

A year earlier I had watched a hacker demonstrate how to break into hotel rooms using an exploit for a digitized key lock he'd built for fifty bucks. Actual thieves had started using the hack to break into hotel rooms and steal laptops. Considering my current situation, this was not comforting. When I asked the sweet lady at the front desk if the hotel had patched its locks for the vulnerability, she looked at me like I was from Mars and assured me that my room was perfectly secure. And yet each evening I still felt compelled to tuck my devices under the couch before lacing up my shoes and bolting outside for air.

Outside, Manhattan's midsummer air was balmy, and Times Square was clogged with tourists. I was desperate for oxygen. For sanity's sake, I made a point of biking up and down the West Side Highway each evening to digest what I had seen that day. My mind had become a maze of NSA acronyms and code names. I was dizzy and disoriented, and it was only at night, pedaling up and down the Hudson River, that I could actually think.

My thoughts were consumed by hundreds of unexplained zero-day references littered throughout NSA documents. Where had they come from? How were they being used? What if they got out? These weren't just backdoors into Russia's systems, or China's, or North Korea's, or Iran's. Two decades ago, we were all using different technology. But that was no longer the case. Backdoors in Apple? Android? Facebook? Windows? Cisco firewalls? This was technology most Americans relied on too. And not just for phone calls and email anymore. We banked on our phones. We monitored our babies on these devices. Our health records were now digitized. The computers that controlled Iran's nuclear plants ran on Windows. Using our iPhones and iPads, we could now tweak the pressure and temperature gauges on oil rigs hundreds of miles offshore. We did this for convenience, for peace of mind, to protect engineers from getting hurt while staving off an explosion. But that same access could be exploited for much darker forces too.

I could not shake the feeling that we were all missing something bigger than the NSA's phone-call metadata collection. That there was another, larger national conversation to be had. Where were these programs taking us? Who else had these capabilities? Where were we getting them from?

Of course, now I look back and am amazed I did not put the pieces together sooner.

After all, six months before I even set foot in that storage closet, hackers had dangled the answer right in front of my face.

CHAPTER 2

The Fucking Salmon

Miami, Florida

S IX MONTHS BEFORE Edward Joseph Snowden became a household name, I found myself flanked by Germany's preeminent industrial security specialist and two Italian hackers at a restaurant in South Beach.

We were all in Miami to attend the same bizarre conference—an annual invite-only gathering of the smartest fifty-plus minds in industrial-control security, a particularly terrifying subset of the industry that examines the myriad ways hackers can break into oil and water pipelines, power grids, and nuclear facilities.

That evening the conference organizer, a former NSA cryptographer, invited some of us out to dinner. Looking back, the invitation had all the makings of a twisted joke: *A reporter, an NSA codebreaker, a German, and two Italian hackers walk into a bar* . . . After only a year on the hacking beat, I was still figuring out my new normal—who was good, who was bad, who was playing it both ways.

Let's just say I stood out. For one, there are not many petite blondes in cybersecurity. To any woman who has ever complained about the ratio of females to males in tech, I say: try going to a hacking conference. With few exceptions, most hackers I met were men who showed very little interest in anything beyond code. And jiujitsu. Hackers love jiujitsu. It is the physical equivalent of solving puzzles, they tell me. I was neither male, a coder, nor particularly interested in getting squashed in a ground fight. So, you see, I had problems.

Growing up, the *New York Times* was my family's bible. I memorized bylines and imagined *Times* reporters being greeted like emissaries from the Lord himself. Not so in cybersecurity. Most people treated me like a child—the less I knew, they told me, the better. Also, as many, many men on Twitter regularly point out to me, nobody in cybersecurity actually uses "cyber" anymore. It's "information security," or preferably "infosec." More than a few times, after introducing myself as a cybersecurity reporter at a hacking conference, I was told to GTFO. (*Dear reader, I leave the deciphering of that code to you.*) As it turns out, introducing yourself as "cyber" anything is the quickest way to the door.

I was learning that this was a small, eerie industry with an intriguing mix of eccentrics. Every bar, at every conference, was reminiscent of the Mos Eisley cantina in *Star Wars*. Ponytailed hackers mingled with lawyers, tech execs, bureaucrats, intelligence operatives, revolutionaries, cryptographers, and the occasional undercover agent.

"Spot the Fed" was a crowd favorite. If you correctly spotted a federal agent at the annual Def Con hacking conference in Las Vegas, you got a T-shirt. For the most part, everyone seemed to know everyone else, at least by reputation. Some despised each other, but there was also a surprising amount of mutual respect, even between opposing sides, so long as you had skills. Incompetence—from a reporter, or especially from those who made their living selling digitized mousetraps—was an unforgiveable offense.

But that night, in Miami, seated between the German with his bespoke suit, ostrich loafers, and neatly combed hair and the two T-shirt-clad Italians with their unkempt 'fros, I felt a tension I had not encountered before.

The German's job title—industrial security specialist—did not quite do him justice. Ralph Langner had dedicated his life to preventing a catastrophic cyber meltdown, the kind of crippling cyberattack that could easily send a Deutsche Bahn train careening off the tracks, take the world's trading hubs offline, blow up a chemical plant, or trigger a tsunami by unlocking the gates of a dam.

The Italians, and a growing number of hackers just like them, were getting in his way. The two had traveled to Miami from the tiny Mediterranean nation of Malta, where they spent their days combing the world's industrial control systems for zero-days, weaponizing them into exploits that could be used for espionage or physical harm, and then selling them off to the highest bidder. I suppose they were invited under the "Keep your friends close and your enemies closer" banner.

That evening I was determined to find out to whom the Italians sold their digital arms and if there were any nation-states, three-letter agencies, or criminal groups to whom they would not. It was a question I'd be asking for years.

I waited until we were two Beaujolais deep.

"So Luigi, Donato, your business model, it's interesting," I stammered.

I directed the words at Luigi Auriemma, whose English proved the better of the two, trying my best to lighten my tone and mask my next question as a casual inquiry, no different from asking about the stock market. "So tell me. Who do you sell to? The U.S.? Who won't you sell to? Iran? China? Russia?"

As much as I attempted to disguise the gravity of the question with a forkful of food, I wasn't fooling anyone. The first rule of the zero-day market was: Nobody talks about the zero-day market. The second rule of the zero-day market was: Nobody talks about the zero-day market. I'd posed this question many times, and I knew it was the one question nobody in his business would answer.

The Luigis and Donatos of the world had rationalized their trade long ago. If companies like Microsoft didn't want them finding zero-day bugs in their software, they shouldn't have written vulnerable code in the first place. Zero-days were critical to national intelligence gathering, and only becoming more so as encryption shrouded the world's communications in secrecy. In a growing number of cases, zero-days were the only way governments could keep from "going dark."

But those rationalizations often ignored the dark side of their business. Nobody was willing to admit that one day these tools could be used in a life-threatening attack, that they were increasingly finding their way to oppressive regimes looking to silence and punish their critics, or infiltrating industrial controls at chemical plants and oil refineries, and that possibly, perhaps inevitably, those who dealt in this trade might one day find blood on their hands.

Everyone at our table knew that I had just casually asked the Italians to stare the shadier side of their business in the face. For a long moment knives and forks went still, and nobody said a word. All eyes turned to Luigi, who would only look down at his plate. I took a long sip of my wine and felt an intense craving for a cigarette. I could sense that Donato understood my question too but was not about to break the silence.

After several tense moments, Luigi finally murmured, "I could answer your question, but I would much rather talk about my salmon."

To my right, I could sense the German fidgeting in his seat. Two years earlier, Ralph Langner had been among the first to crack the code and unveil the plot behind Stuxnet, the most sophisticated and destructive cyberweapon the world had ever seen.

Stuxnet—as the computer worm came to be called—had been discovered in bits and pieces in 2010 as it slithered its way through computers around the globe, using an unheard-of number of zero-day exploits, seven to be precise. Some were clearly designed to infect hard-to-reach—even offline—computers. One Microsoft zero-day allowed the worm to invisibly spread from an infected USB flash drive onto a computer undetected. Others allowed it to crawl across the network from there, climbing ever higher up the digital chain of command in search of its final destination: Iran's Natanz nuclear plant, where it burrowed deep into the offline, or "air-gapped," computers that controlled the rotors that spun Iran's uranium centrifuges. And then, by remote command, Stuxnet silently spun some of Iran's centrifuges out of control, while stopping others from spinning entirely. By the time Iran's nuclear scientists discovered that a computer worm was responsible for the destruction of their centrifuges, Stuxnet had already destroyed a fifth of Tehran's uranium centrifuges and set Iran's nuclear ambitions back years.

Langner had made a name for himself with his analysis of the Stuxnet code—and for having the chutzpah to be the first to call out the weapon's two architects: the United States and Israel. But these days Langner worried endlessly about what might happen when those same capabilities ended up in the wrong hands. The Stuxnet code could work just as well in an attack on a power or nuclear plant, or a water treatment facility in the United States. In fact, Langner had mapped out "target-rich environments"—industrial systems still vulnerable to Stuxnet's code around the globe. The bulk were not in the Middle East. They were in the United States.

It was only a matter of time before our weapons turned on us. "What you end up with is a cyberweapon of mass destruction," Langner told the audience of hundreds that day. "That's the consequence that we have to face. And we better start to prepare right now."

Since his discovery, Langner had been traversing the globe, consulting with the world's largest utility companies, chemical plants, and oil and gas firms, preparing them for what he and others now believed was an impending and inevitable cyberattack of mass destruction. In his eyes, Luigi and Donato were cold-blooded cyber mercenaries, catalysts for our impending doom.

The longer Luigi stared silently at his salmon, the more visibly clenched Langner's jaw became. Finally, after several tense moments, he shifted his chair away from the Italians to face me directly.

"Nicole," he said, loudly for the others to hear. "These men are young. They have no idea what they are doing. All they care about is money. They have no interest in learning how their tools will be used, or how badly this will end."

Then, shifting his gaze back to Luigi, he said, "But go ahead. Tell us. Tell us about your fucking salmon."

SIX WEEKS IN a storage closet in the summer of 2013, and all of them with *fucking salmon* playing and replaying itself in my head. The phrase became my own code word for everything anyone who played in the cyberarms trade refused to tell me.

What had the Italians refused to tell me that night in South Beach?

Who did they sell their zero-days to?

To whom would they not?

Were they—and thousands more free-agent hackers just like them—the missing link in the Snowden documents?

Were there any rules or laws to their trade?

Or were we supposed to put our faith in hackers' own moral fortitude?

How did they rationalize the sale of zero-days in technology they, their mothers, their children—hell, our most critical infrastructure—used?

How did they rationalize the sale of a zero-day to a foreign enemy? Or to governments with gross human rights violations?

Did the United States buy zero-day exploits from hackers?

Did they buy them from foreign hackers?

With American taxpayer dollars?

How does our government rationalize the use of taxpayer dollars to sabotage the world's commercial technology? And global infrastructure?

Aren't we leaving our own citizens vulnerable to cyberespionage, or worse?

Were any targets off-limits?

Are these zero-days being used, or are they rusting away in a stockpile somewhere?

Under what conditions will we use them?

Under what conditions will we not?

How are we protecting them?

What if they get out?

Who else knows about this?

How much money are hackers making?

How are they spending it?

Is anyone trying to stop them?

Who else is asking these questions?

How does anyone sleep at night?

How will I?

It took me seven years to answer my own questions. By then, it was too late. The world's cyber superpower had been hacked, its tools free for anyone to pick up and use. The playing field was leveled.

The real attacks had only just begun.

PART II

The Capitalists

I think we spent our best years fighting on the wrong side.

—Larry McMurtry, *Lonesome Dove*

CHAPTER 3

The Cowboy

Virginia, United States

G ETTING TO THE bottom of the zero-day market was a fool's errand, they told me. When it came to zero-days, governments weren't regulators; they were clients. They had little incentive to disclose a highly secretive program, which dealt in highly secretive goods, to a reporter like me.

"You're going to run into a lot of walls, Nicole," Leon Panetta, the Secretary of Defense, warned me.

Michael Hayden, the former director of both the CIA and the NSA whose tenure oversaw the greatest expansion of digital surveillance in the agency's history, laughed when I told him what I was up to. "Good luck," Hayden told me, with an audible pat on the back.

Word about my quest traveled fast. My competitors told me they didn't envy my mission. Zero-day brokers and sellers prepared for me with bug spray. My calls went unreturned. I was disinvited from hacking conferences. At one point cybercriminals offered good money to anyone who could hack my email or phone. But there came a point when I knew backing off this story would haunt me more than moving forward. I'd glimpsed enough to know where this was all going.

The world's infrastructure was racing online. So was the world's data. The most reliable way to access those systems and data was a zero-day. In the United

States, government hackers and spies hoarded zero-days for the sake of espionage, or in the event they might need to do what the Pentagon calls D5—"deny, degrade, disrupt, deceive, or destroy"—an adversary's critical infrastructure in the event of war one day.

Zero-days had become a critical component of American espionage and war planning. The Snowden leaks made clear that the U.S. was the biggest player in this space, but I knew from my reporting that it was hardly the only player. Oppressive regimes were catching on. And a market was cropping up to meet their demand. There were vulnerabilities everywhere, many of them of our own making, and powerful forces—including our own government— were ensuring that it stayed this way. There were many people and institutions that did not want this story to be told. I came to believe that the only way to contain the spread of the world's most secretive, invisible market was to shine a big fat light on it.

As with most journalistic endeavors, getting started was the hardest part. I moved ahead the only way I knew how. I started with what little was already public and peeled the onion back from there. To do that, I had to travel back more than a decade, to the zero-day market's first public glimmer. I had to track down the men who—in their folly—were under the mistaken impression that they'd pioneered it.

EVERY MARKET STARTS with a small wager. I'd discover that the zero-day market—at least the public face of it—launched with ten bucks.

That's all it took for John P. Watters to buy his first cybersecurity company in the late summer of 2002. That was a heck of a lot less than he paid to engrave his initials—J.P.W.—on the crocodile cowboy boots he wore the evening he marched into iDefense's Chantilly, Virginia, headquarters to figure out whether there was anything left worth salvaging. But Watters figured $10 was a fair price for a company that had been hemorrhaging $1 million a month with no obvious plans to make it back.

And yet, for the two dozen remaining employees gathered at the company's headquarters that August, none of this made any sense. For starters, Watters was nothing like them. Six feet tall and pushing 250 pounds, he did not resemble the emaciated hackers and former intelligence types glued to their computer screens in Chantilly.

When they heard that a mystery millionaire from Texas was buying their shop, they expected a suit to walk through the door. Watters didn't do suits. His standard uniform was a Tommy Bahama shirt, cowboy boots, blade sunglasses if the sun was shining—a little colorful for men who wore black T-shirts and preferred to work in windowless dungeons. Their diet subsisted of sandwiches and Red Bull. Watters preferred Miller Lite and Texas rib eye. He didn't even live in Virginia and had no immediate plans to leave Texas. And, stranger still, he had zero experience with computers. These young men, with their sunken, glowing eyes, lived through their screens.

Watters was a moneyman. He'd spent his early career investing hundreds of millions of dollars for a wealthy family down in Texas. That sputtered out after the family patriarch passed, and his son, a chef, told Watters he planned to join him as co-CEO. For a cowboy used to running his own shop, that was a no-go. So Watters left with some capital commitments from the family in tow, started his own private equity shop, and began looking around for a ripe place to park some cash.

He set his sights on computer security. This was 1999. In just a few decades, the internet had made quantum leaps from the Pentagon's primordial invention—ARPANET—to the clunky commercial web available by dial-up modem, to the web Americans were coming to know through their Netscape browsers. Internet companies like Yahoo and eBay were being marked to absurd valuations.

Watters figured people would be willing to pay good money to lock these systems down. But cybersecurity companies were inefficient as hell. It was the old game of cat and mouse. As soon as virus hunters could immunize their customers from attacks, it was too late. The bad guys had already had their fun. The passwords and credit card data had already left the building. Watters knew the things that had attracted him to this business in the first place—the cops-and-robbers elements—weren't going away anytime soon. The security industry required a new model.

Between 1999 and 2001 Watters patiently waited out the dot-com bubble and burst, all the while keeping his eye on a few cybersecurity companies with potential to become viable businesses—just not at the insane valuations venture capitalists had pegged them to. At one point he pulled his ten-year-old daughter out of school, chartered a private jet, and showed up at iDefense's headquarters in Chantilly to do some good old-fashioned due diligence. iDefense billed itself as an advanced cyber threat alert system for big banks like Citigroup, though the bulk of its clients were government agencies—the Pentagon, the Navy, and the Coast Guard. But when Watters took a hard look at the books, he could see that iDefense was just another noisemaker with no real product to speak of, and no real plans to build one.

He was right. Two years later iDefense filed for bankruptcy, and as fate would have it, the bankruptcy hearing was set for September 11, 2001. If terrorists hadn't hijacked the planes that day, a judge may very well have ordered iDefense to halt operations. Instead, the hearing was postponed. One month later, a judge ruled that, post–9/11, the country was going to need more security companies like iDefense. Rather than forcing the company into liquidation, the judge ruled in favor of a Chapter 11 restructuring.

British investors pledged $600,000, figuring that would buy them enough time to restructure iDefense and sell it for profit. They called Watters and told him the business was for sale: $5 million.

Watters didn't blink.

"Nope," he told them. "I know as well as you do that you're going to run out of money. Call me when you do."

It only took ten months. The Brits told Watters that unless he was willing to pony up and buy their stock, they would close up shop the very next day. Watters told them he would buy iDefense for ten bucks.

They accepted. Watters asked his wife to give him two years to turn the thing around.

THE PARKING LOT outside iDefense's Chantilly headquarters was nearly deserted the evening Watters strolled in for a second time one Tuesday in August 2002. More than two thirds of the company had been let go. The remainder hadn't been paid in six weeks. The two dozen still standing took a hard look at this big bear of a man and his crocodile boots. Their first impression was *Who is this clown?*

Their second reaction was overwhelming relief. Watters was colorful all right, but he was not your typical private equity stiff. Employees hadn't been paid in six weeks. The first thing he did was walk straight into the office kitchen and start writing paychecks to make everyone whole. He gave executives a choice: they could convert their notes to equity, which could translate to bigger profits down the road, or get paid in cash now.

Everyone chose cash—hardly the vote of confidence Watters was hoping for. Employees were willing to give iDefense another few months, but nobody thought this was going to end with bells ringing on a trading floor, or an eight-figure exit. The corporate culture had become one of starvation. And it wasn't as if employees could just walk across the street and ask for another job. Nasdaq reached its lowest point of the dot-com crash the following month. Five trillion dollars in paper wealth vanished. In another two years, more than half of all dot-com companies would disappear.

iDefense's competitive prospects weren't looking so hot either. The day Watters showed up at iDefense, its closest competitor, a start-up called SecurityFocus, was scooped up by Symantec, the security giant, for $75 million in cash. Like iDefense, SecurityFocus offered clients an early cyber threat alert system, in the form of a hacker mailing list called BugTraq.

"Mailing list" never quite did BugTraq justice, though. It was more like a primitive version of Reddit or 4chan—festering grounds for the curious, the self-righteous, and mischief-makers of the early web. The concept behind BugTraq was simple: hackers all over the world posted bugs and exploits to the list. Some did this for fun, curiosity, street cred, or simply to thumb their nose at the tech vendors who ignored or threatened them anytime they tried to call attention to bugs in their code.

In those early days, there were no 1-800 numbers to inform Microsoft or Hewlett-Packard, "Hey, I just turned your server into a massive surveillance tool, and it looks like I may be able to use it to break into NASA." Cold-calling a company rep or software engineer usually resulted in silence, vitriol, or a sternly worded letter from the general counsel. Vendors couldn't be bothered and never made it easy to do the right thing. BugTraq and, later, similar lists like Full Disclosure became the de facto forums for hackers to drop their best finds and slam companies for leaving land mines in their code every day.

Hackers' discoveries formed much of the raw material for iDefense's iAlert product. Where hackers discovered one bug, there were usually more to be found. Its iAlert system gave customers a heads-up to gaping holes in their hardware and software and offered them workarounds. If Symantec closed off BugTraq to iDefense, the company wouldn't have anything to alert its clients to. And if iDefense tried to compete with Symantec's deep pockets, it would fail.

The first few months were bumpy. The more Watters dug into the business, the more he realized just how crude it was. The alert system wasn't distinct from any number of other feeds its customers subscribed to—many of them free.

Watters knew there was something here. He just didn't know yet what it was. He invited a few employees out to lunch, hoping they might offer some ideas, or consolation. Instead, they agreed. There was nothing here left to salvage.

"You bought it. You own it," one of the company's graybeards told Watters.

"Well, yes, I guess I did," was all Watters could muster. "Shit."

By November, Watters told his wife he was going to have to write off the business or move to Virginia full-time. Together, they decided he would give it one last chance. He bought a two-bedroom apartment just outside Chantilly— "the Hacker Hut," he called it—and hunkered down.

Each morning at 5:15 A.M. Watters started firing off emails, hoping to jump-start employees into a new rhythm. He approached every sector of the business as if preparing to defibrillate. His new motto was "Three to one": "We're going to do three years of work in one," he told employees, often to eye rolls. He started replacing the eye rollers with fixers he knew would get the job done. And he began soliciting ideas for new products from the rank and file.

At the time iDefense ran a research lab run by two young hackers in their twenties: David Endler, who spent his early career at the NSA, and Sunil James, who was just a few years out of college. James had joined iDefense in the days after 9/11 with smoke from the Pentagon still wafting into his apartment. He spent his first few nights sleeping at the office.

Endler and James's jobs were to scour software and hardware for zero-days and keep tabs on BugTraq and other hacker forums for vulnerabilities that, in the wrong hands, could hurt iDefense's client base.

"The pool of people who understood vulnerabilities, enough to actually explain them to people in simple terms, was still really small," James told me. Often the two sat in darkness, lit only by the glow of their screens, hunting for bugs. Both men knew their team could only find so many on their own. The vast majority still came from BugTraq. And with BugTraq now in Symantec's hands, they knew their primary feed of threat intel was about to dry up. Unless they found a new pool of vulnerabilities to tap, James told Endler, "We're about to get fucked."

Both men knew that there was a huge, untapped pool of hackers the world over, discovering vulnerabilities at all hours of the day and night. It didn't take a genius to see that vendors were living off the backs of these hackers' free labor, taking their discoveries and using them to make their products more secure. If Symantec closed up BugTraq, James told Endler, tech companies and their customers would lose their primary source of raw threat data. James had an idea: What if we go straight to the hackers and pay them for their bugs?

James knew that inviting hackers to dissect technology products for flaws wasn't exactly de rigueur in this industry. The position most big companies— Hewlett-Packard, Microsoft, Oracle, Sun Microsystems—took at the time was that anyone who drew attention to a flaw in their products should be prosecuted or sued for tampering. Microsoft executives called it "information anarchy" and at one point compared hackers who dropped bugs on BugTraq and at hacking conventions to terrorists "who throw pipe bombs into children's playgrounds." That year, 2002, representatives from the major tech companies convened at the annual Def Con conference in Vegas to lay down the law. Since Def Con was founded in 1993, the conference had become a forum where hackers could present all the ways they could break into companies' products. But recently the companies felt the hackers had gone too far. They were sick of getting called out and shamed on stage. That year, they banded together in Vegas to come up with a new approach to hackers and their bugs. It essentially boiled down to: "Give your bugs to us first. Or we'll sue your ass."

The notion that iDefense would consider paying hackers for an advance look at bugs in their software was not likely to go over well with the Microsofts, Oracles, and Suns of the world. Perhaps it was the Hawaiian shirt and cowboy boots, but there was something about Watters that made James and Endler think he might be open to a different line of thinking.

Endler made the pitch: There are vulnerabilities all over the place, he told Watters. Programmers at Microsoft, Oracle, and the big tech companies were inadvertently writing bugs into their code every single day. By the time one bug was found and patched, developers had already baked new code, with new bugs, into software being used by iDefense's customers all over the globe. The security industry did little to protect their customers from the next attack. They were still focused on mitigating the last one. And besides, Endler told Watters, antivirus software was itself laden with bugs.

Meanwhile, black-hat hackers were exploiting these bugs for profit, espionage, or digital mayhem. White hats who wanted to do the right thing were losing any incentive to report them to vendors. Vendors didn't want to deal with basement dwellers, and were offended by the very notion that their products were flawed in the first place. They much preferred threatening hackers with lawsuits to fixing bugs in their systems. With their belligerence, vendors were just giving white hats more incentive to become black hats. If nobody was going to fix these bugs, they might as well exploit them or blast them out to "script kiddies," low-level hackers who used them to deface a company's website or take it offline. There were rumblings about some kind of underground gray market that paid hackers for their bugs, and their silence. And these days, getting paid was a lot more enticing than getting sued.

Ultimately, the losers in this game, Endler argued, were iDefense's clients: banks and government agencies whose systems were riddled with holes. Doors were being left wide open for attack.

"We could start a program," Endler pitched Watters. "We could pay hackers to turn over their bugs."

iDefense would still turn the bugs over to the vendors, Endler said, but until the vendors were ready to roll out a patch, the iAlert product could function as a legitimate early-warning system for their subscribers. iAlert would give their clients an early heads-up and a workaround so attackers couldn't exploit the not-yet-public flaws. iDefense wouldn't have to pay the hackers much, Endler reasoned. Hackers just wanted a way to turn bugs over that didn't entail getting thrown in prison. "We could use the program to our competitive advantage," Endler told Watters.

Most other CEOs would have balked. The idea that a company with no profits to speak of would use its dwindling funds to pay hackers—pimply thirteen-year-olds in their parents' basements, ponytailed coders from the web's underbelly—to find bugs in other companies' systems would have struck most executives as risky at best. Most lawyers would have shut it down right then and there.

Not Watters. It was the Wild West elements of this industry that had drawn him here in the first place. It also made good business sense: by opening up a legitimate vulnerability market, iDefense would have a first look at technology's defects and holes and finally have something unique and concrete to offer its clients. It would no longer be just another noisy feed.

It also meant Watters might be able to justify raising the costs of the company's alert system.

And so, in 2003, iDefense became the first shop to publicly open its doors to hackers and start paying bounties for zero-day bugs.

THE BOUNTIES STARTED small, $75 a pop. James and Endler didn't know what they were doing. There was no market to mark to, no competitive program—at least none that they were aware of.

The whole effort was a crapshoot. And in those first few months, they knew hackers would be testing them, seeing what they could get away with. Hackers were not about to turn over their best exploits right away; they'd start with the low-hanging fruit. Of the first thousand bugs submitted over the next eighteen months, half were crap. Some were cross-site scripting vulnerabilities—common bugs in web applications that novice hackers could use to deface a website. Or they were bugs that caused Microsoft Word to crash anytime you opened a new Word document. These bugs were a nuisance, but black hats weren't using them to steal intellectual property or customer data. James and Endler contemplated turning these bugs away. But they knew they needed to build trust, so that hackers would come back to them with bigger and better hauls. In those early days, the iDefense team swallowed their pride and shelled out for crud.

The strategy worked, for a time. Among the first to start turning over serious zero-days was a hacker in Turkey by the name of Tamer Sahin. In 1999 Sahin had been arrested for hacking a Turkish web provider. As part of his plea deal, he agreed to help the Turkish government lock up its systems, which meant he was constantly searching the web for vulnerabilities and publishing security advisories for software sold by Microsoft, HP, AOL, and others. Sahin started to make a name for himself in the hacking community, but he had yet to receive a penny for his work.

When iDefense announced its program in 2002, Sahin submitted its first bug, a flaw in an internet protocol that could potentially allow an attacker to intercept passwords and other data between a user and her browser. The bug netted Sahin $75—enough for a month's rent in Turkey. Sahin became a one-man bug factory, turning over fifty bugs and exploits—enough to forgo a day job—for two years.

In Kansas City, a thirteen-year-old named Matthew Murphy was getting hooked, too. Murphy was too young to work legally, but he was getting four-figure checks from iDefense for bugs he found in AOL and antivirus software. Murphy used the money to buy his mom a second computer and a second phone line for dial-up internet.

Eventually he had to tell his mom, a cop, what he was up to—not an easy conversation. She worried that his discoveries could be abused to hurt people. "I told her if the vendors don't want their customers getting hurt, they should have made their software secure in the first place."

iDefense became a go-to shop for hackers looking for quick rewards for their bugs. Every year Watters made a point to throw the biggest parties at Black Hat and Def Con, recruiting more hackers to their program and awarding prizes to hackers who found the most serious flaws. Hackers who barely made it out of their basements would get hammered at the iDefense party, then take their liquid courage to the blackjack tables.

A year into the program, iDefense's two top bug hunters were an Argentine hacker named Cesar Cerrudo and a Kiwi hacker nicknamed Zenith Parsec. Offline, Zenith Parsec was Greg McManus, a New Zealand sheep farmer who much preferred finding bugs in software to shearing sheep. Before long, half of all iDefense's bug bounties were going to McManus.

The Kiwi brought a sophistication to his work that iDefense hadn't seen before. Watters figured it would be cheaper to bring him to Virginia than keep sending thousands of dollars' worth of Western Union payments to New Zealand. So Watters offered McManus a job and a room back at his Hacker Hut.

McManus showed up a few weeks later with his computer, a Rubik's cube, a backpack, and a black T-shirt that read SOMEONE SHOULD DO SOMETHING. As roommates, the gregarious cowboy with a penchant for Miller Lite and the quiet Kiwi with the Rubik's cube were the ultimate odd couple. But they hit it off. And with their nights free, McManus began to teach Watters the art of the hack.

Watters had spent his entire career working for money. Hackers, McManus explained, aren't in it for money. At least, not in the beginning. They are in it for the rush, the one that comes with accessing information never meant to be seen. Some do it for power, knowledge, free speech, anarchy, human rights, "the lulz," privacy, piracy, the puzzle, belonging, connection, or chemistry, but most do it out of pure curiosity. The common thread is that they just can't help themselves. At their core, hackers are just natural tinkerers. They can't see a system and not want to break it down to its very last bit, see where it takes them, and then build it back up for some alternate use. Where Watters saw a computer, a machine, a tool, McManus saw a portal.

Hackers had been around for more than a century. In the 1870s, several teenagers were caught tampering with the country's primitive telephone system. The label *hacker* has a spotted history—one alternately celebrated and condemned—but history's most revered entrepreneurs, scientists, chefs, musicians, and political leaders were all hackers in their own right.

Steve Jobs was a hacker. So is Bill Gates. *The New Hacker's Dictionary*, which offers definitions for just about every bit of hacker jargon you can think of, defines *hacker* as "one who enjoys the intellectual challenge of creatively overcoming or circumventing limitations."

Some say Pablo Picasso hacked art. Alan Turing hacked the Nazi code. Ben Franklin hacked electricity. And three hundred years before them, Leonardo da Vinci was hacking anatomy, machinery, and sculpture. Leonardo famously labeled himself with the Latin phrase *senza lettere*—without letters—because, unlike his Renaissance counterparts, he couldn't read Latin. He learned by tinkering instead, a habit that led many a modern-day hacker to claim him as one of their own. Though society has come to conflate hackers with black hats and criminals the reality is that we owe them much of our progress and, ironically, our digital security.

Everything, even the most secure system on earth, McManus told Watters, can be compromised. And he set to work showing Watters how. The first task was to scan your target's system for any public-facing web servers or applications. Often you found something. If not, patience was critical. There was a reason, years later, that security experts began to call nation-state hacking groups "advanced persistent threats." If they were patient and persistent, they would inevitably find a way to get in and stay in.

Someone inside the organization would inevitably install an application for convenience, and those applications could be exploited for access. The trick was to sit quietly and watch to see what changed on your target's perimeter. An executive might install an application to send and receive phone calls from her landline office phone. As soon as that application is installed, you start poking and prodding it, directing your browser to the application, and jamming it with web traffic to see what the application does. You look for anything strange, signs of quality or signs of crap. When you can't bend it to your will, you hit the support forums, looking for any clues as to how the application was built, what kinds of problems people are reporting, any updates or fixes to the software or hardware. When you find an update, even a snippet of update code, you rip it apart, reverse-engineer and decompile the code.

Like learning any foreign language, reading code comes naturally to some, while others learn over time. Once you read enough code, you get to a point where you might not be able to write it from scratch, but you understand the basic outlines of what you're seeing. Eventually, McManus told Watters, you start to identify patterns. The goal is to get to a point where you can find functions and variables that can be exploited to do something for which they were never intended.

In that one little snippet of a software update, you might be able to inject code into a web server that causes it to turn over the source code to the voice-mail application. Then you start all over again. You tear apart the application's source code, looking for anything odd. You could find nothing and go back to the drawing board. Or you might find the gold mine—a remote code execution bug, the kind of bug that allows a hacker to run code of his choosing on the application from afar. To do that, a password may be required. No problem. You search dark web repositories for stolen emails, usernames, and passwords that match an employee at your target organization. When that fails, you dig further and further into the target's log-in infrastructure, looking for any other signs of weakness. You might find that when someone installed the log-in software, they stored usernames and passwords in plain text in a log file on the server. *Bingo!* You pull the log file, and now you have the full kill chain.

You string your zero-days together. You plug in the plain-text username and password, use the remote execution bug to get into the voice-mail application. And now you're not only in the target's phone system, you're in anything connected to the target's phone system: their computer, any computers on their local network, an IT administrator's computer that has special network privileges to access more computers still. With that level of access, the possibilities for theft and mayhem are endless. You can access data for insider trading. You can sell executives' credentials for a pretty penny on the black market. Or you can do the right thing and tell the voice-mail application company that they have a big, fat problem on their hands, and hope they take you seriously.

Every night McManus walked Watters through these various attack scenarios, pointing out anomalies in the code, identifying weak points and showing him how they could be abused. It was terrifying, but McManus's Kiwi accent made it all sound remarkably pleasant. By day, McManus was like a pig in shit, vetting other hackers' bugs, telling iDefense which weren't worth paying for and weaponizing the killer bugs into exploits. By night he was introducing his boss to a whole new world, run by a language few understood.

McManus didn't ask for much in return. He simply asked Watters for permission to paint the walls of iDefense's lab black. It would make him more comfortable, he said. Watters agreed. Anything to help the Kiwi feel more at home. The guy had a laserlike focus on bugs and exploits, and it was beginning to pay off.

More often than not, iDefense would turn a bug over to a vendor, only to be told that the bug wasn't serious. That's where McManus stepped in. He would furnish a proof-of-concept exploit to show them just how easily the bug could take control of their software or steal their customers' data. For so long, it had been left to vendors to decide which bugs were worth patching. Often there was no rhyme or reason to these decisions. Now McManus was proving to vendors, and their customers, that these bugs could no longer be dismissed. McManus also helped prioritize which bugs required immediate workarounds for customers, and which could reasonably wait.

The program brought something unique to the market, and helped iDefense justify a bump in pricing. That first year Watters doubled iDefense's subscription fee from $18,000 to $38,000. He fired customers unwilling to re-up, and found he could easily replace them with big-budget banks and federal agencies willing to pay top dollar for an intelligence product that came with bug workarounds built in.

To his delight, many of the customers Watters fired came crawling back at higher fees. He fired one customer who was paying $27,000 a year and won them back at $455,000. He fired a government agency that was paying $25,000 and won them back at $1.5 million. By October 2003, Watters had doubled iDefense's revenues and started investing more of his own money in the business.

All this success started to piss off the software vendors. Microsoft, Sun, Oracle—they all hated the program. Their employees accused iDefense of inviting hackers to break into their products. And as the program picked up, iDefense's customers started pestering the tech companies to patch their systems—and fast. Suddenly the biggest tech companies in the world were being forced to work on someone else's timetable. And they went after iDefense with a vengeance.

At the Black Hat convention that year, Endler was accosted by a member of Microsoft's security team who likened iDefense's bug payment program to extortion. All over Vegas, women in short cocktail dresses watched, dumb-founded, as nerds in black T-shirts screamed at other nerds in corporate logo T-shirts about the ethics of bug payments.

Watters heard about it, too. He was invited out to dinner by Oracle's chief security officer, Mary Ann Davidson. This should be interesting, Watters thought. Davidson didn't waste a moment. Before their appetizers arrived, she told Watters that what iDefense was doing, paying hackers for bugs, was "immoral."

Immoral, for fuck's sake? God forbid we find flaws in Oracle's precious code that should never have put there in the first place, Watters thought. Without his program, hackers would still have more incentive to exploit Oracle bugs than report them.

Davidson was so high-minded, so sure of herself, that Watters felt he had no choice but to whip out his cell phone and dial his mother. "Mary Ann, I think you should talk to my mom." He held out the phone. "Maybe you'll at least allow her to counter your position. I think she'll tell you that I'm a pretty moral guy."

For the first time since they'd sat down, Davidson was quiet.

The more bugs iDefense secured, the better business was. Watters found he couldn't sign up customers fast enough. Government agencies were re-upping every year. The more successful its business became, the more vendors like Davidson hemmed and hawed. Watters's business thrived because he was a cowboy on a wide-open range, but that range would start to get fenced in just as his business really started to take off. He would have a few really good years, but the market would shift, other players would move in, and things would get complicated.

THE FIRST SHIFT in the wind was Bill Gates's memo. In 2002, after a series of escalating attacks on Microsoft's software and customers, Gates declared that security would become Microsoft's top priority.

Hackers wrote off the memo that launched Gates's Trustworthy Computing Initiative as a joke. Microsoft had been slapping together sloppy software, riddled with holes, for years. And now everyone was supposed to believe Gates had found religion? Microsoft was the dominant player in the PC market, but ever since two technical prodigies at the University of Illinois by the names of Marc Andreessen and Eric Bina created the first internet browser, Redmond had been playing catch-up. The internet had been around for decades, since 1969 in fact, in an early incarnation cobbled together at the Pentagon. Since then it had evolved considerably. But it was Andreessen and Bina's creation of the first Mosaic internet browser in the mid-1990s that opened the internet up to the masses. The Mosaic browser had color and graphics and made it easy for people to upload their photos, videos, and music. Suddenly the internet started showing up in *Doonesbury* cartoons and in the famous *New Yorker* cartoon featuring two dogs at their PC, one saying to the other, "On the Internet nobody knows you're a dog!" And from that point on, the internet population doubled each year, year over year, until it was too big for the Microsofts of the world to ignore.

Microsoft had been so focused on dominating the PC market that it didn't pay the web much notice until Andreessen moved to Silicon Valley and debuted a commercial version of his Mosaic browser, Netscape Navigator, in 1994. One year later, Netscape was clearing a hundred million hits a day, sending Redmond into full panic mode.

The ensuing battle between Microsoft and Netscape is a thing of legend. Microsoft had dismissed the web as a science experiment and put all its eggs in the personal computer model—independent computers working independently on independent desks or on closed networks. Once they saw the web's market potential, Microsoft hacked together its own web browser—Internet Explorer—cobbled together a quick and dirty web server, and aggressively pushed web providers to choose Microsoft over Netscape. This was before Gates became a philanthropic saint. He was firing off emails to AOL executives, demanding to know: "How much do we have to pay you to screw Netscape?"

In the rush to best Netscape, speed, not security, was the name of the game. More than a decade later, Mark Zuckerberg would coin a name for this approach at Facebook with his motto "Move fast and break things."

Just as soon as these products hit the market, hackers began unspooling them with glee. They wanted to see how far the bugs in their new internet toys could take them—which, as it turned out, was quite far. Hackers found they could tunnel through Microsoft's systems to customers all over the web. They tried to point out errors to Microsoft, but rarely, if ever, was their work taken seriously. Part of the problem was that code—not diplomacy—was their strong suit.

It wasn't until Uncle Sam stepped in—after a series of destructive attacks—that things started to change. In 2001 attackers unleashed Code Red, a computer worm that turned hundreds of thousands of computers running Microsoft software into useless paperweights. Attackers later used some of the bugs from the Code Red worm to launch a massive attack that took the computers of hundreds of thousands of Microsoft customers offline, and attempted to take one notable Microsoft customer offline: The White House.

Code Red came on the heels of a series of other embarrassing Microsoft-related attacks. One computer virus, named Melissa by its author after a stripper in Florida, exploited Microsoft flaws to shut down the servers at some three hundred corporations and government agencies, leaving behind a cool $80 million in damages. Another virus born out of the Philippines, ILOVEYOU, wiped files and infected victims at a rate of some 45 million a day, forcing major Microsoft customers like Ford Motor Company to shut off email.

And then there was Nimda, an attack that slowed the internet to a crawl. Nimda took advantage of an unpatched Microsoft bug to infect everything in its reach—email, servers, hard drives—and then happily reinfected everything it had already hit. It had only taken twenty-two minutes for Nimda to become the worst cyberattack of its time. The tech-research firm Gartner warned Microsoft customers to "run, don't walk, away" from Microsoft's web server software.

Nimda's timing—just one week after 9/11—caused government officials to suspect cyberterrorists. A line in the code—"R.P. China"—pointed to China. Or had it been planted there to throw off responders? Why RPC and not PRC? Was it the work of a Chinese speaker who didn't know English grammar conventions? Or terrorists planting a false flag? Nobody ever found out. But the mere suspicion that the attack had been the work of cyberterrorists made Microsoft's security woes too alarming for the government to ignore.

Before 9/11, there were so many holes in Microsoft's products that the value of a single Microsoft exploit was virtually nothing. After 9/11, the government could no longer let Microsoft's security issues slide. Officials at the FBI and the Pentagon started ringing up Microsoft executives to rip them a new one.

Nimda was the least of it. Officials were growing concerned about a new class of Microsoft flaws that allowed attackers to take invisible control of customers' machines. They wanted to know who, at Redmond, was taking this seriously. And they made clear that if Microsoft continued to stick its head in the sand, Uncle Sam was prepared to take its business elsewhere.

ON JANUARY 15, 2002, just as iDefense was getting going, Gates fired off the cybersecurity equivalent of the "shot heard round the world." From that point on, Gates said, security would be the company's "highest priority,"

"Trustworthy Computing is more important than any other part of our work," Gates wrote in his now infamous memo. "Computing is already an important part of many people's lives. Within 10 years, it will be an integral and indispensable part of almost everything we do. Microsoft and the computer industry will only succeed in that world if [chief information officers], consumers and everyone else sees that Microsoft has created a platform for Trustworthy Computing."

What the security community wrote off as a stunt became an economic force. Microsoft froze new products and dredged up existing ones, ripping its software apart and training nearly ten thousand developers to build it back up again with security principles at the core. For the first time, procedures were put in place to embrace the hacking community. Microsoft set up a customer service line for hackers, tracked each caller and even logged their psychological quirks, noting which hackers needed to be handled with kid gloves, which had rock-star status, and which were just trolls. It instituted a regular system for rolling out software patches, releasing them on the second Tuesday of every month—"Patch Tuesday"—and offered customers free security tools.

And while plenty of zero-day bugs were still discovered, the frequency and severity of Microsoft bugs started to dry up. By the time I joined the security beat eight years later, I would always make a point of asking hackers, "I know you hate the vendors, but of all of them, who do you hate least?"

The answer was always the same. "Microsoft," they would tell me. "They turned their shit around."

The ripple effect of Gates's memo could be seen far from Redmond, in underground dark web forums and in hotel rooms at the big security conferences. There, in the shadows, a growing number of defense contractors, intelligence analysts, and cybercriminals started doling out higher rewards to hackers who promised to keep their bug discoveries secret.

In these subterranean circles, people started assigning a far higher value to Microsoft zero-day exploits than what iDefense was paying. "In 2000, the market was saturated with Microsoft exploits," Jeff Forristal, an early hacker, told me. "Today a remote Windows exploit would cost you six, maybe seven figures. Back then, you could get those for pennies on the dollar."

The same exploits hackers had once happily traded for free, or dumped online to shame vendors into releasing a patch, started taking on higher monetary values as a new group of mystery buyers began creating a market for their finds and giving hackers far more reasons—much more profitable reasons—to quietly sell the holes they found than turn them over to vendors to be sealed shut.

IT DID NOT take long before Watters started getting the calls. At first they were few and far between, but as iDefense's program took off in 2003 and 2004, the calls became more frequent; the callers more desperate. The men on the other end wanted to know if Watters would consider withholding some of hackers' bug submissions from vendors and clients in exchange for higher profits.

The same bugs that Watters was paying hackers $400 for? These mystery callers were willing to pay iDefense $150,000 for a single bug, so long as iDefense kept the sale quiet and didn't tip anyone else off.

The men said they worked for government contractors that Watters had never heard of. He'd heard rumblings of a gray market for zero-days, but the fact that these contractors were willing to pay so much blew him away.

When Watters turned them down, they switched tack and tried to appeal to his patriotism. These bugs would be used to spy on America's enemies and terrorists, they told him. Oh, the irony, Watters thought. Vendors had called him criminal. Now here were government contractors telling him he would be doing his country a service.

Watters was a patriot. But he was also a businessman. "It would have killed us," he told me. "If you're coconspiring with the government to leave gaping holes in core technology used by your customers, you're inherently working against your customers."

His callers eventually got the message. But beyond iDefense, something was shifting. Other market forces were at work. Hackers started hoarding bugs. Bug submissions dropped as hackers got greedy, often asking for significantly more than iDefense was willing to pay. Some alluded to having "other options."

It was clear that there were new competitors in this market. In 2005 a mysterious new outfit called Digital Armaments started announcing five-figure bounties for bugs in widely used Oracle, Microsoft, and VMWare systems. Beyond a bare-bones website registered in Tokyo, little was known about Digital Armaments' customers, or its backers. They solicited "exclusive rights" to these bugs and said only that they planned to notify the vendors "eventually."

The market Endler and James believed they had created was morphing. A new tier of hackers had emerged. Hackers the company had never worked with started to approach iDefense with the holy grail of software bugs: flaws in Microsoft's Internet Explorer browser that could be used to remotely take over someone's machine. But they were only willing to hand them over for six figures. The most iDefense had paid for a similar bug was just shy of $10,000.

"There was no way we were even going to pay close to that," James recalls.

Just three years into the iDefense program, hackers were demanding $4,000 for bugs that earned just $400 three years earlier. In another five years, a single bug would command $50,000. Watters told me the first thousand bugs iDefense paid $200,000 for in the first eighteen months of the program would have cost $10 million today.

iDefense was getting priced out of the very market it helped spawn. Others had caught on to what Endler had known all along: there was more to be gained by embracing hackers and their discoveries than pretending the holes did not exist. But these newer players were entering the market for very different reasons. And they had much bigger pockets.

Watters read the writing on the wall. By 2005 he had put $7 million of his own money into iDefense. He had told his wife it would take him two years to turn the company around. In the end, it took three. That July, nearly three years to the day he scooped up iDefense for ten bucks, Watters sold the company to Verisign for $40 million, left the Beltway, and moved back to Dallas.

It was time to let the market run where it would.

CHAPTER 4

The First Broker

The Beltway

I T WOULD HAVE been a huge business," one of the very first zero-day brokers told me of his pitch to Watters one rainy day, in between bites of his enchilada.

Years before Watters green-lighted the cash-for-bugs program, the market for zero-day bugs and exploits had already taken on a secret life of its own. While James and Endler were busy drawing up their quaint price lists—$75 for this bug, $500 for that—a handful of government brokers and defense contractors within ten miles of their lab were offering hackers as much as $150,000 for their discoveries, so long as they kept quiet.

Indeed, the most coveted bugs required it. Secrecy was a necessary precondition for zero-days. The moment a zero-day bug is no longer a secret, digital authorities painstakingly assign the bug a name and a score—ranging from "Eh, this can wait" to "Patch this like your life depends on it." Known bugs are then fed into a publicly accessible national vulnerability database. When the bug gets patched, hackers lose access. And spies were learning that the world's growing mass of data was only useful if they could access it.

Government spies determined the best way to guarantee long-term access to data was a zero-day exploit. They were willing to pay hackers far more for that access than the pitiful amounts iDefense was paying. And once they shelled out six figures for those zero-days, they weren't about to blow their investment and access by disclosing the flaw's existence to anyone—especially a journalist from the *New York Times*.

This made the far more lucrative government market for zero-days even harder to crack. It took me years to find an underground broker from the market's earliest days who would talk, and it wasn't for lack of trying. Every time I caught a lead, I would reach out only to never hear a peep. Many never responded. Most denied any involvement in the underground cash-for-bugs market.

Some would tell me they had gotten out years ago, refusing to offer anything further. Some just hung up. And then there was the one who told me that not only would he not talk to me about the market, but he'd already warned everyone he knew not to talk to me. If I continued on this thread, he told me, I would only be putting myself in danger.

Was this frightening? Yes. But I knew much of it was smoke and mirrors. Most just feared for their bottom line. In their line of work, keeping your mouth shut was essential.

Every deal required trust and discretion, and most deals were wrapped in nondisclosure agreements and, increasingly, classified. The most profitable brokers kept their zero-day business, the sheer fact that there was a business, a secret. The more discreet the broker, the more buyers coveted his business. A broker's quickest road to bankruptcy was to talk to the media. Still is.

This was not a matter of paranoia. Underground brokers have a perfect case study in the perils of talking to a reporter about the zero-day market. The case that comes up again and again is that of a well-known South African exploit broker, based in Bangkok, aka the Grugq. The Grugq just couldn't help himself. Unlike most zero-day brokers, who avoid any platform that leaves a digital trace, the Grugq is on Twitter—he has more than 100,000 followers—and in 2012, he made the fatal mistake of openly discussing his business with a reporter.

He would later claim that he thought he was speaking off the record, but he was also happy to pose for a photo next to a large bag of cash. As soon as the story appeared in *Forbes* magazine, the Grugq immediately became persona non grata. He received a visit from Thai security officials. Governments stopped buying from him. His brokerage business, several people close to him told me, plummeted by more than half.

Nobody was looking to follow in the Grugq's footsteps, to forsake their fortune and reputation for fame, or transparency.

IT WASN'T UNTIL the fall of 2015—after two years of trying and failing to convince someone to talk—that one of the market's first brokers agreed, against his better judgment, to sit down with me face-to-face.

That October I flew to Dulles to meet with a man I'll have to call Jimmy Sabien. Twelve years earlier, Sabien had been the first man bold enough to call up Watters to try to convince him to sell zero-days to him on the sly, instead of handing them over to iDefense's customers and the big tech companies. Sabien and I arranged to meet at a Mexican restaurant in Ballston—just a few miles from several of his former customers—and over enchiladas, he relayed what many hackers and government agencies have long known, but kept quiet.

Sabien had been out of the market for years, but in the late 1990s he was recruited to one of three boutique government contractors that first started buying zero-day bugs on behalf of U.S. intelligence agencies. Back then the deals were not yet classified, which meant he wasn't breaking any laws by talking to me. Even so, Sabien now worked in a different capacity with many of the same government clients and security researchers that he did back then, which is why he would only meet on the condition that I not use his real name.

Sabien was the first to pitch Watters on what would have been a hugely profitable side business. "You couldn't dream up better profit margins," he told me. He was willing to pay iDefense $150,000 to turn over some of the very same exploits Watters's team had paid hackers $400 for. When that pitch didn't work, Sabien pivoted to patriotism. "You'll be doing your country a service," Sabien recalled telling Watters.

The two spoke on and off for months, until Watters made it clear the answer was a hard no. Twelve years later, Sabien still shook his head at the slight. "It could have been huge."

Before he helped pioneer the exploit business, Sabien was in the military, protecting and managing military computer networks around the world, and he looked and played the part. Tall, broad-shouldered, hair cut high-and-tight, he was jovial, fast-talking, and punctual. On the rainy day we were scheduled to meet, I showed up a few minutes late to find him schmoozing with a man he introduced as a former client at a government agency. Sabien introduced me as a reporter. The man gave Sabien a suspicious glance that said *What the hell are you doing talking to her?* As Sabien and I made our way to our table, the man yelled, "Just remember, 'Congressman, I have no recollection of said programs.'"

Sabien glanced at me and laughed nervously. We both knew talking to me could land him in hot water.

Protecting computer networks for the military, Sabien told me, had left him intimately acquainted with technology's flaws. In the military, secure communications mean the difference between life and death, but the big technology companies didn't seem to grasp that. "People were clearly designing these systems for functionality, not for security. They weren't thinking about how they could be manipulated."

Manipulating computer systems was pretty much all Sabien thought about when he left the military for the private sector. He joined a boutique contractor just up the road from where we sat, where he managed a twenty-five-person team that researched and developed cyberweapons and intrusion tools for military and intelligence agencies and, to a lesser extent, law enforcement.

Sabien quickly learned that the sophisticated cyberweapons they were building were useless without a way to deploy them. Access to a target's computer system was critical. "You could be the best jewelry thief in the world, but unless you know how to bypass the Bulgari store alarm system, it doesn't get you anywhere," he told me.

"Access," he said, "is king."

In the mid-1990s Sabien's team started trafficking in digital access, searching for bugs and exploiting them for customers. The bulk of his company's revenues—more than 80 percent—came from the Pentagon and intelligence agencies, with the remainder from law enforcement and other U.S. government agencies. The goal was to deliver their government customers secret tried-and-tested ways into every system used by the adversary, be it nation-states, terrorists, or low-level criminals.

Some of their work was opportunistic. If they could find a bug in a widely used Microsoft Windows system, they'd develop it into an exploit and sell it to as many customers as possible. But much of their work was targeted. Government agencies would come to Sabien's team looking for a way to monitor workers at the Russian embassy in Kyiv or the Pakistani Consulate in Jalalabad. In those cases, Sabien's team would have to do reconnaissance to decipher which computers the targets were running and what kind of operating environment they were working with, and note every application connected to it. Then they would have to find a way inside.

There was always a way. Alas, humans are not perfect. So long as humans were responsible for writing computer code and designing, building, and configuring machines, Sabien's team knew there would always be mistakes. Finding those flaws was only half the battle. The other half was writing and honing the exploit code that would give government agencies a reliable, clean beachhead.

And Sabien's government clients didn't just want a way in; they wanted a way to crawl through their adversary's systems undetected; a way to plant invisible backdoors that kept them in even after their breach was discovered; and ways to pull the adversary's data back to their command-and-control servers without tripping up any alarms.

"They wanted the entire kill chain—a way in, a way to beacon out to their command-and-control server, an exfiltration capability, an obfuscation capability," Sabien told me, using military-speak. "It makes sense when you think of the Special Forces and SEAL Team Six. They have snipers, sweepers, exfil specialists, and people who break down the doors."

This is what Sabien's team provided in the digital realm. But their work wasn't about shock-and-awe. Quite the opposite: every step had to be stealthy and invisible. The harder it was for their adversary to discover their code, and their presence, the better. The trifecta was a string of zero-day exploits that offered reliability, invisibility, and persistence. Rarely did you get all three. But when you did: "*Ka-ching!*" Sabien said.

When I asked Sabien to discuss specific exploits, he recalled some with the same affection with which others might recall their first love. His favorite was a particularly stubborn zero-day exploit in a video memory card. The memory card ran on the computer's firmware—the software closest to the bare metal of the machine—making the exploit nearly impossible to find and harder still to eradicate. Even if someone wiped the machine clean and restored all of its software, the video memory card exploit stuck. The only way for a target to be sure they had rid their systems of spies was to throw their computer in the dumpster. "That exploit was the best," Sabien recalled with a twinkle.

The first thing spies would do after successfully breaking into a machine, Sabien told me, was listen in for other spies. If they found evidence that the infected machine was beaconing out to another command-and-control center, they would scrape whatever the others were catching. "And if they were really selfish," Sabien told me, "they'd patch the system and kick everyone else out."

It wasn't abnormal, Sabien said, to find multiple nation-states listening in on the same machine—especially in the case of high-profile targets, diplomats, government shell companies, arms dealers, and the like. There was one well-known exploit in HP printers that for years, Sabien told me, was utilized by "government agencies all over the world." The exploit allowed anyone with knowledge of its existence to scrape any files that passed through HP's printers and offered spies a foothold in their target's network, where IT administrators would least suspect.

The day the printer exploit was discovered and patched by Hewlett-Packard, Sabien said, "I just remember thinking to myself, 'A lot of people are having a very bad day.'"

THE SHORT LIST of government agencies looking to acquire their own zero-day arsenals did not stay short for long.

The NSA boasted the largest and brightest army of cyberwarriors of the intelligence community, and in those early days, the agency didn't require much outside help.

But in the mid-1990s, as the masses took to the web and email, sharing a fine-grain record of their daily lives, relationships, inner thoughts, and deepest secrets, a growing number of intelligence agencies began to worry that they were not prepared to exploit the rapid adoption of the internet and the intelligence gold mine it presented. In late 1995 the CIA created a special working group to assess the agency's readiness to utilize the web as an intelligence tool. The group's principal finding was that the CIA was woefully underprepared for this brave new world. The same was true at the other intelligence agencies, who were even further behind, with significantly smaller budgets and few people on staff with the skills to find zero-days and code them into reliable exploits. A growing number of agencies started looking to buy their way into these capabilities.

Amassing those stockpiles became a competitive enterprise. Cyber was one of the only bright spots in an otherwise dismal decade for defense spending. In the 1990s, the Pentagon's military budgets were chopped by a third, with cyber being the one exception. Congress continued to approve vague "cybersecurity" budgets, without much grasp of how dollars funneled into offense or defense or even what cyber conflict necessarily entailed. Policymakers' thinking on cyber conflict was, as former commander of U.S. Strategic Command James Ellis put it, "like the Rio Grande, a mile wide and an inch deep." But inside each agency, officials were learning that the best zero-days netted the best intelligence, which in turn translated to bigger cyber budgets down the road.

And there was Sabien, right in the middle of all of it.

Sabien told me his team couldn't churn out exploits fast enough. Different agencies all wanted ways into the same systems, which played well from a bottom-line perspective, but not so much from the American taxpayer's. His company was selling the same zero-day exploits two, three, four, times over to different agencies. The overlap and waste, Sabien recalls, became too much to stomach.

The government has a name for this problem—duplication—and it wastes millions in taxpayer dollars every year. But duplication is even worse in the digital world, where contracts for bug-and-exploits are sealed up in nondisclosure agreements and often classified. The prevailing wisdom among intelligence agencies is that once word of an exploit gets out, it's only a matter of time before it gets patched, and its value plummets. So little gets shared across agencies, let alone discussed.

"Each of the agencies wants the win," Sabien told me. "They want to up their budgets so they can do the more advanced offensive operations."

The duplication waste became so problematic that Sabien finally called up his contacts at four of the intel agencies he sold to. "I said, listen, as a contractor I'm not supposed to talk about this, but as an American taxpayer I'm going to need you guys to go to lunch. You have common interests that you should probably discuss."

The overlap and waste was only exacerbated after 9/11. As defense and intelligence spending ballooned by more than 50 percent over the next five years, there was a virtual stampede from the Pentagon and intelligence community to Beltway contractors that specialized in digital espionage.

But bugs and exploits took time to find and develop, and Sabien came to the same conclusions as Watters's team at iDefense. His twenty-five-man team could scour for bugs and develop and test exploits nine-to-five, but it would be far easier to outsource that work to the thousands of hackers around the world who spent their days and nights glued to their computer screens.

"We knew we couldn't find them all, but we also knew there was a low barrier to entry," Sabien recalled. "Anyone with two thousand dollars to buy a Dell is in the game."

And so the underground market for zero-day bugs began, the same one that would quietly undermine iDefense's business, and eventually consume it—and the rest of us.

SABIEN'S STORIES FROM those early days landed like a spy novel, complete with cloak-and-dagger meetings, bags of cash, and murky middlemen—only in this case none of it was literary or imagined. It all checked out.

In the beginning, Sabien's team started scanning BugTraq, taking the bug discoveries that hackers were volunteering for free and tweaking them slightly before baking them into their own exploit. But eventually they started reaching out to hackers on the forums directly, inquiring whether they'd be willing to develop something unique for Sabien's customers and never tell a soul.

The money provided plenty of incentive. In the mid-1990s, government agencies paid contractors roughly $1 million for a set of ten zero-day exploits. Sabien's team would budget half that to buy bugs and then develop them into exploits themselves. A decent bug in a widely used system like Windows might fetch $50,000; a bug in an obscure system used by a key adversary might fetch twice as much. A bug that allowed the government's spies to burrow deep into an adversary's system, undetected, and stay awhile? That might net a hacker $150,000.

Sabien's team avoided idealists and whiners. And because there were no rules to this market, the bulk of their suppliers were hackers in Eastern Europe.

"With the breakup of the Soviet Union, you had a lot of people with skills, without jobs," Sabien explained. But the most talented hackers, he told me, were based in Israel, many of them veterans of Israel's Unit 8200. I asked Sabien how old his youngest supplier was, and he recalled a transaction with a sixteen-year-old kid in Israel.

It was a secretive business, and mind-blowingly convoluted. Sabien's team couldn't exactly call up hackers, ask them to send their exploit by email, and mail them back a check. Bugs-and-exploits had to be carefully tested and retested across multiple computers and environments. In some cases hackers could get away with demonstrating their exploits over video. But in most cases, deals had to be done face-to-face, often in hotel rooms at hacker conventions. Sabien's team would have to understand the exploit well enough to take it back and re-create it for their government clients, usually before anyone got paid. If the exploit didn't work reliably, nobody got paid.

Increasingly, Sabien's team had to rely on murky middlemen. For years, he said, his employer would dispatch an Israeli middleman with duffel bags stuffed full of half a million dollars in cash to buy zero-day bugs from hackers in Eastern Europe. Again, these weren't weapons. They were gaping security holes that could be exploited to break into hardware and software, and the American taxpayer was being asked to bankroll the entire supply chain.

It was not exactly a well-oiled operation. Every step in this insanely complex deal-making structure was filled with shady characters and relied on *omertà*. Every interaction necessitated a startling amount of trust: government clients had to trust their cyberarms dealers to deliver a zero-day that would work when they needed it to. Contractors had to trust hackers not to blow the exploit by using it themselves, or reselling it. Hackers had to trust that contractors would pay them after their demonstration, not just take what they'd gleaned and develop their own variation.

Each deal was shrouded in secrecy—it was nearly impossible to know if the six-figure bug you purchased from an Israeli teenager hadn't been resold to your worst enemies. And then there was the issue of payment. This was pre-Bitcoin. Some payments were done via Western Union, but the bulk of exploits were paid in cold, hard cash—a necessary byproduct, Sabien told me, of a business that couldn't afford to leave a trace. You couldn't dream up a less efficient market if you tried.

Which is why, in 2003, Sabien began to take note of a little company called iDefense that began openly paying hackers for their bugs. When Sabien first called up Watters and told him he was willing to pay six figures for the very same exploits Watters's team had acquired for three, the first words out of Watters's mouth were "Why on earth are you paying this much?"

To a businessman like Watters, who was busy trying to push the market out into the open, what the contractors were doing was idiotic, dangerous even.

"Nobody wanted to talk openly about what they were doing," Watters recalled. "There was this whole air of mystery to it. But the darker the market, the less efficient it is. The more open the market, the more it matures, the more buyers are in charge. Instead they chose to work out of Pandora's box, and the prices just kept going up."

As Watters started getting calls from more and more contractors, the offers only continued to climb. And it wasn't just U.S. government agencies who were buying anymore; there was new demand from other governments, and government front companies, all of whom kept driving up the price of exploits and making it difficult for iDefense to compete. As the market started to spread, what really troubled Watters wasn't the effect the market would have on iDefense; it was the increasing potential for an all-out cyberwar.

"It's like having cyber nukes in an unregulated market that can be bought and sold anywhere in the world without discretion," he told me.

THE CLIENTELE WAS shifting along with the earth beneath their feet. The certainty of the Cold War era—with its chilling equilibrium and clarity—was giving way to a vast uncharted digital wilderness where adversaries were no longer defined by national borders but by cultures and religion. You weren't quite sure where they would pop up, or when.

In this new world order, enemies were seemingly everywhere. In the United States, intelligence agencies began relying on cyberespionage to collect as much data about as many people as possible. It also began developing a cyberweapons arsenal in the event it would have to disrupt enemies' networks, or infrastructure, one day. And an entire army of Beltway contractors was more than willing to supply the digital weaponry, the reconnaissance tools, and all the requisite parts.

In the beginning, Sabien told me, he was one of only three contractors—that he knew of—dealing in the cyberespionage and weapons trade. But as more government agencies and foreign nations started their own cyber offense programs, the cost of exploits—and the number of contractors eager to traffic in them—began to double every year.

The big defense contractors—Lockheed Martin, Raytheon, BAE Systems, Northrop Grumman, Boeing—couldn't hire cyber specialists fast enough. They began poaching from the intel agencies and acquiring boutique contractors like Sabien's.

By the time Sabien had agreed to meet with me, he'd been out of the market for more than a decade, but the market was hard to avoid these days.

"In the nineties, there was just a small community of people working on exploits and selling them. These days it's so commoditized. It's blown up. Now"—he swirled his finger in a wide circle in the air around the Beltway—"we're surrounded. There are more than a hundred contractors in this business, probably only a dozen that know what they're doing."

The Drug Enforcement Agency, the U.S. Air Force and Navy, and agencies most of us have never heard of all had their own reasons for acquiring zero-days. Ever heard of the Missile Defense Agency? Me neither, until a former Pentagon analyst told me that the agency—which is responsible for defending the nation from missile attacks—acquires zero-day exploits. "I couldn't even tell you if there's anyone at that agency that knows how to use them," he said.

The market's spread to U.S. agencies didn't bother Sabien. It was its spread abroad that had him rattled.

"Everyone has their enemies," he told me. For the first time since we'd sat down, his countenance was no longer jovial. "Even countries you would never suspect are stockpiling exploits for a rainy day. Most do it to protect themselves."

"But one day soon," he added as we got up to leave, "they know they might have to reach out and touch someone."

Before we parted ways, Sabien told me he had something he wanted to show me. He passed me his phone. On the screen was a quote attributed to Nathaniel Borenstein, who I vaguely recalled as one of two men who invented the email attachment, the invention so many nation-states now used to deliver their spyware.

"The most likely way for the world to be destroyed," it read, "most experts agree, is by accident. That's where we come in; we're computer professionals. We cause accidents."

I passed the phone back to Sabien.

"Keep going," he told me. "You're onto something. This will not end well."

And with that, he was gone.

CHAPTER 5

Zero-Day Charlie

St. Louis, Missouri

I f his old bosses from the NSA had just showed up with the $50,000 in cash that afternoon—as Charlie expected they would—the government might have been able to keep the fact of his bug, of the entire underground bug-and-exploit market, its dirty little secret.

Indeed, when Charlie Miller bid goodbye to his wife and rode the I-170 toward the St. Louis airport hotel that morning in 2007, he was sure that by the time he returned home, they would have the cash they needed for their kitchen renovation. Why else would the NSA insist on coming all the way to St. Louis to discuss his bug in person?

Charlie pulled off the highway and into the front entrance of St. Louis's Renaissance Airport Hotel, an imposing tall, black-mirrored structure that may as well have been the miniature version of his old stomping grounds at Fort Meade.

But the suits from the NSA had something different in mind.

IT HAD ONLY been a year since Charlie left the agency. The decision had not been an easy one. The NSA had hired him in 2001 as a promising young math PhD to join the ranks of the agency's top cryptographers. But once he'd finished the agency's three-year training program, he decided that hacking—not math—was what he wanted to do with his life. He was an obsessive tinkerer. Anything you put before him—cars, computers, phones—he pulled apart, simply because he relished the puzzle of taking something built for one purpose and bending it to his will.

Charlie would go on to become one of the NSA's top "global network exploitation and vulnerability analysts," a fancy title that meant he spent most of his time hunting for vulnerabilities that granted the agency entry into the most sensitive networks in the world. "You can do things at the agency you can't do anywhere else," Charlie told me.

Outside the agency, he had become something of a celebrity on the hacker circuit. His exploits made headlines and won hacking contests. He'd especially made a name for himself poking holes in Apple's devices. Apple famously took a black-box approach to its security. Its security defenses were considered top-secret. To this day, Apple employees are forbidden from even disclosing the number of people who work on security. Its Cupertino headquarters is surrounded by vertical slabs—the same ones Trump chose for his border wall—in part because they can't easily be scaled.

Apple has always insisted that its strict vetting procedures keep malware, spyware, and spam out of its iTunes store. Charlie famously trashed that myth when he submitted a fake stock-ticker app with a glaring security hole that allowed it to infect other apps on the iPhone, just to see if Apple would notice. Apple's vetters missed the hole, and when Apple learned from news articles that Charlie's app was a Trojan horse, they blacklisted him. The episode earned Charlie infamy in hacker circles and a nickname, Zero-Day Charlie. He relished it.

In February 2016 I flew to St. Louis to see Charlie in person. (He had to postpone our meeting twice because he was too busy appearing on *CSI: Cyber*.) We'd first met years earlier at a rooftop hacker party in Vegas, and he was just how I'd remembered him: tall, skinny, with sharp features and serious eyes, his mouth dripping with sarcasm. That first night we met in Vegas, he was dressed in an all-white hip-hop tracksuit. It was hard envisioning him as a math PhD.

The last time Charlie and I had spoken by phone was after he and another researcher discovered a zero-day exploit in the Jeep Cherokee that allowed them to seize control of the steering wheel, disable the brakes, screw with the headlights, indicators, wipers, and radio and even cut the engine from a remote computer thousands of miles away. Eight months later, the automaker was still dealing with the fallout.

On this ice-cold day, I met Charlie at his "office," the basement of his home in the St. Louis suburbs, where he parked himself day and night in front of multiple computer monitors beside his pet hedgehog, Hacker. You had to watch your step to avoid the mangled car parts strewn over the floor. At the time Charlie was working for Uber, doing security on what the company hoped would one day be its driverless car. But he spent most of his free time these days trying to one-up his latest Jeep hack.

When I landed in St. Louis and picked up my rental car, his exploit was still fresh in my mind. "Do not, under any circumstance, give me a Jeep," I told the Budget salesman. When I recounted this to Charlie, he asked where I was staying. When I told him, he delightedly explained that my hotel elevator ran on the same vulnerable platform as the hacked Jeep. I started taking the stairs.

Charlie had left the agency in 2005, and he still couldn't discuss his work there apart from vague mentions of computer-network exploitation. Not even his wife knew what kind of work he'd done at Fort Meade. "We would go out with coworkers," he said, "and nobody would have any idea what to talk about in front of her because we couldn't talk about work."

I mentioned that I was somewhat familiar with the NSA's exploitation programs from my stint in the Snowden closet. "You probably came across some of my work!" he told me.

In the decade since he left the agency, Charlie's career had taken some intriguing twists and turns. He'd made enemies at some of the country's top technology companies—Apple and Google—and been hired by others— Twitter and Uber. The NSA was never a big fan of those who departed its ranks preretirement. The agency's graybeards considered anyone who left to be traitors. If, after leaving, you discussed anything even remotely related to the work you did at the agency, you were Benedict Arnold.

But when Charlie left the NSA, he had good reason. His wife was pregnant with their second child, and she'd accepted a job teaching anthropology at Washington University in St. Louis, where their families still lived. Charlie took a gig doing security for a broker-trader firm, which basically entailed reminding people to constantly change their passwords. Compared to breaking into foreign governments for a living, it was mind-numbing work. But at night he would sit in the dark, glued to his computer screen, hunting for zero-days beside a giant poster from the 1995 Angelina Jolie film *Hackers*.

After his second was born, paternity leave made it possible for him to turn his moonlighting into a full-time gig. In between diaper changes and bottle feeds, he would dig deep into the web for bugs and unforeseen ways to bend them to his will. "That's why the Europeans are so good at writing exploits," he says. "After babies, European parents get like a year to hack."

Hunting for bugs and writing exploits had become an addiction, the digital equivalent of smoking crack. He could disassemble programs and applications for hours without so much as a glance at the clock. Exploits were, for Charlie, just like a mathematical proof. Once he had turned a bug into an exploit, there was no room for debate about the bug's severity.

What Charlie found late one evening in 2006 was the type of bug most could spend a lifetime searching for and never find—the kind of zero-day that could have allowed him to run amok through NASA's computer systems or hijack the password to a Russian oligarch's trading account.

It was an exploitable bug in the Linux program called Samba that allowed Charlie to take over a target's machine unnoticed. And from the moment he discovered it, he knew he'd struck gold. The way he saw it—the way any high-level hacker would see it—he had five options. One, he could tip off vendors to the bug, and hope he didn't get threatened or sued in the process. Two, he could keep the zero-day to himself and use his newfound invisible powers for his own purposes. Three, he could take his zero-day to the press, or dump it on a mailing list like BugTraq, get street cred for his discovery, and shame the vendors into patching it.

But none of those options would get Charlie anything in the way of a financial reward. And he was not about to turn over this bug for free. This was worth something.

He briefly considered selling the bug to iDefense, which would give him credit for the discovery and see to it that the bug got patched. But Charlie knew this bug was worth more than the crummy couple grand iDefense was willing to spend. "I knew the price would be much higher with other sellers," he told me.

I asked him just how he knew.

"There was a small community of us who just knew," is all he would offer.

That left option five: sell his zero-day on the underground market, directly to a government agency, or indirectly through a broker. The problem was, he would not be able to talk about the bug, nor get credit for its discovery. It also meant he might have to worry about how it was being used.

But Charlie didn't want to just sell his bug—he wanted to go public about the market.

Ever since he'd ditched cryptography for hacking, he'd been struck by the dichotomy between how the government and the private sector treated hackers. At the NSA, "network exploitation" was considered a prized skill that required infinite patience, creativity, and a keen knowledge, honed over years, of computers and the networks they operated in. Outside the NSA, hackers were treated like low-level criminals. Most hackers didn't realize there was legitimate value—in some cases six-figure value—in what they were doing. They were too focused on avoiding lawsuits. "I just thought it was time people should know there was an underground market."

And then there were the problems with the market itself. If efficient markets require high levels of transparency and free flows of information, then the zero-day market was just about the least efficient model you could imagine.

Sellers were sworn to never speak a word about their zero-day sale. Without data, it was impossible to know whether they had achieved a fair price. And it was often impossible for sellers to find buyers without cold-calling multiple interested parties. If they described their zero-day or handed it over for evaluation, a buyer might simply feign disinterest and use it anyway.

The time lag between a hacker's zero-day demo and when he got paid was brutally long. Zero-days took weeks, if not months, to vet—all the more time for the vulnerability to be found and patched. Six-figure zero-days could turn to dust in seconds, leaving sellers up a creek.

And just as with blood diamonds, there was the fairly significant issue of one's conscience. As more buyers made their way into the market—foreign governments, front companies, shady middlemen, cybercriminals—it was becoming impossible for hackers to know how zero-days would be used. Would their code be used for traditional state-to-state espionage, or to track down dissidents and activists and make their lives a living hell? There was no way to tell.

From the buyer's perspective, the underground market was equally frustrating. A government agency couldn't openly advertise that they needed a way into a Lebanese bank or an arms dealer's phone. When they did find someone advertising a way into their target's system, they could never guarantee that seller wouldn't turn around and sell it to someone less discreet. There was always the risk that another buyer would come along and blow a six-figure exploit in a sloppy operation. If the market weren't bankrolled by blissfully ignorant taxpayers the world over, it might never have been able to reach the volume, or dollar amount, it has today.

It was the academic in Charlie that decided he would use his Linux zero-day to write a white paper about the market itself. He knew nobody would take him seriously unless he actually stepped into the market. "To have credibility in this space, you had to be doing what you said you were doing."

And so Charlie began shopping his zero-day around.

But first, he had to run his discovery by his former employer. The NSA has a strict prepublication review policy: anything former employees publish, for as long as they shall live, must be run by the agency's minders. Until an NSA review board has cleared them for publication, anything a former NSA employee wishes to publish will remain classified.

And in Charlie's case, the minders weren't having it. The agency denied his request to make his zero-day public.

Not one to take no for an answer, Charlie appealed. There was nothing about his zero-day that was classified. He had found it as a private citizen. Any number of other high-level hackers could have found it.

It took another nine months, but the agency ultimately ruled in his favor. He was free to do whatever he wanted with his zero-day.

HIS FIRST STOP was a buddy at a now-extinct company called Transversal Technologies, who maintained a Rolodex of government contacts at relevant intelligence agencies. For a 10 percent cut, his friend agreed to shop Charlie's zero-day around to various U.S. agencies.

One government agency—Charlie never did disclose which—came back quickly with an offer of $10,000. It was a nice premium over what iDefense and others were offering, but a lowball compared to what some agencies were rumored to have paid. When a second government agency expressed interest, Charlie pulled a number out of the air and asked for $80,000—nearly twice his annual salary at the NSA. He assumed that would be the start of some larger negotiation, but instead the agency agreed, "Too quickly," he noted, "which likely means I had probably not asked for enough."

The deal came with one caveat, though: the agency was only interested in Charlie's zero-day exploit if it worked with a particular flavor of Linux used by one of its targets. As for who that target was, Charlie would never know. There was the moral trade-off. Once his zero-day was in the agency's hands, they could use it to spy on whomever they chose. In the United States, the likeliest targets were terrorists, foreign adversaries, or drug cartels, but there were never any guarantees that very same zero-day wouldn't come back to haunt you.

Charlie handed over his zero-day for evaluation. For a month he was racked with anxiety as he waited to hear back. He became convinced that somebody else would find his bug, or worse, that the people he had given it to would claim credit for his work.

Five weeks later, the agency came back with bad news. His exploit did not work on the system it needed to pierce—but they were willing to purchase it at a cheaper price, $50,000. Charlie agreed, and two weeks later he received his check. As part of the arrangement, the agency would license his zero-day—and mandate his silence—for two years.

Charlie used that time to dig further into the market, compiling what little data he could find for eventual publication. To round out his own experience, he agreed to help another researcher sell an exploit in an older version of Microsoft PowerPoint. But just when they had settled on a $12,000 deal with a foreign seller, the exploit's value plummeted to zero; Microsoft had patched the underlying bug. It was proof, Charlie believed, of the market's underlying inefficiencies and failures.

He kept going. He spent much of his free time over the course of those two years trying to learn as much as he possibly could about exploit pricing. What he found was that the pricing data was all over the place. One government official told him some agencies were willing to pay as much as $250,000 for a single exploit. A broker told him the going rate for a single reliable exploit was more like $125,000. Some hackers said they'd been offered between $60,000 and $120,000 for a way into the Internet Explorer browser. There seemed no rhyme or reason to it all. The way Charlie saw it, hackers were getting screwed left and right. The only way to restore sanity to the exploit market was to crack it wide open.

In 2007, as his two-year agreement to stay mum about his zero-day expired, Charlie began putting the finishing touches on his white paper—academically titled "The Legitimate Vulnerability Market: Inside the Secretive World of Zero-Day Exploit Sales." It was then that he received the call from Fort Meade.

THE VOICE ON the other end of the line did not offer much in the way of details. They needed a face-to-face. The agency would be sending some of its people out to St. Louis to meet with him directly.

As he hung up, Charlie ran through all the possible reasons the NSA's higher-ups would fly all the way to St. Louis to meet in person. There seemed only one plausible explanation: perhaps, having failed to keep his exploit a secret through the agency's bureaucratic channels, the NSA was now willing to buy his silence and purchase it outright. *Ka-ching!* "I was sure they were going to show up with a big bag of cash," Charlie recalled.

That evening he told his wife that their dream kitchen would soon be a reality.

A couple days later, Charlie drove to the Renaissance St. Louis Airport Hotel. As he walked through the lobby and rode the elevator to the top floor, he recalled the mythology, the stories recounted time and again, about hackers' rendezvous in hotels with government officials, just like this, who had demonstrated their zero-days and left with cold hard cash. The stories always sounded too cinematic, too cloak-and-dagger, to be true, but as Charlie rode the elevator to the twelfth floor, he couldn't help but grin. He was about to live the tale.

As Charlie exited the elevator and made his way to the hotel conference room, something seemed off. The room did not strike Charlie as the set of a clandestine rendezvous. Light poured in through floor-to-ceiling windows that overlooked an airstrip twelve floors below. Cheap watercolors adorned the wall, and the floors were covered with bright-red wall-to-wall carpeting. Waiting to greet him were four suits from the NSA. Noticeably absent was a duffel bag of cash.

Charlie assumed he was in for a long financial negotiation. Instead the meeting took less than fifteen minutes. The suits, it quickly became evident, were not there to buy his exploit. They were here to keep him quiet. "Think about your country," one of the agents told Charlie. "You can't be talking about this."

They wanted Charlie to abandon his white paper, to never speak a word of his zero-day, or its sale, or the mere fact of the marketplace to a soul.

Charlie half listened as he watched planes take off and touch down outside. He couldn't believe the agency had dispatched four of its brass all the way to St. Louis to make the do-it-for-your-country pitch. He wasn't interested in their moral probity, and he'd already fulfilled his patriotic duty. At a certain point, he tuned them out altogether.

It was never about the money. The first time I asked Charlie why he sold his exploit to the government was over beers at a dive bar in San Francisco's Tenderloin neighborhood. Back then he'd answered sarcastically: "For the green! Money money money!" But it was hardly convincing, and we both knew it. The more I pressed, the more evident it was that it was about the principle, about getting paid fairly for the work.

For too long vendors had been free-riding off the work of hackers who had taken the time to discover and notify them of gaping security vulnerabilities in their products, and often made a point of threatening the hackers in the process. And the payouts iDefense and others were willing to pay were, in Charlie's mind, a joke.

It was time for hackers' work to be taken seriously and the only way Charlie could foresee that happening was to make sure they got paid. If that meant exposing the zero-day market, so be it.

He didn't even bother explaining this to the suits. They would never understand.

When it became clear that no cash would be exchanged that day, Charlie got up to leave. "Sorry you had to fly all the way out here for nothing," he told them. "But the answer is no. I'm going public."

The suits looked at Charlie with a mix of exasperation and contempt.

"Enjoy the arch," he told them as he headed for the door.

*   *   *

A FEW MONTHS later, Charlie Miller stepped to a lectern at Carnegie Mellon beneath a giant projected image of a $50,000 check made out in his name.

The account holder's information—name, address, bank, even signature—had been redacted to protect the unnamed agency Charlie had sold his exploit to. But over the course of half an hour, Charlie relayed to an audience of economists and academics how he had gone about selling his zero-day to Uncle Sam.

For the first time, the secret was out: the U.S. government was willing to pay hackers—quite a bit, as it turned out—to turn over vulnerabilities in the products and leave their customers—including American citizens—vulnerable in the process. And the government was doing so with money from taxpayers, the very people the government was charged with protecting.

Inside the lecture room, the academics' reaction to Charlie's presentation that day was muted in comparison to the outrage Charlie's white paper generated 250 miles southeast. At Fort Meade, and all around the Beltway, government officials were apoplectic. Not only had Charlie ignored the government's plea for *omertà* but his white paper included a list of the payments—as high as $250,000—that government agencies were giving hackers for their exploits. Not only would this practice be hard to defend, but Charlie's paper was sure to drive up prices for zero-days.

From Redmond to Silicon Valley, technology executives at Microsoft, Adobe, Google, Oracle, and Cisco combed through Charlie's paper with a mix of alarm and agitation. He had only confirmed what executives had long suspected: their own government was perfectly willing to throw them, and their customers, under the bus in the name of national security.

The revelation had all the makings of a public relations nightmare. Executives shuddered to think of the potential fallout for their market share. What would foreign customers think when they learned that the U.S. government was actively paying hackers to spy via their products? And then there were the hackers. No doubt they would use Charlie's white paper to start demanding bounties for their bugs. How could they compete with the five- and six-figure payouts the government was offering? Considering how much market share they stood to lose as customers stopped trusting the security of their products, how could they not?

In the hacking community, Charlie's paper was alternately celebrated and condemned. Some cast him as an unethical researcher who, by selling his zero-day to the government and waiting so long to come forward with it, had put millions of Linux users at risk. Some pushed to have his cybersecurity license stripped.

But some hackers believed Charlie had done yeoman's work, even if it had netted him $50,000 in the process. For nearly two decades they had been treated as crooks and lowlifes by companies they chose to help for free. Now they knew there was a lucrative government market for the very work vendors had shamed them for doing.

THAT COULD HAVE been the end of Charlie's story. He had opened the proverbial kimono; the market was now exposed. He told himself the bug he'd sold Uncle Sam would be the last bug he ever sold.

But that's not how he would be remembered. One month after unleashing his white paper on the world, Charlie made an even bigger name for himself with the first remote hack of an iPhone.

The conventional wisdom had always been that the iPhone—with its sleek design and closely held code—was more secure than the alternatives. But Charlie blew a hole right through that theory. He demonstrated before an audience of hundreds how easily he could remotely control anyone's iPhone simply by steering their browser to a malicious website he created.

The iPhone exploit—had Charlie sold it—would have easily netted a six-figure payout on the underground market. But he wasn't interested in the cash. Charlie was in it for the intellectual curiosity and the street cred. This time, he'd tipped off Apple to what he'd found, and helped the company's engineers come up with a patch. Another eight months later, he did it again, hacking Apple's MacBook Air in less than two minutes. He changed his Twitter bio to "I'm that Apple oday guy."

When Google unveiled a beta-version of its Android operating system that year, Charlie couldn't help himself. He broke it almost immediately with an exploit that made it possible to remotely capture an Android user's every keystroke, text, password, email, anything they did on their phone.

As with his Apple exploits before, Charlie went straight to Android's developers with his zero-day and offered to help them patch it. Now that—as he genuinely believed—his white paper had changed the conversation between hackers and vendors, surely Google would welcome his discovery.

And Google did appreciate his efforts, as far as Charlie could tell—at least, until the day his boss at the security consultancy where he worked asked him, "How's that Google exploit fix coming?"

"Great!" Charlie replied. "We're working together on a fix."

"I know," his boss said. "I've been BCC'd on all the emails."

Unbeknownst to Charlie, Google had gone behind his back. Executives had called his boss and informed him that his employee was illegally breaking into Google's newfangled mobile system. They were now BCC'ing Charlie's boss on every email they traded with him.

Had Charlie's boss not been familiar with the brutal dynamic between hackers and vendors, he might have fired Charlie. Instead, he took Charlie's side, telling Google that his employee had done nothing wrong; that in fact he was doing the company a favor by not allowing bad actors to find the Android hole first.

For Charlie, the episode was a dark one. His white paper had changed nothing. The big vendors—Google, for Christ's sake—still had it wrong. They would rather bury their head in the sand and threaten hackers than work with them to secure their products.

Charlie cut off communications with Google and took his exploit to the *New York Times*, which wrote about his discovery. And he vowed never to give Google, or anyone else for that matter, another free bug. Google's Android executives had just started a movement, even if they didn't know it yet.

"YOU WANT TO hear something fucked up?"

It was late. Charlie was drunk at a dive bar in Manhattan with two other security researchers, Dino Dai Zovi and Alexander Sotirov. Between the three of them, they had hacked Apple, Microsoft, and the biggest names in cybersecurity.

When Charlie relayed his interaction with Google and the lengths Android executives had gone to try to get him fired, Dai Zovi and Sotirov nodded in between long sips of their beers. The tale was a familiar one, but the duplicity of Charlie's story, combined with the alcohol, propelled them into new territory. "That's fucked up, man," Dai Zovi told Charlie.

The most moral option for any hacker—going straight to the vendor—still generated the worst outcomes. How could that be? The vendors didn't appreciate quality assurance? The three hackers were so worked up, they barely noticed a prostitute approaching their table.

Particularly in the wake of Charlie's foray into the underground bug-and-exploit market, it offended him to think that he could have just as easily sold the bug for five, six figures underground. And yet here he was, being punished for giving Google a critical bug for free.

The three talked late into the night. The vendors needed to be taught a lesson. Together, they agreed to fight back, and they gave their campaign a name: No More Free Bugs.

THAT MARCH, 2009, the three took the stage before hundreds of hackers at a security conference in Vancouver.

Clad in all black, Dai Zovi and Sotirov held up a giant sign that read NO MORE FREE BUGS. It might as well have been translated as NO MORE MISTER NICE GUY.

Earlier that day Charlie had won the conference's hacking contest for the second year, breaking into a MacBook Pro via a Safari browser exploit—once again. He netted himself a free computer and a nice $5,000 prize in the process. But this time—true to his new campaign—Charlie hadn't given Apple a courtesy call beforehand. He would let Apple reckon with the fallout. And now he took the stage to make the case that, starting now, others should do the same.

"This is it," Charlie told the crowd. "From now on, stop giving away your bugs for free. We do all this work, and all we get is threats and intimidation."

"This is the moment," he said. "Stop. *No more free bugs!*"

Hackers aren't exactly known for their enthusiasm, but as Charlie spoke, they rose to their feet, cheered, and applauded. "NO–MORE–FREE–BUGS," a few yelled. Their chants began to trickle out online. Several took to Twitter to tweet "No more free bugs!" #NMFB started trending as their campaign gained momentum.

At the back of the conference hall, the government men—khaki-clad, shirts tucked, hair high and tellingly tight—were not smiling. At those back tables, nobody stood. Nobody clapped. Their mouths stayed tight. One gave his coworker at the agency a wink. They'd been buying hackers' exploits for years. With the vendors out of the picture, there'd be more hackers selling on the underground market. Fewer patches. Unfettered espionage.

But one man in the back of the room shook his head. He had been at the NSA for years. Now he scanned the foreign faces in the room—French, Chinese, Russian, Korean, Algerian, Argentine hackers. Who would these frustrated hackers sell their bugs to? Not all these exploits would go to Uncle Sam.

How would those exploits be used?

This, the man told himself, would only end badly.

PART III

The Spies

The enemy is a very good teacher.

—THE DALAI LAMA

CHAPTER 6

Project Gunman

Moscow, Russia

A S IT TURNED out, the NSA's fixation with zero-days did not start in the
1990s. It started in the predigital era, with an enemy's attack on our
analog systems—an attack so clever it subverted the spy world order. Had it
not stayed classified for so long, we might have paused for breath before diving
headfirst into the digital abyss.

In 1983, among the darkest years of the Cold War, American embassy
workers in Moscow began to suspect that everything they said and did—even
messages that they carefully encrypted—was leaking to the Soviets.

Employees knew that they were under constant Soviet surveillance, even
in their personal lives. Their apartments were bugged, but that was the least
of it. They would often arrive home to find clothing missing from their closets
and dirty liquor glasses in their sink. But this was different. It seemed that
everything that transpired inside the embassy, even their unspoken communications, were being relayed to the Soviets. American spies became convinced
the embassy had a mole.

And were it not for a vague tip from the French, they may have never
discovered that the mole was in their machines. In 1983, the French embassy
in Moscow discovered the KGB planted bugs in its teleprinters, which had
relayed all their incoming and outgoing telegrams to the Soviets for six years.

The Italian embassy in Moscow discovered the same. French diplomats were adamant: Washington should assume the Soviets had bugged the equipment at the American embassy too.

American officials knew the bugs could be anywhere: their printers, copiers, typewriters, computers, crypto gear—anything plugged into a wall. The Soviets had proven themselves creative geniuses when it came to eavesdropping.

At the end of World War II, as Soviet-American relations fell apart, the Soviets stepped up their surveillance of their former ally. A U.S. sweep of the Moscow embassy in 1945 had turned up a staggering 120 hidden microphones, planted in the legs of newly delivered tables and chairs, plaster, everywhere. The discovery forced the Soviets to get more creative. In 1945, Soviet schoolchildren had presented the American ambassador with an elaborate hand-carving of the Great Seal of the United States. The carving hung in the ambassador's home office for seven years, until U.S. officials discovered a tiny bug—called "Golden Mouth"—hidden deep in the wood that allowed the Soviets to eavesdrop on the ambassador at will. The furniture had been switched out over time, but the bug kept its foothold in the wall. The bug survived four ambassadors, until its discovery in 1952. The ambassador at the time, George Kennan, recalled feeling "acutely conscious of the unseen presence" in his office. Back at the Oval Office, intelligence officials briefed President Reagan on their options, and they were few. Picking up and leaving the Moscow embassy was not one of them. They were already four years into the construction of a brand-new $23 million embassy in Moscow, and it was becoming an intelligence embarrassment. The Reagan administration was already spending more than twice that figure on experimental X-ray machines and specially trained staff who, day after day, were finding more and more eavesdropping equipment embedded in the concrete of the construction. The new embassy was on track to become an eight-story listening device. The likelihood that embassy workers would eventually move into the building was growing slimmer by the day.

The White House knew that their only move was to match the creativity and cleverness of the Soviets in finding the bug and replacing embassy equipment—all under the watchful gaze of the adversary. It was a long shot, but they didn't have a choice. America was unlikely to win the Cold War with the Soviets anticipating their every move.

And so, in February 1984, President Reagan personally approved what would come to be known as Project Gunman, a classified six-month NSA effort to remove every single piece of electrical equipment from the U.S. embassy in Moscow, bring it back to Fort Meade for examination, and replace it with equipment the agency could guarantee was not bugged.

BACK AT FORT MEADE, Walter G. Deeley, the agency's strong-willed deputy director of communications security, tapped himself for the job. He'd gone over his bosses at NSA and the Pentagon and personally made the case to Reagan for why he should run Gunman.

For years Deeley had taken a personal interest—some called it an obsession—in rooting out leaks. He had been with NSA for thirty-four years, most of that time intercepting foreign communications, but for the last chapter of his career he'd found himself on the opposite side of the equation, figuring out how to protect American communications from foreign spies.

Having tasked analysts with finding ways into the NSA's own systems, Deeley was increasingly distraught by what they found. If his own analysts could find ways in, he knew the Soviets were likely already there. He'd been impressed with the ingenuity of Soviet eavesdropping—when U.S. officials had managed to uncover it—and he knew they had a huge advantage on the Americans, in that they were uninhibited by red tape. Unless the United States started taking security seriously, he worried that it would lose the Cold War.

Nobody at the agency doubted that Deeley could get things done. During the Vietnam War it was Deeley who created the NSA's 24/7 nerve center, known as the Signals Intelligence Operations Center, the predecessor to what would eventually become the National Security Operations Center, which handles the NSA's most time-sensitive operations.

The same year Deeley stepped into his post as chief of the operations center, a U.S. Navy EC-121 patrol plane was shot down over the Sea of Japan. Over the next few hours NSA analysts scurried from office to office all over Fort Meade, scrambling to collect the intelligence necessary for a coordinated response. Deeley deemed the response sloppy, inefficient, and unacceptable. Under his tenure, he created a dedicated twenty-four-hour watch center to collect near-real-time intelligence, connect dots, and triage crises popping up around the globe. It was Deeley who insisted that analysts deliver daily briefings to the head of the agency.

Now nearing retirement, Deeley had set off on one final mission: hunt down the source of the embassy's leaks, neuter it, and outfit the embassy with spyproof equipment.

DEELEY PULLED TOGETHER the NSA's best people for the job. They had one hundred days to complete phase 1, which entailed swapping out every piece of machinery at the embassy with new equipment.

That in itself proved a challenge. The agency found it nearly impossible to find replacements. The agency's typewriters, IBM Selectrics—the "Cadillac Supreme" of typewriters in those days—had been whisked off store shelves. And IBM Selectrics that worked on Russian voltage were even harder to find. The NSA scoured its inventory and called in favors at IBM, but ultimately was only able to pull together 50 of the necessary 250 typewriters to install in the most sensitive areas of the embassy. The rest would have to wait.

It took Deeley's team two months just to prepare the new equipment for shipping. To make sure each piece of new equipment hadn't been bugged, NSA analysts disassembled, reassembled, and scanned every single piece of machinery with X-ray machines, noting any irregularities, before installing antitamper sensors and tags on the inside and outside of each typewriter.

When it came time to ship the typewriters to Moscow, every piece of equipment had to be carefully stored in tamperproof bags that were not available in the Soviet Union, lest Soviet spies attempt to break through the bags and replace them with new ones.

Armed guards protected ten tons of equipment as it made its way from top-secret trailers at Fort Meade to Dover Air Force Base, to Frankfurt, Germany, and eventually to the Moscow embassy, where, conveniently, after the first ton of equipment went up, the Soviets shut down the embassy elevator for "maintenance." NSA workers had to manually cart the remaining nine tons of equipment up the embassy stairs, and all ten tons had to be manually carted back down.

Even the U.S. ambassador was kept in the dark. The day the NSA's agents arrived, they passed him a curt handwritten note carefully instructing him to tell embassy workers that the government had generously decided to swap out their old machines.

Convinced the KGB would try to compromise the new equipment, NSA technicians ran wires from the antitamper sensors on each box down to a Marine guard station on a different floor, to be monitored around the clock by armed guards.

Over the next ten days NSA technicians systematically emptied each box, filled it with the corresponding older equipment, and then reengaged the tamper sensors, lest the KGB try to intercept the bugged equipment and remove whatever was inside. Armed guards escorted the old equipment as it made its way from Russia's Sheremetyevo Airport to Frankfurt, Frankfurt to Dover, and back to Fort Meade, Maryland.

It took the full hundred days just to get the embassy's equipment back to the Fort. The race to find the Soviet's exploit was on.

DEELEY TOLD THE NSA's best twenty-five analysts to meet him in a trailer in a corner of the NSA parking lot. He'd neglected to mention that the entrance to the trailer was four feet off the ground. The analysts scavenged the lot for cinder blocks and empty wire spools to get themselves up and in.

Deeley didn't bother with niceties. "We're meeting here in this fucking shithole because I don't want any rubberneckers in OPS3 [the main information security building] getting curious. You've all been told this project is VRK [very restricted knowledge], right?"

The analysts nodded and murmured "yes" in agreement. Their supervisors had instructed them to mention their task to no one, not their colleagues, not their spouses, not even their dogs.

"You know what VRK really means?" Deeley didn't waste a breath. "It means I'll cut your fucking balls off if you breathe a word about what you're doing to anyone, and I mean anyone." He motioned to the door. "You do this my way or it's the highway."

For years Deeley had been ringing alarm bells that the agency needed to do more to protect its communications from Soviet interception. Now he was in a rare position to prove as much. With his retirement looming, this would likely end up being the final chapter of his long career. And despite his feats, he knew the nation had a short memory when it came to success. His legacy was on the line.

"Okay, so here's the deal," he continued. "We're dividing the gear into two groups; the accountable stuff [the crypto gear] will be looked at in our labs in OPS3, while the rest [the teleprinters, copiers, and typewriters] will be examined here. Each of you has been given a specific assignment by your supervisor, and work starts the second I leave."

Then he offered up a carrot and a stick. "For better or worse, NSA's reputation will be riding on how well each of you does your job. And we don't have forever," he warned. "The longer we take finding whatever exploits the other side has slipped in, the more chance that we're gonna get fucked by assholes at State, at Langley, or the Russians."

Deeley paused to take a drag of his cigarette and blew it back in the analysts' faces. To speed things up, he offered a five-thousand-dollar bonus to the first analyst who could find the smoking gun.

For the first time, the analysts perked up: $5,000—the equivalent of $12,000 in today's dollars—was roughly a quarter of their salaries.

"Any questions?"

There were none.

"Okay," Deeley said, "Get to work." And with that, he was gone.

THE AGENTS WORKED nights and weekends, ignoring anyone who questioned what they were doing out in the trailer in the parking lot. The likeliest place for the Soviet exploit was in the crypto gear. How else would the Soviets be capturing their encrypted communications? They started there, carefully unpacking each piece of equipment, disassembling each part, and running it through X-ray machines in search of anything unusual. They spent weeks searching every component, photographing the slightest anomaly, but nothing turned up.

Deeley would stop by once a week to hover, chain-smoking at a faster pace than usual, watching as they painstakingly sorted through tons of embassy equipment. He was sure the crypto gear was the culprit. When the gear was cleared, he began to panic. He had gone over the heads of the NSA, the secretary of defense, the CIA director, and the national security advisor to get to

Reagan. He'd been slamming his fists on tables for years, warning that the NSA was no match for Soviet eavesdropping. He had personally convinced Reagan to let him run Gunman. And now he was facing the ugly realization that he might have nothing to show for it.

At night he would go home, grab a beer, and storm to his study in silence, where he'd blast Gershwin's *Rhapsody in Blue* on repeat—code to his wife and eight kids to leave him the hell alone. He'd worked his way up the Pentagon from a lowly army sergeant to a deputy director of the agency—and this would be his final chapter? By week ten, with only a few pieces of equipment left to examine, Deeley was coming to the realization that Gunman had failed. The damned Soviets must have intercepted the equipment. They must have found a way to remove or disable their bugs en route back to the Fort. Listening to Gershwin's burst of brass, Deeley began to confront the terrible magnitude of the missing bug, not just for his own legacy, but for the republic.

Back inside the trailer that summer, analysts were sweating through their shirts. They had worked through the faxes, the teletypes, and scanners. That left just the typewriters. But embassy inspectors had carefully scanned the typewriters before and nobody had ever noticed anything out of the ordinary. Until now, they had kept an open mind. Now they wondered whether Deeley was losing his.

But late one evening, on July 23, 1984, one of the analysts working alone that night noticed an extra coil on the power switch of a Selectric typewriter. That was hardly unexpected—newer typewriter models contained additional memory that would explain additional circuitry and coils. But the analyst decided to run the typewriter through an X-ray machine, top to bottom, just to be sure.

"When I saw those X-rays, my response was 'holy fuck.' They really were bugging our equipment," he recalled.

There on the X-ray film, he could see that inside an unassuming metal bar that ran the length of the typewriter was the most sophisticated exploit he, hell, anyone at the agency, had ever seen. The bar contained a tiny magnetometer, a device that measures the slightest distortions in the earth's magnetic field. The magnetometer was converting the mechanical energy from each typewritten keystroke into a magnetic disturbance. And next to it was a tiny electronic unit recording every disturbance, cataloging the underlying data, and transmitting the results in short bursts via radio to a nearby Soviet listening post. The entire implant could be controlled by remote control and had been specifically designed to allow the Soviets to turn it off when American inspectors were in the area.

Deeley's team couldn't help but admire the mastery involved. All the encryption gear in the world would not have kept the Soviets from reading the embassy's messages. The Soviets had found a way to collect each and every keystroke before anyone had a chance to scramble them. It was masterful work, and a lesson the NSA would never forget. Years later, the NSA would turn the same trick on iPhones, computers, and on America's biggest technology companies, capturing data as it flowed between Google and Yahoo's data centers in unencrypted form.

The agency had always assumed the Soviets used recording devices and bugs to monitor the embassy's activities. Now, for the first time, the Soviets had shown themselves adept at electromechanical penetration too.

For Deeley, this was complete vindication. For years he had argued that encryption was not enough. To truly thwart government interception, the NSA would have to lock down anything that plugged into an outlet. Now he had proof.

He briefed the NSA director, Lincoln "Linc" Faurer, and handpicked the team to accompany him to the White House to personally inform President Reagan.

If anything, their discovery only added to the urgency of their mission. Initially they determined that six of the embassy's typewriters were bugged, but they knew there had to be more. Deeley's team started briefing intelligence officials and agents headed to the Soviet Union in soundproof, anti-echo chambers, where their reactions to the Soviet's implant ranged from astonishment to admiration to anger.

Deeley's team trained other NSA agents to look for telltale signs of the typewriter attack—the modified power switch, the metal bar—and showed them how to x-ray the typewriters. Agents would eventually locate the implant in seven more typewriters used by the embassy's highest-ranking officials and their secretaries, and three more in the U.S. consulate in Leningrad.

It was clear that the Soviets were pouring significant resources into their analog exploits. They were intercepting electric pulses and offering the NSA a playbook to do the same. Years later, with the switch from analog to digital technologies, they would use the same techniques to intercept information as it translated into ones and zeroes.

The NSA would eventually discover more than five different variants of the Soviet implant. Some were designed specifically for typewriters that plugged into the wall. More sophisticated versions worked on newer battery-powered typewriter models.

When Deeley's team searched back through their inventory, they discovered that the first implant had been installed in a typewriter shipped to the embassy in 1976. That meant that by the time Gunman was complete, the Soviets had been siphoning American secrets off their typewriters for eight years.

Routine embassy inspections had missed the implants. American inspectors had come close after finding an antenna in the embassy chimney at one point, but they'd never discovered its purpose. And analysts never wondered why the Soviets treated their own typewriters with such paranoia. The Soviets banned their own staff from using electric typewriters for classified information, forcing them to use manual models for top-secret information. When Soviet typewriters weren't being used at the country's own embassies, the Soviets stored them in tamperproof containers. And yet the Americans hadn't bothered to ask why.

"I think people tend to fall into the trap of being disdainful too often of their adversaries," Faurer would later recall. "We tended to think that in technical matters we were ahead of the Soviet Union—for example, in computers, aircraft engines, cars. In recent years, we have encountered surprise after surprise and are more respectful. Most folks would now concede that they have enormously narrowed the gap and have caught us in a number of places."

The Americans never did find out how the Soviets got their implants into their typewriters. Some suspected they'd been intercepted in shipment; others during maintenance; and still others suspected a mole. However they got there, Gunman was a whole new level of spycraft, and it irrevocably altered the game. Almost forty years later, few computers do not speak with others. Each is linked to a network that is linked to a larger, more complex network still, forming an intricate and invisible web that zigzags the globe and reaches out to the farthest corners of our galaxy, where an orbiter now dispatches more data than ever before from the desolate landscapes of Mars. Gunman opened the door to what was possible. Now, everywhere you look are endless opportunities for espionage—and destruction.

CHAPTER 7

The Godfather

Las Vegas, Nevada

T HAT WAS OUR big wake-up call," James R. Gosler, the godfather of American cyberwar, told me one afternoon in late 2015. "We were lucky beyond belief to discover we were being had. Or we would still be using those damn typewriters."

If any single technologist can be credited with spurring the United States to scramble, catch up, and take the lead as the world's most advanced digital superpower, it is a recently retired man in his late sixties who bears an uncanny resemblance to Santa Claus, now living on the outskirts of Las Vegas.

There, the only sign of Gosler's long, redacted intelligence career is a box filled with intelligence awards, all of them doled out in mostly private, classified ceremonies for accomplishments the public will likely never know about.

It was Gosler who pushed to declassify Project Gunman early.

"I was the irritant," he said, chuckling.

Irritant would be a generous understatement. I asked nearly every single one of the men who guided the CIA and NSA through the turn of the century to name the father of American cyberwar, and none hesitated: "Jim Gosler."

And yet in hacker circles Gosler remains an unknown.

Even the thousands of hackers who flock to Vegas every year for Black Hat and Def Con to watch the biggest names in hacking break into iPhones,

ATMs, and elevators are clueless to the fact that the real wizard of their trade lives just a few miles away. When Gosler and I first met in person, we met at The Venetian hotel during the Black Hat hacking conference. It was the first time in the conference's almost twenty-year history that he had stepped anywhere near it.

"This is a terrible place to recruit," Gosler informed me. The crème de la crème of nation-state hackers, he told me, weren't showing their skills off at conferences. They were laboring in stealth in university labs and secure operations centers.

Over the course of Gosler's career, this unassuming man served as the central catalyst for the United States government's vulnerability discovery and exploitation programs as society made the transition to digital. And if he were any less humble, Gosler would probably concede as much.

Instead, he credits his colleagues and bosses in the intel world and a host of New Age management gurus. Gosler frequently cites Malcolm Gladwell—"The Outlier is fantastic!" he told me, more than once. Gordon Moore and Andy Grove, two former chief executives of Intel, were his heroes. Grove's book *Only the Paranoid Survive* is his bible. But his all-time favorite is Price Pritchett, the organizational management guru.

For years, anytime intelligence officials visited Gosler's office at CIA headquarters in Langley, they were greeted with the following Pritchett quote on the wall:

> Organizations can't stop the world from changing. The best they can do is adapt. The smart ones change before they have to. The lucky ones manage to scramble and adjust, when push comes to shove. The rest are losers, and they become history.

The quote perfectly encapsulates Gosler's take on American intelligence agencies' belated response to advances in technology and its endless potential for exploitation, espionage, and destruction.

From its founding in 1952, the nation's preeminent spy agency, the NSA—"No Such Agency" or "Never Say Anything," in the old joke—was America's chief eavesdropper and codebreaker. For the NSA's first three decades, the agency's sole mission was snatching intelligence as it flew through the atmosphere. At Fort Meade thousands of brilliant PhDs, mathematicians, and codebreakers would cull through messages, decrypting, translating, and analyzing them for critical nuggets that informed America's next move in the Cold War.

But as the world's data migrated to typewriters, then mainframe computers, desktops, laptops, and printers, closed networks and then the internet, the NSA's old passive model of sitting back and waiting for Soviet communications to hit its global collection system would no longer suffice. Unimaginable volumes of nation-state secrets—previously relegated to locked file cabinets—were suddenly being transmitted in ones and zeroes and freely available to anyone with the creativity and skill to find them. Miniature cameras used by CIA spies to photograph top-secret documents in file folders had given way to something else entirely.

And Gosler, more than most, saw to it that America seized every last exploitable digital opportunity.

GUNMAN WAS HIS proof. In our conversations, he would refer back to it ad nauseam—partly because Gunman was declassified, whereas everything else Gosler touched was not. But it was also because it was proof that we had reason to be paranoid. It was evidence that America's enemies were mastering their digital interception capabilities—and already had a wide lead.

The Americans boasted intelligence feats of their own, to be sure. In the mid-fifties, the CIA and their British counterpart, the MI6, undertook a monumental effort known as Operation Regal to intercept a buried Soviet cable in East Berlin. They managed to covertly build a 1,400-foot tunnel underneath Berlin, where they accessed East European and Soviet communications for more than a year before the Russians discovered it. Later, in a joint NSA-CIA-Navy operation in the seventies, Operation Ivy Bells, American divers successfully tapped a Soviet communications cable on the sea floor, just north of Japan. Believing the cable was out of American reach, the Soviets barely bothered to encrypt anything that funneled through it. For years the NSA was able to filch Soviet secrets off their cable tap, until a double agent—one of the NSA's own analysts—tipped the KGB off to the American's deep-sea intelligence program in 1981.

But Gunman was different. "A-plus. Technically brilliant," Gosler would tell me. The Soviets weren't relying on a well-placed bug or cable tap; they had found a way to embed in the typewriters themselves, pilfering every keystroke before Americans even had a chance to encrypt their messages. In intelligence jargon, this would become known as "hacking the end points," and Gunman set a new bar. In this new era, the NSA would take note as the Apples, Googles, Facebooks, and Microsofts of the world moved to encrypt the world's communications. They too would have to master the art of hacking the end points—the phones and PCs that contained information in unencrypted, plain text.

"That kind of technical collection wasn't invented with the advent of the computer. The Soviets had been doing this since the 1970s, but Project Gunman made it real," Gosler told me. "You couldn't pretend it away anymore."

In Gosler's lexicon, there's BG—Before Gunman—and AG. BG, the Americans were "fundamentally clueless," he told me. "We were in la-la land."

Three decades AG, we were breaking into anything with a pulse.

BEYOND DATES AND essentially meaningless job titles, Gosler won't say much about what transpired between the day he walked into Sandia National Laboratories in 1979 as a bright-eyed twenty-seven-year-old and the day he retired as a fellow there in 2013.

For the most part, it is all highly classified. You have to press him for basic details. At dinner parties, when others inquired, Gosler would say only that he worked for the federal government.

"You had to be very careful about what you say, especially when abroad, for personal safety reasons," he told me with a whisper.

We were seated at a restaurant. Like so many others I would meet, Gosler made a point of arriving early, finding a table near an exit, and sizing up everyone inside. He had taken the seat facing the entrance—the best position for survival.

Over the course of our conversations throughout 2016 and 2019, the father of American cyberwar began to piece together his career and, through it, the United States' own evolution as the world's most skilled exploiter in the digital realm. He was careful to redact vast, classified stretches of his résumé.

Those, I had to fill in for myself.

Gosler spent his first five years at the Energy Department's Sandia National Labs in the—"Oh, I'll just call it the computer department," he told me. His first gig involved digging through the innards of mainframe computers and the operating systems that ran Sandia's back-end administrative systems, like payroll.

Of New Mexico's two national nuclear laboratories—Los Alamos National Laboratory in Santa Fe and Sandia in Albuquerque—Los Alamos is the one hallowed in our national memory. Los Alamos gave birth to the Manhattan Project and has been the crucible of America's nuclear weapons research and development. But much of the real weaponization of America's nuclear weapons program is conducted at Sandia, which oversees the production and security of 97 percent of the non-nuclear components that make up America's nuclear arsenal. Five years into his gig at Sandia, Gosler transitioned into the team charged with making sure that each nuclear component would work exactly how it was supposed to work whenever the president authorized them to, and, critically, not work under any other circumstance. Accidents and malfunctions were more common than one would think. One Sandia study found that between 1950 and 1968, at least twelve hundred nuclear weapons had been involved in "significant" accidents.

Even bombs that would live in infamy didn't quite work as intended. Little Boy—the very first nuclear weapon America dropped in war—killed eighty thousand people on Hiroshima. But the destruction could have been much worse—only 1.38 percent of its nuclear core fissioned. Three days later, when Americans dropped their second bomb—codename "Fat Man"—on Nagasaki, it accidentally detonated one mile off target, though it still managed to kill forty thousand. A 1954 test of a hydrogen bomb in the Bikini atoll produced a yield of fifteen megatons—triple the amount America's nuclear scientists anticipated—blanketing hundreds of square miles in the Pacific—and, as a result, America's own weapons observers—with lethal radioactive fallout.

These were precisely the scenarios that Sandia's scientists were charged with avoiding. But Gosler's team was less concerned with accidents than they were with intentional sabotage by the enemy. By the mid-1980s, U.S. scientists were exploring ways to reliably deny and disrupt the Soviets' communication networks and nuclear weapons systems in the event of armed conflict. Sandia's scientists could only assume the Soviets were doing the same.

The only way to ensure that America's weapons arsenal was not subverted was to be the first to find and fix any security holes in the components themselves. And from the moment Gosler stepped foot in Sandia's bureaucratically titled Adversarial Analysis Group, in 1984—the same year Gunman turned up the typewriter bugs—he made a name for himself finding critical vulnerabilities in nuclear components, in how those components fit together, and in the applications layered on top of them.

"I'd find problems and . . . fix them," he told me one day.

"You mean exploit them and use them on the enemy?" I asked.

He laughed nervously. "You'll have to ask someone else about that."

And so I did.

IT TOOK ME the better part of two years, but I was able to fill in some of the blanks in Gosler's résumé through documents, a lawsuit by a former Sandia employee, and the oral and written accounts of Gosler's underlings and superiors at the NSA and the CIA—all of whom agreed that without the wise, bearded, bespectacled man I would come to know, America's offensive cyber program would never be where it is today.

By 1985 Gosler was only a year into his new gig hunting for vulnerabilities, and he could already see that his work was about to get a lot harder, impossible even. Like so much else, nuclear weapons design was evolving away from discrete electronic control systems to more complex microchips. In dissecting those chips down to the very bit, Gosler could see that these advancements—and the complexity they introduced—would only create more room for error, malfunction, and eventually enemy subversion and attack.

The previous year, Gosler had heard a famous lecture by Ken Thompson. Thompson, who had won the 1983 Turing Award for cocreating the Unix operating system, used his turn at the lectern to share his concerns on where technology was headed. He'd titled his lecture "Reflections on Trusting Trust," and his conclusion was this: unless you wrote the source code yourself, you could never be confident that a computer program wasn't a Trojan horse.

Thompson had perfectly articulated what Gosler knew to be true. But by the time Gosler listened to Thompson's lecture, he could see that the predicament was getting exponentially worse. Very soon, he knew, they would not be able to guarantee the security of the country's nuclear weapons arsenal.

"You could still find vulnerabilities, sure, but your ability to guarantee no other vulnerabilities existed was becoming impossible." He paused for emphasis. "That's important, Nicole. You could no longer make the statement that any of these microcontrolled systems were vulnerability-free."

Others might have thrown up their hands. Many did. But Gosler was never one to shy away from a challenge, and it was there, deep in those microchips, that he discovered his life's purpose and the foolishness of the human condition—all wired together.

These chips were at once a hacker's paradise and a national security nightmare. Each carried endless potential for exploitation and subversion, espionage and destruction.

And for the next three decades of his career, Gosler proved as much.

GOSLER BEGAN WITH two experiments. That year, 1985, he convinced his bosses at Sandia to sponsor a study. They called it Chaperon, and its premise was simple: Could anyone design a truly secure computer application? And could someone subvert that application with a malicious implant that could not be detected, even through a detailed forensic investigation? In other words, a zero-day.

Sandia divided its top technical brass into bad guys and good guys: the subverters and the evaluators. The former would plant vulnerabilities in a computer application. The latter would have to find them.

Gosler still spent most of his evenings away from work breaking hardware and software for the fun of it. But professionally, he had only ever played the role of evaluator. Now, he relished the chance to play subverter. He designed two implants and was sure the Evaluators would discover his first subversion.

"I was immersed in a fantasy world back then," Gosler told me. When he wasn't breaking software, he was playing the 1980s computer game Zork, popular with some of the techies he worked with.

For his first trick, he inserted a few familiar lines from the Zork game into the security application's code. The Zork text effectively fooled Sandia's application into revealing secret variables that could be used by an attacker to take over the application—and any data the application secured. Gosler was sure his colleagues would pick up on it quickly.

For his second subversion, Gosler inserted a vulnerability that he and others would later only describe as a "groundbreaking technical achievement."

THE EVALUATORS NEVER did find Gosler's two implants. Even Gosler's Zork subversion proved maddeningly difficult to track down. Sandia's evaluators still describe the study as one of the most frustrating experiments of their career. They spent months looking for his implants before they finally threw up their hands and demanded that he tell them what he had done.

It took Gosler three eight-hour briefings, pacing in front of a whiteboard covered in notation, which he attacked in bursts, to painstakingly explain his implant. His peers nodded along, but clearly they were baffled.

Initially Gosler thought the second implant could be useful as a Sandia training exercise, but seeing employees' frustration, his bosses rejected the idea outright. They worried that the exercise would only compel new recruits to quit.

Instead, his bosses decided to start over and put together a new study: Chaperon 2. This time, they chose someone other than Gosler to lead the subversion. Some one hundred Sandia engineers spent weeks and months hunting for the implant. While others came close, only one—Gosler—discovered the subversion and presented it in a detailed hours-long briefing.

Word of the studies and Sandia's techno wizard got back to the intelligence chiefs at the NSA—"the big dogs back East," as Gosler called them—who called Sandia and asked for Gosler by name.

RICK PROTO AND Robert Morris Sr., the respective chiefs of research and science at the NSA's National Computer Security Center, thought Gosler could teach their analysts a thing or two.

This was 1987. Proto was a giant at the agency. Morris Sr., the government's most senior computer scientist at the time, would earn infamy one year later as the father of Robert Tappan Morris, the Cornell student who unleashed the "Morris worm" from MIT, which would brick thousands of computers at a cost of tens of millions of dollars. Gosler had worked with some of the government's top computer scientists before, but nothing prepared him for "the Fort."

Walking into Fort Meade, his first impression was simply, "This is a different league."

At their first meeting, Gosler asked Morris Sr. the question that had been troubling him for some time now. "How complex can software be for you to have total knowledge of what it could do?"

Morris Sr. knew this was a loaded question. Hulking mainframe computers were ceding to smaller, cheaper devices, with microelectronics and microcontrollers baked inside. Computer applications were now incorporating more and more lines of code, creating more and more room for error and incorporating them into bigger and bigger—and more critical—attack surfaces. And these applications were being loaded into airplanes, naval ships, and, perhaps most critically of all, America's nuclear weapons.

Despite the security implications, there appeared to be no sign of turning back. The first complete version of the Linux operating system contained 176,000 lines of code. Five years later, it would contain 2 million. By 2011 Linux would contain more than 15 million lines of code. Today, the Pentagon's Joint Strike Fighter aircraft contains more than 8 million lines of onboard software code, while Microsoft's Vista operating system contains an estimated 50 million lines.

Each line of that code contains instructions that can potentially be subverted for any number of means. The more code, the more difficult it is to discover errors, typos, or anything suspicious. Finding Gunman's typewriter implant was an intelligence feat. Finding the equivalent of that implant in a next-generation fighter jet? Good luck.

Morris Sr. told Gosler that, off the top of his head, he would have "100 percent confidence" in an application that contained 10,000 lines of code or less, and zero confidence in an application that contained more than 100,000 lines of code. Gosler took that as his cue to share with Morris Sr. the more complicated of the subversion tactics he had developed for Sandia's Chaperon 1 study. Turns out it was an application with fewer than 3,000 lines of code.

Morris Sr. invited an elite NSA squad of PhDs, cryptographers, and electrical engineers to take a look. Not one discovered Gosler's implant, nor could any replicate the subversion once Gosler pointed them to it. Clearly the NSA had overestimated its own ability to detect vulnerabilities in the nation's most sensitive computer systems. Suddenly anything with more than a couple thousand lines of code looked suspicious. Once they had seen Gosler's demo, all sorts of mischief, leaks, and national security disasters that they had never even imagined seemed possible.

"Even if you found something, you could never be confident you found everything," Gosler said. "That's the awful nature of this business."

By 1989 the intelligence community was still reeling from the ingenuity the Soviets had demonstrated in Gunman. The internet was just around the corner and would present an entirely new attack surface. The NSA had been lucky to find the typewriter implant, but God only knew how many other implants they had left to find.

They needed Gosler's help. Proto asked Gosler to stay. Over the next two years, Gosler would serve as the agency's first "visiting scientist," schooling the NSA's defense-oriented analysts in all the ways modern software and hardware could be defeated and subverted. His mission was to help the nation's top defenders track down Soviet implants and anticipate those of every other adversary who sought to do the United States harm.

Wherever they might be.

"I WAS LIKE a kid in a candy shop," Gosler told me of his two-year stint at the NSA.

Everything about the agency—the people, the culture, the mission—intimidated him. Everything and everyone, it seemed, was on a strictly need-to-know basis. "This was a whole new level of secrecy," Gosler recalled. "People at the agency spent a lot of time measuring you up, making sure you were trustworthy and bringing something to the table, before they would say anything to you. You had to talk to people multiple times before they would trust you with an assignment. And once you got it, all you could think was 'Don't screw up.'"

Gosler spent most of those two years in the NSA's defense division, what is now known as Information Assurance. But it wasn't long before he was exposed to what he will only call the agency's "dark side"—the disparate skunk works that would one day mature and combine to form the agency's elite hacking division, known until very recently as Tailored Access Operations (TAO).

The NSA's offensive operation was still in its infancy then—nowhere near the cadre of thousands of NSA hackers that operate at Fort Meade and across the country today—but the agency had been warned as far back as the late 1960s that the technology they increasingly relied on could be exploited for both breaches and eavesdropping.

In 1967—nine years before the first email would traverse the internet—a computer pioneer by the name of Willis H. Ware outlined the myriad vulnerabilities in modern computer systems and all the ways they could lead to classified information leaks or be exploited for espionage. His so-called Ware Report proved a catalyst for the Pentagon to convene a Defense Science Board task force to study computer security. That task force came to several ominous conclusions, chief among them: "Contemporary technology cannot provide a secure system in an open environment."

The report was the first to advance the notion that computers were leading humanity, and with it the nation's intelligence apparatus, down a dangerous path. But it offered little in the way of solutions. So over the next few years, the U.S. government culled some of the report's authors, as well as top contributors at the NSA and CIA, to analyze the security risks posed by computers and offer recommendations.

The sum of their work—which would come to be known as the Anderson Report, after the report's primary author, James P. Anderson—would set the U.S. government's cybersecurity research agenda for the next decades, and lay the groundwork for the United States' cyberwar operations.

The Anderson Report concluded that computers provided would-be attackers with a "unique opportunity for attempting to subvert" their systems and access their underlying data. "Coupled with the concentration of the application (data control systems, etc.) in one place" (the computer system), that capability "makes computers a uniquely attractive target for malicious (hostile) action." The design of hardware and software systems was "totally inadequate to withstand attack," the report concluded; and if one malicious user could control a single computer node, "the entire network may be compromised." The only limits to attack were attackers' own imagination and skill.

The possibilities were endless. An attacker could gain "unauthorized access to classified data by exploiting a pre-programmed weakness due to careless design or implementation" or plant "a 'trap door' in the computer application or in the programming and operating systems supporting the application."

So long as computer operating systems accepted software updates without question, the report concluded, computers would be manipulated to accept trapdoors.

The authors had analyzed Honeywell's computer operating system, among the first developed with security in mind, and discovered a number of serious flaws that allowed them to seize control of any computers it touched, and any data stored inside. The same was true, they found, for most every other contemporary computer systems they probed.

In the absence of grave government interference, the Anderson Report determined, there was "little hope" of keeping the nation's most sensitive secrets—military plans, weapons, intelligence, and spycraft—safe from foreign adversaries. The threat would only become more acute in the future, the authors concluded, once America's political enemies recognized just how many national security secrets could be stolen from U.S. government databases, and how little effort is required to subvert them. And all this, before the internet.

"Don't you think it odd that we started getting these warnings in the late sixties?" Gosler would ask me decades later. "And yet look at where we are now. What is it going to take?"

GOSLER HAD TAKEN one look at what the NSA's early hackers were up to, and he knew there was nothing else he would rather do with his life than join them.

This was 1989. The Pentagon's precursor to the internet, ARPANET, had become an artifact. It was now a slow, tiny, and antiquated part of a much larger and faster internet, and the Pentagon decided the time was right to shut ARPANET down for good.

ARPANET's successor, the internet, was quietly growing to 100,000 host machines, each with multiple users, and was nearing its tipping point. As Netscape Navigator and Internet Explorer made their way onto PCs, and the world's infatuation with the web grew, so did the NSA's.

"Why did Willie Sutton rob banks?" Gosler would repeatedly ask his bosses and underlings at the intel agencies. "Because that's where the money is!"

There was still money in brick-and-mortar banks, but the motherlode was moving to the internet, along with everything else. Pulling out critical intelligence would require a complete transformation of the way the NSA did business. The alternative, Gosler argued—the status quo—would only guarantee that the United States would become, as his guru Pritchett put it, "losers."

"We could not just go with the flow. We had to be proactive," Gosler told me. "We had no other choice."

Americans could no longer rely on the old models of spycraft—a game in which they would wait for adversaries to dispatch messages along radio signals, microwave transmissions, and telephone lines. They would have to chase it to the source: the hardware, software, imaging, sensors, satellite systems, electronic switches, PCs, and networks it was moving to. With so much data blossoming on so many millions of machines, Operation Ivy Bells looked quaint.

The NSA would have to go out and penetrate not only undersea fiber-optic cables but also the networks, and networks inside those networks still. They would have to map out the machines all these networks connected to, and find those that stored the most critical data, all the while exploiting vulnerabilities in hardware, software—and people—to extract crucial intelligence. And there would be no point in doing this unless they could successfully achieve it on a massive scale.

As Gosler's two-year stint at the NSA came to an end in 1990, he saw clearly the challenges—and opportunities—the intelligence community faced.

Either the intelligence community would grow and adapt, or the internet would eat us alive.

GOSLER PROBABLY WOULD have happily stayed at Fort Meade, had it not been for his promise to return to Sandia. He was of the old-school mindset that one owes loyalty to those who train them, and he knew his loyalties were in Albuquerque.

But before he set off for New Mexico, Gosler struck a deal with the NSA's then director (and the soon-to-be deputy CIA director) William O. Studeman. Gosler would use his vacation days at Sandia probing the hardware and software that America's enemies now depended on. In return, Admiral

Studeman would let Gosler work directly on some of the NSA's most closely guarded offensive missions—"a classified, really neat project" is all Gosler would give me.

So in 1990 Gosler officially returned to Sandia, while continuing his unofficial classified work for the NSA. "I would go to work, break software and hardware, come home at night, eat dinner, and break software and hardware," he recalled decades later. It was the beginning of a strategic relationship between the NSA and Sandia that would only grow closer over the next few years, then decades.

Gosler will not discuss the exploit work he did for intelligence agencies over that time. It is all still highly classified. The whole story will be told only when secret U.S. documents are declassified, probably in the second half of this century.

But all one needs to do is look at the source of his department's funding to see how critical his work became for the nation's intelligence apparatus. When Gosler first returned to Sandia in 1990, his department was running on $500,000 from the Department of Energy's National Nuclear Security Administration. Five years later, Gosler's department was plush with $50 million in intelligence funding. This investment in digital exploitation was in stark contrast to post–Cold War intelligence budget cuts, which saw billions slashed from intelligence budgets and a complete freeze on new hires at the NSA by the mid-1990s.

The only glimpse the world has into the work the NSA was outsourcing to Sandia, beyond classified leaks and Gosler's generalized commentary, comes in the form of an employment suit filed by one of Gosler's former underlings at Sandia. In the suit, a Sandia employee accused the lab and fifteen of its employees, including Gosler, of firing him for refusing to be conscripted into an NSA "infowar." The suit claims that Gosler told Sandia employees about work his team was doing for the NSA on a "covert channel" that essentially entailed "virusing computer software and hardware" and "spiking" the machinery and encryption algorithms used by America's foreign adversaries to make them easier to crack. Sandia's official justification for firing the employee was that he had demonstrated a lax approach to classified intelligence, and for his "flagrant attack on a valued Sandia customer"—the NSA.

The lawsuit never says it outright, but it insinuates that Gosler's team may have contributed to some of the NSA's most stunning intelligence coups in modern history.

THE SAME YEAR Gosler allegedly admitted to his Sandia coworkers that he was aiding the NSA's exploitation work on a covert channel, a Swiss national by the name of Hans Buehler was arrested in Tehran for espionage.

Buehler, a top salesman for a Swiss encryption company called Crypto AG, would spend the next nine months in an Iranian prison, much of it in solitary confinement. "I was questioned for five hours a day for nine months," Mr. Buehler told reporters after his release. "I was never beaten, but I was strapped to wooden benches and told I would be beaten. I was told Crypto was a spy center."

Crypto AG, the German wing of it anyway, paid Tehran $1 million to release Buehler, even though, as far as Buehler knew, there was no truth to Tehran's allegations. Not until three years later would two reporters at the *Baltimore Sun*—Scott Shane, who would later join the *Times*, and Tom Bowman, who joined NPR as a Pentagon correspondent—break the story of why Tehran would have reason to be suspicious.

For years—as far back as World War II—the NSA had, with the CIA and Crypto AG's blessing, and perhaps a helping hand from Sandia's premier exploitation specialists, been spiking Crypto AG's encryption machines to make any messages that passed through them easily decipherable to American decoders and analysts.

Crypto AG was a perfect foil for U.S. intelligence agencies. Among its high-profile clients were some of America's biggest adversaries in Iran, Iraq, Libya, and Yugoslavia, all of which had entrusted their most sensitive military and diplomatic secrets to the Swiss machines. They never imagined that the Swiss—known for their secrecy and neutrality—would agree to a deal that guaranteed American spies could unscramble their data.

The agency accomplished this, in part, through its own version of Gunman: NSA agents worked with Crypto AG executives to introduce trapdoors into Crypto AG's encryption machines, allowing NSA decoders to easily decrypt their contents.

Gosler never confirmed this. When I asked him whether that "really neat classified project" he worked on for Admiral Studeman was code for Crypto AG or a project just like it, he laughed. Over the course of our conversations, I came to know that particular laugh well. It meant "Nice try."

Gosler never did discuss any classified operation he participated in or was privy to, or even any he was not. What he would tell me was this: in the decades after Gunman was discovered, the intelligence community—with Gosler's help—created a taxonomy of adversaries that could exploit technology for surveillance. And they made damn sure the United States was on top.

At the bottom of this pyramid were the barely competent Tier I and Tier II adversaries, the script kiddies of nation-states, the ones who couldn't find a zero-day if their lives depended on it. Instead, they relied on sheer purchasing power to buy click-and-shoot exploits from hackers off sites like BugTraq, or from contractors on the underground market.

Right above those states in this taxonomy were the Tier III and IV adversaries, who trained their own teams of hackers but also relied on outside contractors to discover zero-day vulnerabilities, write exploits, deploy them on a target, and "make hay," as Gosler put it.

And above those states were the "big dogs," as Gosler called them, the Tier V and VI nation-states who spent years and billions of dollars finding mission-critical zero-days, developing them into exploits, and—in a feat of glory—inserting them into the global supply chain. The only difference between Tier V states and Tier VIs, Gosler told me, was that the VIs were doing all of this on a massive, push-button scale. At that time, at least, the only countries capable of that level of sabotage were Russia, China, and—though he would never say it—the United States.

"Think about it," he told me one day. "Nothing is American-made anymore. Do you really know what's in your phone, or in your laptop?"

I looked down at my iPhone with a renewed sense of intrigue, the kind of look you might give a beautiful stranger.

"I do not."

INSIDE THAT SLEEK black glass sandwich was a universe of hardware—circuitry, encryption chips, flash memory, cameras, logic boards, battery cells, speakers, sensors, and mystery chips—pieced together by faceless, haggard workers on a factory floor somewhere far beyond my reach.

And yet here we were, entrusting our entire digital lives—passwords, texts, love letters, banking records, health records, credit cards, sources, and deepest thoughts—to this mystery box, whose inner circuitry most of us would never vet, run by code written in a language most of us will never fully understand.

As Gosler spoke, my mind went straight to Apple's haggard, faceless factory workers in China. In my mind, that factory worker now had a face, and his dormitory now had a mattress stuffed with all the cash foreign spies had been paying him in bribes to swap in their spiked encryption chip—the one with the weak crypto that cryptographers back at Fort Meade, or Cheltenham, or Moscow, or Beijing, or Tel Aviv, could easily crack. Or maybe it was the factory worker's supervisor? Or maybe the C-level execs? Or maybe the CEO himself? Or maybe that factory worker wasn't bribed, but blackmailed? Or maybe he was a CIA line officer all along?

The opportunities to sabotage the global supply chain were endless, Gosler told me. My mind also darted back to *Times* headquarters, to Sulzberger's closet, to one of those two classified NSA documents Glenn Greenwald was so hesitant to give up—the one that laid out, in intelligence jargon, how the NSA was spiking the global supply chain.

The document was a 2013 NSA intelligence budget request, outlining all the ways the agency was circumventing encryption on the web. The NSA called it the SIGINT Enabling Project, and the vast reach of the agency's meddling and incursion into the world's digital privacy was disguised in typical agency lingo:

> The SIGINT Enabling Project actively engages the US and foreign IT industries to covertly influence and/or overtly leverage their commercial products' designs. These design changes make the systems in question exploitable through SIGINT collection (e.g., Endpoint, MidPoint etc.) with foreknowledge of the modification. To the consumer and other adversaries, however, the systems' security remains intact. In this way, the SIGINIT Enabling approach uses commercial technology and insight to manage the increasing cost and technical challenges of

discovering and successfully exploiting systems of interest within the ever-more integrated and security-focused global communications environment.

Portions of the NSA's budget request were even more explicit. In an attempt to solicit more funds, the agency boasted that some of these "enabling" projects were "at, or near total completion." That year, 2013, the NSA claimed that it planned to "reach full operating capability for SIGINT access to a major Internet Peer-to-Peer voice and text communications system." The Snowden leaks did not say which system, but Skype was the obvious suspect. It also claimed to have "complete enabling for the two leading encryption chip makers used in Virtual Private Network and Web encryption devices."

In other words, the NSA was making fools of anyone who believed they could thwart spies using off-the-shelf encryption tools like VPNs—virtual private networks that route a person's web activities through a protected, encrypted tunnel. Theoretically, the whole point of using a VPN is to shield your data from would-be snoopers and spies.

I thought of Gosler's laugh. *Nice try.*

THE NSA DID not achieve this scale of spycraft alone. It got a big assist from its partners across the river in Langley. And there too, like a cyber Forrest Gump, Gosler was uncannily present for the agency's biggest leap into digital exploitation.

In December 1991, as Gosler's team was busy spiking hardware and software in New Mexico, the CIA's Soviet operatives were clinking champagne glasses at their annual holiday fete. That year, the mood was especially festive. Affixed to suits were campaign-style buttons featuring the red Soviet hammer and sickle with the words THE PARTY'S OVER.

With the CIA agents' hangover headaches still pounding days later, Russian soldiers marched into the Kremlin and swapped out the Soviet flag with the Russian flag not seen since 1917. The Cold War was over, but new enemies were on the horizon, and the champagne would not flow for long. One year later R. James Woolsey, President Clinton's new pick for CIA chief, would tell senators, "Yes, we have slain a large dragon. But we live now in a jungle filled with a bewildering variety of poisonous snakes. And in many ways, the dragon was easier to keep track of."

Beyond long-standing American adversaries in Russia, China, North Korea, Cuba, Russia, Iran, and Iraq, the United States was now up against a growing list of complex national security threats: the proliferation of nuclear, biological, and chemical weapons; criminal groups and drug cartels; regional instability in the Middle East and Africa; and new and unforeseen terrorist threats.

Woolsey's Senate testimony was tragically prescient. Just three weeks later, Islamic fundamentalists detonated twelve hundred pounds of explosives in a van beneath the World Trade Center. And eight months after that, the mutilated bodies of American soldiers would be dragged through the streets of Mogadishu after Somalis downed two U.S. Black Hawk helicopters.

At Fort Meade, NSA leadership was struggling to reconcile its place in this dynamic national security realm, just as the internet was changing spycraft forever. The CIA was at the same crossroads. The secrets American spies were after were now passing at high speed through a maze of computer servers, routers, firewalls, and personal computers. To do their job, America's intelligence agencies would have to acquire every last possible scrap of digital information—the "whole haystack," as Keith Alexander, the NSA's director from 2005 to 2014, would call it—and to get it, they would have to hack it.

Back in 1993, the challenge was daunting, but as Michael Hayden, who would go on to lead both the NSA and CIA, put it, "We also knew that if we did this even half well, it would be the golden age of signals intelligence."

THE GOLDEN AGE of signals intelligence inevitably pitted the codebreakers at the NSA against their counterparts forty miles away at Langley.

In its infancy, the NSA had always looked to the CIA as a sort of older brother. Officials at the Fort sought out Langley's guidance on budgeting, what intelligence it should collect, and how it should be produced. But as the NSA's budget doubled, tripled, then quadrupled, it became a force of its own. By the 1970s, the NSA no longer felt the need to filter intelligence through Langley's middlemen, when it could just as easily send its reports directly to the White House, the secretary of state, and the National Security Council.

The CIA came to resent what officials saw as NSA overreach. The two agencies had reached a tentative truce decades ago. The NSA would stick to its "rice bowl" and collect data "in transit," while the CIA targeted intelligence at the source, using spies to break into their targets' homes, briefcases, computers, floppy discs, and file cabinets. But as NSA moved from traditional SIGINT—the passive, dynamic atmospheric interception it had been doing for decades—to proactively hacking the end points, or what NSA, in a slick vocab flip, was now calling "SIGINT at rest," it began encroaching on CIA turf. Langley knew that if it did not quickly redefine its role in this emerging digital landscape, it would be elbowed out for good. Already the CIA leadership was busy pushing back on policymakers advocating that—in the post–Cold War era—the CIA should be abolished, with its primary functions turned over to the State Department.

The CIA was having a hard time coming up with rebuttals. The agency set up a series of working groups to figure out how the hell the CIA was going to justify its budgets, when so much was moving away from file cabinets and locked safes.

By 1995 each group had come back with the same demoralizing finding: the CIA was frankly not set up to seize the internet's new collection opportunities. A small information warfare team of twelve people was created. Half would work on defensive analysis, the other half on offensive zero-day exploits and hacking tools that they loaded onto giant floppy discs and passed around for sensitive operations.

But it was clear that the agency needed more than an ad hoc team. A Special Projects Staff group recommended that an entirely new office be created to tackle the new battleground of information technology.

They called it the Clandestine Information Technology Office (CITO). And as the CIA's then director, John Deutsch, scanned the field for possible candidates to lead the organization in 1995, one name came up again and again.

BY THEN GOSLER was a legend—in highly classified circles at least. The spectacled Sandia scientist was now widely considered to be the country's preeminent digital exploitation specialist, the man the government went to "to solve the impossible," as one of his employees later put it.

When the CIA called, Gosler hesitated at first. He had found his life's work at Sandia. But a senior intelligence officer told him he'd be nuts to turn down the job. So in 1996 he took over directorship of CITO—the predecessor to the CIA's Information Operations Center—and started reporting directly to the agency's spy service and to its science and technology division. The CIA's equivalent of Q Branch in the James Bond films, the agency's Directorate of Science and Technology developed surveillance devices that imitated flying insects and gave birth to the lithium-iodine battery. The CIA developed the battery to improve its surveillance operations, but eventually it would be used in smartphones, electric vehicles, even pacemakers.

Gosler started pulling in the division's best technical experts. But he knew that for his mission to succeed, he needed to pull in the spies, too. He would have to convince everyone from the powers on high to the least techno-savvy of CIA spies that from now on, digital exploitation would play a powerful role in most, if not all, future spy operations.

He began to preach digital espionage to the spies at every opportunity. In trainings, in the hallways, at the water cooler, he tried to convince the CIA's line officers, many of them technophobes, that he needed their help. He reassured them that his division was not out to replace them. He was just looking to supplement their spycraft in revolutionary ways. Digital exploitation could be a powerful tool, he told them, in blackmailing and recruiting spies, unleashing previously unimaginable foreign secrets, even embedding in the enemies' weapons themselves.

Gosler showed a knack for spycraft, which played well with many of his Langley recruits. Sometimes digital exploitation was as easy as sending a CIA officer into the C-suite of a major technology supplier and asking them, point-blank, to incorporate spiked NSA hardware or microchips into their supply chain—the old do-it-for-your-country approach. But more often than not Gosler explained how CIA officers could use traditional modes of spycraft to recruit a well-placed insider at a foreign hardware or arms manufacturer, shipping hub, or even a hotel employee to spike systems covertly. With so much personal data flowing online, the time required to identify CIA assets—and understand their soft underbellies—took minutes now, versus days, weeks, months even, of casing their homes, employers, job-change patterns, love lives, personal debts, travel patterns, vices, and favorite haunts.

All of this was now instantly available with a few clicks.

With the internet, CIA officers could now turn to online databases to pinpoint those with intimate access to the technologies the NSA sought to infiltrate. Some of that data proved useful as blackmail. But it also helped them weed out those whose spending habits, gambling and other addictions, or extramarital affairs put them at a higher risk for babbling or becoming double agents.

Other times, the job required that CIA operatives go undercover as systems engineers, designers, shipping couriers, logistics and maintenance men, or hotel or cleaning staff to booby-trap computers as they made their way from manufacturers to assembly lines, shipment centers, and warehouses en route to a hostile leader, nuclear scientist, drug trafficker, or terrorist.

"People were the primary access points, given that somebody held the data room combination, the encryption codes, the passwords, and the firewall manuals," Gosler told his CIA trainees. "People had written the software. People managed the data systems. Ops officers should recruit computer hackers, systems administrators, fiber-optic techs, and even the janitor if he could get you into the right data-storage area or fiber-optic cable."

From 1996 until 2001, when he returned home to Sandia, Gosler's shop partnered with the NSA and other intel agencies to determine which technologies and weapons systems were worth targeting. CITO's technicians helped with the technicalities, but it was up to the CIA officers to run field operations and figure out how best to get hardware implants or software modifications into adversaries' systems.

"Gosler preached that we would be the pathfinders for the CIA's cyber offensive against an array of targets. As adversaries grew to acquire and use digital data, and as the internet expanded, so would the CIA's cyber operations. We would be in the forefront," Henry Crumpton, one of Gosler's trainees, would later write in his autobiography. "To reassure us, Gosler stressed that while this was new terrain, he would support us and we, the operations officers, should focus on operations. We did not need degrees in computer science. We just needed to understand the relationship between foreign intelligence in digital form and human nature. We would exploit the relationship. That I understood.

That I could do. The advent of espionage in cyberspace was nearly instantaneous. Its rapid growth and impact on our operations was stunning, even revolutionary."

It had taken nine years for one CIA operative to secretly photograph twenty-five thousand pages of classified Soviet and Polish military documents at the height of the Cold War. Suddenly a well-placed implant could siphon terabytes upon terabytes of intelligence booty in hours, in some cases minutes. "You begin to understand both the opportunity and the challenge," Gosler told me, when you stop to consider that one terabyte is equivalent to a thirty-one-mile-high stack of paper, each sheet packed with single-spaced data.

Once Gosler helped the NSA and CIA turn on the spigot, there would be no turning it off. Five years earlier, the American intelligence community's greatest fear was that the change in information flows would cause them to go blind or deaf. Now their greatest fear was drowning.

Pulling out critical, credible, actionable intelligence was getting to be nearly impossible as unprecedented flows of noisy, seemingly unrelated data made its way through an endless maze of digital pipes back to the Fort. Solving for Big Data would consume U.S. intelligence agencies for decades.

FIVE YEARS INTO his time at the CIA, Gosler figured that he had given all he had to give to what he still affectionately calls the "greatest HUMINT organization on earth."

It would take someone with a different combination of skills to take America's exploitation programs to the next level, to make sense of the fire hose of data flowing back to the Beltway. Gosler could not discuss much, if any, of his work at CIA, but in that short time he had earned the National Intelligence Medal of Achievement, the William J. Donovan Award, the Intelligence Medal of Merit, the CIA Director's Award, and the Clandestine Service Medallion. He remains the most decorated person to ever work in cyber in the American intelligence community.

His handiwork now infused nearly every clandestine operation the CIA undertook. Digital espionage and traditional spycraft had become symbiotic. Exploitation was now core to the CIA's mission as it tracked, captured, and killed terrorists all over the world. It was clear that the opportunities to exploit the digital domain were only growing exponentially with the CIA's growing reliance on unmanned aerial drones, which now carried their own video cameras, sensors, and interception equipment.

It was time for someone else to take the helm. And after five muggy Washington summers, Gosler was looking forward to getting back to the dry heat of the New Mexican desert. On his last day at Langley, he packed up his medals, management books, and the IBM Selectric typewriter bar he kept as a memento from Gunman and wished the men and women of the CIA well. They were the only ones who would ever know the impact of his work. Some had become his personal heroes; others his family. In the five years Gosler spent at Langley, his daughter had grown up, and he could never tell her exactly what he had been doing while he was away.

This was May 2001. By the time the godfather of American cyberwar walked out to the parking lot, got in his Jeep, and drove away, American intelligence agencies were siphoning unprecedented amounts of data from more than a hundred strategically placed implants in Iran, China, Russia, North Korea, Afghanistan, Pakistan, Yemen, Iraq, Somalia, and terrorism safe havens around the world. But even with all that data gushing back to Fort Meade that May, America's exploitation programs were still—at least by modern standards—highly targeted.

Four months after Gosler moved back west, the planes hit the buildings. Little would ever be "highly targeted" again.

CHAPTER 8

The Omnivore

Fort Meade, Maryland

I N THE DARKEST months of the post 9/11 inferno, shuttle buses packed with young recruits streamed toward unmarked NSA facilities around Washington.

On board, nobody said a word. Nobody knew why they were there, exactly, or what they would be interviewing for. The men on the bus—and for the most part they were all male engineers, hackers, and codebreakers—had been told, in the vaguest of terms, that they each possessed a unique set of skills that could help their country. With images of the planes streaking toward their targets, the towers collapsing, the Pentagon burning, and the charred wreckage smoldering in a lonely Pennsylvania field seared into their minds, they felt compelled to show up for the bus.

Their shuttles eventually stopped before an unremarkable building off campus from the Fort. As each disembarked, he was handed a red badge. The badge triggered a blinking red light anywhere he walked—a signal to anyone who saw it that these men lacked security clearance.

"Welcome to your operational interview," a coordinator told them. It was an oddly warm greeting for what was perhaps the country's most intense professional assessment.

Each recruit was handed an agenda that included several hours' worth of tests to assess their trustworthiness, discretion, ambition, skill, and any "derailers"—management-speak for signs of a dark side. There would be a technical interview, a polygraph, a drug test, and a psychological evaluation. Those who passed would receive a formal job offer in the mail. Their salaries started at $40,000—less than half what their engineering classmates could expect to make in Silicon Valley—for a secret job that, for all they knew, could have been cleaning toilets. It would be many more months before they were cleared to learn their mission. And then, of course, they were forbidden from telling anyone what exactly it was that they were doing here. As a reminder, the agency hung up a giant sign in the cafeteria: SHHH! DON'T TALK ABOUT WORK.

These men would be joining the NSA's elite, highly classified team of TAO hackers—a unit deemed so critical in post–9/11 intelligence gathering that for years the government tried to deny its very existence.

THE U.S. NATIONAL SECURITY apparatus, which had spent the decade leading to 9/11 probing and exploiting vulnerabilities the world over, was now face-to-face with its own dark shadow. It was collecting more data on more targets around the globe than it ever had in history, and yet it had missed critical intelligence. It had failed to connect the dots.

When American intelligence agencies rewound the tapes to the moments before the planes hit, they discovered that they'd had everything they would have needed to prevent the attacks. Intelligence officials had sounded the alarm bells on al-Qaeda no less than forty times in President Bush's daily intelligence briefing. All nineteen of the 9/11 hijackers were trained in CIA-monitored al-Qaeda camps in Afghanistan. Two of the hijackers had attended an al-Qaeda summit in Kuala Lumpur the previous year, yet had still been granted U.S. entry visas. A July 2001 dispatch from an FBI agent in Phoenix, warning colleagues that Osama bin Laden may have sent recruits to American flight schools for possible terrorist attacks, had carelessly slipped through the cracks. And just weeks before the attacks, FBI agents had even picked up a thirty-three-year-old Islamic radical by the name of Zacarias Moussaoui—the so-called twentieth hijacker—for suspicious behavior at a Minnesota flight school. Among his various belongings, agents discovered knives, binoculars, a handheld aviation radio, a laptop, and a notebook. Agents were still waiting for a judge to grant them access to Moussaoui's notebook and laptop when the towers came crashing down.

In the blame game that followed, the 9/11 Commission and other lawmakers—many of whom had voted to slash intelligence budgets over the previous decade—would all agree: intelligence had been at fault. The intelligence community needed more resources, more legal authorities, more data, more machines, and more people to ensure that nothing like 9/11 ever happened again. The Patriot Act was signed, and later the Foreign Intelligence Surveillance Act was amended to expand the government's ability to conduct electronic surveillance without court orders. Annual intelligence budgets surged to $75 billion from just a few billion dollars. The Office of the Director of National Intelligence, the National Counterterrorism Center, and the Department of Homeland Security were established to coordinate intelligence from disparate agencies and head off future threats.

In 2002 the Pentagon announced its Total Information Awareness project to ingest as much data as possible. Even after Congress cut off spending one year later—and the public wrote TIA off for dead—NSA continued to mine calls, emails, telephone conversations, financial transactions, web searches, and more as part of a classified project code-named Stellar Wind that would not be fully revealed until years later. Leads from Stellar Wind were so vague and voluminous that agents called them "Pizza Hut cases"; seemingly suspicious calls frequently turned out to be takeout orders.

The goal was to track down every lead and watch every terrorist, would-be terrorist, terrorist sponsor, and foreign adversary. The U.S. government wanted to know who they knew, where they slept, who they slept with, who paid them, what they bought, where they traveled, when they ate, what they ate, what they said, and what they thought in an expansive effort to anticipate terrorist plots well before things went *boom*. "If we didn't know where they got their hair cut, we weren't doing our jobs," one former NSA employee told me.

In the thirty months after 9/11, as al-Qaeda attacks picked up, NSA lawyers set to work, aggressively reinterpreting the Patriot Act to collect Americans' phone records in bulk, and later weakening the Foreign Intelligence Surveillance Act to allow for warrantless wiretapping. The agency began intercepting foreigners' phone calls, including Americans' long-distance phone calls abroad. NSA analysts mined phone calls in Iran, Iraq, North Korea, Afghanistan, and Russia. They scooped up calls from officials at Mexico's cartel-tracking agencies. Even America's closest allies—Israeli Air Force officials, German chancellor Gerhard Schroeder, and his successor Angela Merkel—made the NSA's target list. The agency devoured data from fiber-optic cables and telephone switches and mandated that America's biggest telecoms turn over metadata for every single call made into, out of, and wholly within the United States. As the world migrated from phone calls to web calls, email, text messages, and, later, encrypted messaging channels like WhatsApp, Signal, and ISIS's own messaging app, the Amaq Agency, NSA was never far behind. The agency had become what my colleague Scott Shane called "an electronic omnivore of staggering capabilities."

No morsel was too small for America's ever-expanding, well-funded cyberespionage machine. More knowledge would save us from the next terrorist attack, went the thinking; or, as Chairman Mao put it, "the only real defense is active defense." Of course, those in the know also recognized that with a few taps on a keyboard, digital espionage could be redirected for attacks. In an intravenous delivery system, once the needle is in, it can be used to replenish fluids or to drop in a lethal dose of medication. So too with exploits. Those same holes the NSA was scouting, stockpiling, and exploiting for reconnaissance could be injected with a payload that could destroy machines on the other end.

In fact, it was only a matter of time.

EVERY TAO EMPLOYEE remembers the first time they entered the Fort, passed through electric fences guarded by antitank barriers, motion detectors, and rotating cameras, and arrived in the small city of fifty buildings, each its own Faraday cage, its walls and windows shielded with copper mesh to keep any signals from leaving the building. At the center of the Fort was a bank, a drugstore, and a post office. Beyond those were the NSA's very own police force and fire brigade. And farther in still, in a segregated complex barricaded by steel doors and armed guards, was the heart of TAO's Remote Operations Center, known as the ROC—one of the rare government offices where jeans and T-shirts outnumbered suits.

Nobody inside the ROC called themselves hackers, but for all intents and purposes, that is what they were. Few beyond its walls knew what this unit did. Its work was so closely protected that at one point its bosses considered installing iris scanners outside their doors. They quickly abandoned this idea, concluding that the scanners only offered a veneer of security; in reality they just introduced more complexity—more ways to get hacked. Inside, hundreds of military and civilian computer specialists worked in rotating shifts, twenty-four hours a day, seven days a week, at desks cluttered with cans of Diet Coke and Red Bull. It was not unusual for analysts to get paged in the middle of the night, beaconing them to call in on a one-way classified phone for a critical mission. In the aftermath of 9/11, these hundreds turned to thousands as TAO accelerated its breaking-and-entering mission around the globe, through a combination of brute-force hacking, cracking passwords and algorithms, finding zero-days, writing exploits, and developing implants and malware that bent hardware and software to their will. Their job was to find every crack in every layer of the digital universe and plant themselves there for as long as possible.

In the post 9/11 decade, implants once, under Gosler's tenure, reserved for a few hundred terrorists and foreign officials in China, Russia, Pakistan, Iran, and Afghanistan now numbered in the tens of thousands and eventually grew into the millions. As iDefense was drawing up its modest price lists for bugs in Chantilly—and getting pilloried for it by the big tech companies—TAO hackers, fifty miles east, were mining for bugs on BugTraq, scouring obscure hacker magazines, and ripping apart any new hardware and software that hit the market in search of bugs to add to the agency's zero-day arsenal. Those backdoors I'd first glimpsed in the closet of Snowden's leaked classified secrets? Those were just the tip of the iceberg, former TAO hackers told me. Compared to the outsize role Snowden would come to play in the public's imagination, his role, and access, inside NSA was actually quite limited.

"Snowden was a low-level admin," one former TAO hacker told me. "The NSA's capabilities were far, far more expansive than what Snowden revealed."

Beyond Snowden's reach—up another several rungs of access—was an arsenal of exploits available only to the agency's elite TAO hackers. There, TAO vaults contained a catalog of vulnerabilities and exploits that granted entry into most nooks and crannies in the digital universe. The agency could hardly keep track of all the hacking tools at its disposal. It had to turn to—what else?—computer algorithms to name its various exploits.

"Originally they had us targeting terrorist channels, then the operating systems," a former TAO operator, whose stint at the NSA included the run-up to and aftermath of 9/11, told me. "Then we started going after the browsers and third-party applications. Eventually, there was a big shift and we started targeting the metal with kernel-level exploits."

The "kernel" is the nerve center of any computer system. It manages communications between a computer's hardware and software. In a machine's pecking order, the kernel is at the very top, allowing anyone with secret access to it to take full control of the device. The kernel also forms a powerful blind spot for most security software, allowing attackers to do what they want, unnoticed, and remain there for months, years even, regardless of how vigilant their victim is at installing software patches and updates. Spies coined a name for these attacks: "The race to the bare metal." The closer to the bare metal of a computer that TAO's hackers got, the deeper and more resilient their access. The NSA began recruiting hackers who specialized in kernel exploitation. Within a decade TAO's hackers invisibly lurked inside the kernel of thousands of machines, with access so deep they could capture every morsel of their target's digital life. Like an addict, the unit could never get enough.

In a nod to just how deep into the bare metal TAO had ventured, the unit's hackers developed their own logo—a spoof on Intel's ubiquitous Intel Inside logo, which alerts computer users that Intel processors are baked inside. TAO crafted its own TAO Inside logo, a tongue-and-cheek reminder that their unit was now inside everything.

TAO became a digital Ford assembly line for cyberespionage. One unit searched for vulnerabilities and developed exploits. Another honed and sharpened the implants and payloads they used once TAO hackers had their beachhead. When a terrorist, Iranian general, or arms dealer could not be easily monitored, a separate elite TAO team was tasked with finding a way to intercept him. Sometimes this entailed a coding breakthrough. Other times, it entailed hacking his mistress or maid, and leapfrogging into his home or office from there. Yet another TAO unit, known as the Transgression Branch, oversaw the NSA's "fourth-party collection" efforts—jargon for piggybacking on another country's hacking operation. This branch was considered especially sensitive because it frequently involved hacking American allies, like South Korea or Japan, to net intelligence on notoriously hard-to-reach targets like

North Korea. Other units managed the back-end infrastructure used to collect and analyze the fire hose of data gushing in from TAO's implants to strategically placed NSA servers around the globe, many of them operated by front companies in China or the tiny, geographically well-situated nation of Cyprus.

Separate TAO units worked closely with the CIA and FBI to reach unreachable offline targets and networks. In some cases American agents would spend months grooming someone in a target's inner circle to get them to physically place a TAO implant on the target's computer. Other times, TAO would closely monitor their target's purchase history, tipping off agents to any opportunities to seize a target's package and plant an implant in transit. Sometimes it was as simple as a CIA officer throwing on a hard hat, dressing up as a construction worker, and walking into a target's offices. "It's amazing what people let you do when you're wearing a hard hat," one former CIA officer told me.

Once inside, CIA agents could place the implant themselves or disguise it as a thumb drive and leave it on a secretary's desk. As soon as someone plugged the thumb drive into the target's network—*Bingo!*—TAO could beacon out to other devices in the building, and digitally crawl toward their target from there. The NSA wasn't the only one using this schtick. In 2008 Pentagon officials were horrified to discover Russian hackers inside their classified networks. When analysts traced back to the source of the breach, they discovered that Russian spies had scattered infected USB drives around the parking lot of a U.S. army base in the Middle East. Someone had picked one up and plugged the compromised thumb drive into a classified network shared by the military, intelligence agencies, and senior officials at the White House. (The Pentagon later sealed its USB ports with superglue.)

The War on Terror and the wars in Afghanistan and Iraq only drove demand for TAO's tradecraft. In the rush to gather as much data as possible for the NSA's "customers"—another bit of agency jargon—at the White House, Pentagon, FBI, CIA, and across the Departments of State and Energy, Homeland Security and Commerce, everything became fair game for interception.

The post–9/11 decade also marked the golden age of spying: Google became a common verb. Its ubiquity and utility offered spies a fine-grained record of a target's life, rich with mundane and compromising details, stored in a permanent archive, accessible from anywhere on earth and protected, in most cases, by nothing more than a single password. TAO hackers suddenly had access to every place their targets went, what they did, who they talked to. Using a target's GPS coordinates alone, TAO hackers could track visits to substance abuse or mental health clinics or one-night-stand motels, all useful grist for blackmail. Google search histories gave spies an intimate window into targets' twisted curiosities. "One of the things that eventually happens . . . is that we don't need you to type at all," Google's then chief executive Eric Schmidt said in 2010. "Because we know where you are. We know where you've been. We can more or less guess what you're thinking about." And so, too, could the NSA.

With the arrival of Facebook in 2004, it was often hard to see where the NSA's efforts ended and Facebook's platform began. Suddenly people were gleefully uploading reams of personal data—photos, whereabouts, connections, even their inner monologues—to the web. The agency could now read the musings of Islamic fundamentalists, capture vacation photos of Russian oligarchs skiing in Val d'Isère and gambling in St. Moritz. Using an automated NSA program called Snacks—short for Social Network Analysis Collaboration Knowledge Services—analysts could visualize their target's entire social network. Every friend, family member, or work acquaintance could lead to still further TAO targeting.

But nothing changed the surveillance game more than Apple's unveiling of the first iPhone in 2007. TAO hackers developed ways to track an iPhone user's every keystroke, text message, email, purchase, contact, calendar appointment, location, and search, and even capture live audio and video feeds of her life by hijacking her phone camera or hot-miking her microphone. The NSA swallowed up mobile alerts from travel companies—flight confirmations, delays, and cancellations—and cross-matched them with itineraries of other targets. An automated NSA program called Where's My Node? sent analysts an email anytime an overseas target moved from one cell tower to another. TAO was now conducting instant, total invasion of privacy with minimal effort. And, until Snowden, iPhone users seemed blissfully oblivious to their invisible NSA ankle bracelet.

Meanwhile, TAO became a vast archipelago of tranches, each with its own incentives to glean and analyze as much intelligence as possible. TAO's operations were scattered and compartmentalized across eight embassies around the world and satellite offices across the country. In Aurora, Colorado, TAO employees partnered with the Air Force to hack spacecrafts and satellites. In Oahu, they worked with the navy to intercept threats to American warships. At an NSA hub in Augusta, Georgia—appropriately code-named Sweet Tea— TAO hackers intercepted communications from Europe, the Middle East, and North Africa. And at an old Sony chip factory in San Antonio, TAO hackers monitored drug cartels and officials in Mexico, Cuba, Colombia, Venezuela, and occasionally the Middle East.

All this took place under utmost secrecy. Americans only caught their first glimmer of what the NSA was up to when San Antonians started complaining on neighborhood forums that their garage doors were opening and closing at random. Some filed police reports, believing neighborhood thieves were to blame. But the cops were at a loss. The incidents forced the NSA to make a rare admission that a rogue agency antenna was to blame. It was inadvertently interacting with old makes of garage door openers.

THERE WAS A necessary detachment to this work. Breaking algorithms, deciphering ones and zeroes, and probing hardware and software to find vulnerabilities and weaponize exploits was all just a day's work. It became harder and harder to follow the itinerary of a single NSA exploit from discovery of a flaw through its life cycle as an instrument for surveillance or attack.

And so, in the years following 9/11, the NSA decided to give its top analysts a glimpse into the fruits of their labors. Senior officials organized a briefing that, two people who were present told me, would be seared into their memories for years to come. In a secure room at Fort Meade, the officials projected more than a dozen faces onto a bright screen. Each man on the screen, the analysts were told, was dead—thanks to their digital exploits. Half the men in the room felt an intense pride in knowing that their work had been used to kill terrorists. "The other half had an allergic reaction," one former TAO analyst told me. "It was 'Here's your work. Here's the death count. Good job. Keep up the great work.' Until then, it was all about breaking an algorithm. It was about math. Suddenly it was about killing people. That's when things changed. There was no going back."

If TAO hackers had any misgivings about their work, often all they had to do was glance over at what their Chinese counterparts were doing to feel better about their mission. Chinese hackers were not just engaged in traditional state espionage; they were pilfering intellectual property from every major company in the Fortune 500, American research laboratories, and think tanks. No longer content to be the world's cheap manufacturing hub, Beijing had dispatched its country's hackers to steal trade secrets from innovators abroad, the vast majority of them in the United States, and were now passing billions, by some estimates trillions, of dollars' worth of American research and development to China's state-owned enterprises.

Mike McConnell, the former director of national intelligence, would later tell me, "In looking at any computers of consequence—in government, in Congress, at the Department of Defense, aerospace, companies with valuable trade secrets—we've not examined one yet that has not been infected," by China.

TAO may have been putting its hands into everything, but at least employees could tell themselves they weren't engaged in theft for profit. The agency had a tendency to recast its mission as a noble calling. "Sigint professionals must hold the moral high ground, even as terrorists or dictators seek to exploit our freedoms," one classified NSA memo declared. "Some of our adversaries will say or do anything to advance their cause; we will not."

As for the American data getting snagged in their expansive dragnet, TAO hackers I spoke with were quick to point out that the agency had strict protocols that prohibited them from querying for any American data picked up in what the agency called "incidental collection." TAO employees' queries were closely monitored by a team of NSA auditors, who reported up to another oversight team, the inspector general, the agency's lawyers, and general counsel. "Querying for Americans' data would have landed you in jail," one former TAO operator told me.

But that was not entirely accurate. In the years following 9/11, a dozen NSA employees were caught trying to use the agency's vast eavesdropping apparatus to spy on their exes and love interests. The incidents were by no means common, but the agency had coined a name for the practice, nonetheless: LOVEINT, a new twist on SIGINT, "signals intelligence," and HUMINT, "human intelligence." In each case, NSA auditors caught the offenders within days, demoted them, cut their pay, and revoked their clearances, which in many cases left them with little choice but to leave the agency. But not one was criminally prosecuted.

And as much as agency officials would later cite the NSA's oversight by the Foreign Intelligence Surveillance Court, which by law had to approve any surveillance operation that might target Americans, the court had become something of a rubber stamp. The NSA's arguments were heard without any opposing counsel, and when the court—under public pressure from the Snowden revelations—finally released numbers, it showed that of the 1,789 applications it received to surveil Americans in 2012, it approved 1,748 without any changes. Only one case had to be withdrawn.

Sometimes the NSA's efforts bore immediate fruit. In one case they successful thwarted a plot to kill the Swedish artist who had drawn pictures of the Prophet Muhammad. In another, they were able to tip off workers at New York's JFK airport to the names and flight numbers of a Chinese human smuggling ring. Eavesdropping equipment aboard a Pentagon plane, flying sixty thousand feet over Colombia, tracked the whereabouts and plots of Colombia's FARC rebels.

"You have to understand," one of the NSA's TAO analysts told me, "we were collecting crazy intelligence. You couldn't believe the shit you were looking at. Our work was landing directly in presidential briefings. You felt your work was saving countless lives."

By 2008 the NSA feverishly began removing human decision-making—and with it any complicated moral calculus—from their work. A highly classified NSA software program code-named Genie began aggressively embedding implants not just in foreign adversaries' systems but in nearly every major make and model of internet router, switch, firewall, encryption device, and computer on the market.

By 2013 Genie was managing 85,000 implants—four times the number of implants it had managed five years earlier—according to U.S. intelligence budgets, with plans to push that number into the millions. While three-quarters of those implants still prioritized targets in Iran, Russia, China, and North Korea, TAO had become far less discriminate.

In a secret post to an internal NSA message board, leaked by Snowden, an NSA operative described the new high-priority target: foreign IT systems administrators whose administrative credentials gave them broad access to hundreds of thousands, if not millions, of more potential targets still. Each NSA implant began pulling back huge volumes of foreign secrets in the form of text messages, emails, and voice recordings, previously unimaginable to digital exploitation pioneers like Gosler.

THE CLOSEST I would ever get to the NSA's large-scale SIGINT operations was Operation Shotgiant. For years American officials blackballed Huawei of China—the world's largest manufacturer of telecom equipment—from American business dealings. More recently, the United States has been on a crusade to pressure allies to ban Huawei's equipment from new high-speed 5G wireless networks, citing suspected ties between the company and China's Community Party. The Trump administration has gone as far as threatening to hold back intelligence if allies move forward with Huawei.

American officials routinely point out that Huawei's founder, Ren Zhengfei, "China's Steve Jobs," was a former Chinese PLA officer, and warn that Huawei's equipment is riddled with Chinese backdoors. Chinese intelligence could use that access to intercept high-level communications, vacuum up intelligence, wage cyberwar, or shut down critical services in times of national emergency.

That all may very well have been true. But it is also certainly true in reverse. Even as American officials were publicly accusing China of embedding trap-doors in Huawei's products, my *Times* colleague David Sanger and I learned from leaked classified documents that the NSA had pried its way into Huawei's headquarters in Shenzhen, years ago, stolen its source code, and planted its own backdoors in the company's routers, switches, and smartphones.

Conceived as far back as 2007, Shotgiant was initially designed to sniff out any ties between Huawei and the PLA, which had been hacking American companies and government agencies at will for years. But it wasn't long before the NSA started using its foothold to penetrate Huawei's customers, particularly countries like Iran, Cuba, Sudan, Syria, and North Korea that studiously avoided American technology.

"Many of our targets communicate over Huawei-produced products," one classified NSA document said. "We want to make sure that we know how to exploit these products," it added, to "gain access to networks of interest."

But the NSA did not stop at Huawei. Shotgiant expanded into the hacking of two of China's largest cellular networks, which were now riddled with NSA implants too. At the time our story broke in 2014, classified documents made clear that the NSA was still working on new implants and malware—tools that could sniff out voices of interest on Chinese cell networks, capture select cuts of their conversations, and siphon them back to Fort Meade, where teams of NSA translators, decoders, and analysts broke them down for critical intelligence. In short, the NSA was doing everything it accused Beijing of doing, and then some.

By 2017 the agency's voice recognition and selection tools had been widely deployed across Chinese mobile networks. And the United States was not just hacking into China. Hundreds of thousands of NSA implants were deeply embedded in other foreign networks, routers, switches, firewalls, computers, and phones around the globe. Many were actively siphoning texts, emails, and conversations back to the agency's server farms every day.

Many others were sleeper cells, dormant until called upon for a rainy day or some future shutdown—or all-out cyberwar.

IN THE POST–9/11 urgency to capture and analyze as much data as humanly possible, leaked classified documents and my interviews with intelligence officials made it clear that few had stopped to question what the potential implications might be if word of their digital escapades ever got out.

No one asked what this might one day mean for the American tech companies they were breaking into, who were now servicing more customers abroad than in the United States. During the Cold War, the NSA did not have to reckon with this dilemma: Americans spied on Russian technology, while Russians backdoored American typewriters. But that was no longer the case. The world was now using the same Microsoft operating systems, Oracle databases, Gmail, iPhones, and microprocessors to power our daily lives. Increasingly, NSA's work was riddled with conflicts of interest and moral hazards. Nobody seemed to be asking what all this breaking and entering and digital exploitation might mean for the NSA's sponsors—American taxpayers—who now relied on NSA-compromised technology not only for communication but for banking, commerce, transportation, and health care.

And nobody apparently stopped to ask whether in their zeal to poke a hole and implant themselves in the world's digital systems, they were rendering America's critical infrastructure—hospitals, cities, transportation, agriculture, manufacturing, oil and gas, defense; in short, everything that undergirds our modern lives—vulnerable to foreign attacks. There are no patents on vulnerabilities, exploits, and malware. If NSA found a way to exploit a digital system, there was a good chance that one day, months or years down the road, other bad actors would discover and exploit those very same weaknesses.

The NSA's answer to this moral hazard was more secrecy. So long as its tradecraft was highly classified and invisible, the agency could keep kicking the can down the road. Its critics argued that those classification levels did little to make Americans more secure; they only shielded the agency from further accountability, raised the stakes when its programs and tradecraft inevitably leaked into the public domain, and inspired others—not just the elite cyber powers—to join the game.

"The NSA's fatal flaw is that it came to believe it was smarter than everyone else," Peter G. Neumann, one of America's sages of cybersecurity, told me one day.

Neumann, now in his late eighties, is one of the few computer scientists—if not the only one—who can brag that he discussed the nation's inherent vulnerabilities with Albert Einstein. And over the years he became a voice in the wilderness, warning officials at NSA, the Pentagon, and everyone down the chain that security flaws would one day make for disastrous consequences.

There was an arrogance to the NSA's work, Neumann told me. By inserting backdoors into any piece of technology it could get its hands on, the NSA assumed—to the country's detriment—that all the flaws it was uncovering in global computer systems would not be discovered by someone else. "They were dumbing everything down so they could break it. They thought, 'Only we have backdoors that we can use,' without realizing that these were backdoors that every nation in the world wanted too."

"It's the arms race all over again," Neumann told me. "In the race to exploit everything and anything we could, we painted ourselves into a dead end where there is no way out. It's going to be a disaster for the rest of the country."

The same implants TAO employees relied on for espionage could also be used for future cyberattacks. They simultaneously functioned as sleeper cells that could flip from benign to devastating. Malware used for espionage could be swapped out or amended to destroy data on the other side of a machine. It could shut down foreign networks, or even take out physical infrastructure, at the stroke of a key. Any industrial systems used by our enemies—even if they were the very same systems used by Americans—were being targeted for a rainy day.

By the rules of spycraft, this was entirely fair play. America spies, China spies, Russia spies. But in 2009, without any debate at all, in the cordoned-off copper walls of Fort Meade, the United States set new rules for cyberwar.

Starting that year, it was not only acceptable to implant code in a foreign nation's critical infrastructure; now the United States made it perfectly okay to reach across a border and take out another nation's nuclear program.

So long as nobody ever uttered a word about it. And so long as it did so with code.

CHAPTER 9

The Rubicon

Natanz Nuclear Facility, Iran

G ET ME A third option," Bush told aides that June.

It was 2007, and the United States, under pressure from Israel, had to deal with Iran and its weapons-grade uranium. For almost a decade Iran had concealed the construction of its Natanz nuclear enrichment plant, burying two cavernous halls roughly half the size of the Pentagon beneath some thirty feet of rock, dirt, and concrete.

Diplomacy hadn't stopped it. Option 1 was out. Pentagon analysts had started war-gaming how Tehran might respond to an Israeli strike, and what it would mean for U.S. troops in the region. An incursion would almost certainly send oil prices skyrocketing. American troops, if they got involved, were already overstretched throughout the rest of the Middle East. Option 2, gone.

President George W. Bush needed something to get the Israelis off his back that didn't entail starting World War III.

It was Keith Alexander, the NSA's techno wizard, who proposed the Hail Mary, option 3. As NSA director, General Alexander had always been something of an anomaly. Unlike his predecessors at the agency, he was a hacker in his own right. At West Point he'd toiled on computers in the electrical engineering and physics departments. In the 1980s, at the Naval Postgraduate

School in Monterey, he'd built his own computer and developed his own programs to move the army's clunky index-card system onto automated databases. On his first assignment to the Army Intelligence Center in Fort Huachuca, Arizona, he'd made it a point to memorize the technical specifications for each and every army computer, then mapped out the center's first intelligence and digital warfare data programs. As he worked his way up the chain of command, he'd picked up master's degrees in electronic warfare, physics, national security strategy, and business.

Just before he was picked to be the sixteenth director of the NSA, Alexander served as director and then chief of an army intelligence organization in Fort Belvoir, Virginia, where he had worked out of a *Star Trek*–themed starship *Enterprise*, complete with a captain's chair and doors that went *whoosh* anytime they opened and closed. Some called him Emperor Alexander—a nod to his *Star Trek* fixation, but also because he had a way of using his geeky charm to get what he wanted. Military types called him Alexander the Geek, but Alexander's predecessor at the agency, Michael Hayden, called him a cowboy because he had a reputation for acting first, apologizing later. Hayden's other derisive nickname for Alexander was The Swoosh, after Nike's swoosh logo and "Just Do It!" slogan.

By the time Bush was demanding a "third option," the NSA and the national energy labs had already been engaged in a years-long effort to map out Iran's nuclear facilities. At TAO, hackers had conducted nonstop reconnaissance to pull back any nuclear blueprints. They'd deployed viruses designed specifically to look for so-called AutoCAD files—software that maps out the computer networks on manufacturing—or, in this case, nuclear enrichment plant floors. They noted any common Iranian operating systems, applications, functions, features, and code, and began stockpiling zero-days in nearly every make and model of machine that Iran's nuclear workers and contractors used, in search of more data. All of this had been done for espionage, but Alexander knew that the very same access could be used for a cyberattack of a different nature, one that could sabotage the infrastructure on the other end. Those "computer network attacks"—CNA in intel jargon—required specific presidential approval by law. And until 2008, they had been fairly basic and limited. At one point the Pentagon tried to take out al-Qaeda's communications in

Iraq, for example—but that was child's play compared to what Alexander proposed next.

IN AN EPIC feat of translation, Alexander briefed Bush on what a destructive cyberattack on Iran's nuclear facilities might look like.

At the Energy Department's Oak Ridge National Laboratory in Tennessee, engineers and nuclear experts had built a near replica of Natanz's nuclear facility, complete with Iran's P-1 centrifuges. The engineers understood that to break Iran's program, they had to break its centrifuges—devices that spin at supersonic speeds, more than 100,000 revolutions per minute—and separate the isotopes that make things go boom.

They also understood that the weak links in all centrifuges are the spinning rotors, which are fragile and fickle. They must be light, but strong and well-balanced, with ball bearings to reduce friction. Move the rotors too fast, and you risk blowing up the centrifuges. Hit the brakes too suddenly, and you risk spinning a giant six-foot-tall centrifuge off its axis, like a tornado that demolishes anything in its path. Even under normal conditions, it's not uncommon for centrifuges to break or blow up. The United States has blown up its fair share of centrifuges over the years. In Iran, engineers regularly replaced about 10 percent of their centrifuges each year due to natural accidents.

What Alexander proposed in 2008 was a cyberattack that could mimic and accelerate those accidents using weaponized code. The rotors that spun Natanz's centrifuges were controlled by specialized computers called programmable logic controllers (PLCs). PLCs made it possible for Natanz's nuclear technicians to monitor centrifuges remotely, check their speeds, and diagnose any problems. If, Alexander explained, NSA could get far enough inside the Natanz computers that accessed those PLCs, its hackers could control the rotor speeds and spin the centrifuges out of control, or stop them from spinning entirely.

Best, the process could happen gradually, so that Iran's technicians might write off the incidents as normal technical malfunctions. Yet the destruction could set Iran's nuclear ambitions back years. It would, no doubt, be the riskiest gamble the United States had ever taken in the cyber realm. But if worked, it might just force Tehran to the negotiating table.

Alexander's pitch always reminded me of the quote that Sabien, the original zero-day broker, shared with me years later: "The most likely way for the world to be destroyed, most experts agree, is by accident. That's where we came in; we're computer professionals. We cause accidents."

There were no guarantees. But in 2008, the Israelis moved to a maximum-pressure campaign to convince the Bush White House to either hand over bunker busters capable of obliterating Iran's nuclear facilities, or get out of its way. That June, the Israeli Air Force dispatched more than a hundred F-15 and F-16 fighter jets, and accompanying refueling tankers and helicopters, to Greece. The White House didn't have to guess why. The distance between Tel Aviv and the Acropolis was almost the same distance from Tel Aviv to Natanz. "It was Israel's way of telling us it was time to shit or get off the pot," a Pentagon official told me later.

The Israelis did not bluff. In a solo operation one year earlier, Israeli fighter jets, under cover of darkness, had decimated a Syrian nuclear reactor on the banks of the Euphrates after Bush made clear the United States would not bomb it. Israel had done the same back in 1981, when they laid waste to Iraq's Osirak nuclear reactor. They were now planning an imminent attack on Natanz. And if Israel went through with it, Pentagon simulations showed that the United States would be dragged into World War III. A record number of American soldiers had died in Iraq the previous year and Bush's political capital, and approval ratings, had reached a new nadir.

With no other good options on the table, Bush went for Alexander's Hail Mary.

There was just one caveat: they would have to bring in the Israelis. They'd have to show them there was another way. The Israelis also had the clearest view of anyone into how Iran's nuclear systems worked. And their cyber skills started to rival even those of TAO.

OVER THE NEXT few weeks, Olympic Games came together.

Some say it was an NSA computer algorithm that came up with the name. Others said it was purposely chosen for the five Olympic rings, symbolizing the unprecedented five-way cooperation between NSA, Israel's Unit 8200, the CIA, Mossad, and the national energy labs. For months, teams of hackers, spies, and nuclear physicists traveled back and forth between Fort

Meade, Langley, Tel Aviv, Oak Ridge, and Israel's own nuclear test site, Dimona, where the Israelis had built their own giant replica of Natanz. They planned out their entire mission like a SEAL Team Six operation: navigation, entry and exit strategies, delivery vehicles, and customized weapons equipment.

Iran's leaders inadvertently gave the operation a major opening. On April 8, 2008—Iran's "national nuclear day"—Mahmoud Ahmadinejad invited reporters and photographers on a personal tour of Natanz. With cameras snapping along the way, Ahmadinejad walked reporters past Natanz's three thousand spinning P-1 centrifuges and then, like a proud new father, showed off Iran's glistening new second-generation centrifuges, which, he bragged, could enrich uranium "more than five times" as efficiently as the P-1s. The photographs offered the clearest glimpse yet into a world previously known only to the Iranians and a few nuclear inspectors.

"This is intel to die for," a British nuclear proliferation expert said at the time, unaware that at that very moment NSA analysts were mining every single photograph for a digital entry.

With those images, designs, and centrifuges in hand, the Americans and Israelis drew up a list of everything they would need for their cyberweapon. They needed a list of every person, contractor, and maintenance person who walked into Natanz. They needed a way to bypass any antivirus or security protections in place. They needed to know everything about each computer inside the building: operating systems, features, printers, how they connected to one another and, most importantly, to the PLCs. They needed a way to invisibly spread their code from machine to machine. And the NSA lawyers needed guarantees that when it came time to drop their payload, the attack would be highly targeted, with minimal collateral damage. The lawyers were rightly apprehensive: PLCs were used for everything from spinning cotton candy to the brake system on roller coasters and in auto and chemical plants all over the world. They needed to be sure that the attack would only work on the exact PLCs that spun Iran's centrifuges. The Americans and Israelis then needed to design a payload, the actual instructions that would spin the rotors and destabilize the centrifuges. And they would need a way to convince Natanz's technicians that all was well, even as their centrifuges spun into oblivion. There could be no fingerprints. No accidental misfires. And no impulsive strikes. The code would need to lie dormant, undetected, over time, so as not to blow their cover.

Meeting even one of these needs was a feat. Meeting them all in tandem, over months and years, under cover of darkness, was an espionage coup of such unprecedented digital magnitude that it would later draw comparisons to the Manhattan Project.

Eight months later, a busted P-1 centrifuge was transported by private jet from Oak Ridge, Tennessee, to the White House Situation Room.

Bush gave Olympic Games the final go.

WE STILL DON'T know exactly who brought the worm in. Some suspect a Mossad spy, a CIA officer, a Dutch mole, a well-paid insider, or an unwitting contractor at one of the five Iranian companies that Olympic Games targeted in the lead-up to the first attack. And we may only find out in 2039, when Olympic Games is set to be declassified. For now, all we know is that it had to have been a human with an infected thumb drive.

Natanz's computers were "air-gapped," specifically to keep the Americans and Israelis out. Years earlier, the Americans allegedly tried to sabotage the centrifuges with a far more rudimentary attack. U.S. spies had intercepted Natanz's power supplies as they traveled from Turkey to Iran, and when the equipment was plugged in, a powerful electrical surge ripped through the frequency converters that controlled the centrifuge motors, causing them to blow up. The surge was so obviously an outside attack that Iran switched out its suppliers and ensured that no machine inside the facility would ever touch the internet.

Yes, that means that for all the advancements in digital exploitation over the previous three decades, sometimes technology only got you so far.

But once the human touch triggered the attack, the rest could be accomplished using the magic of seven zero-day exploits, four in Microsoft software and three in the German Siemens software inside the PLCs.

We still do not know where—with two glaring exceptions—these zero-days came from, whether they were developed "in-house" by TAO or Israel's Unit 8200 or procured off the underground market. What we do know is that the worm—in its final form, 500 kilobytes—was fifty times bigger than anything discovered before it. It was one hundred times the kilobytes required to send *Apollo 11* to the moon. And it was pricey, easily worth millions of dollars. But, held against a single $2 billion B-2 bomber, it was a Costco bargain. Each of the seven zero-days played a critical role in getting their worm into the Natanz computers and crawling its way into the centrifuges.

The first was a zero-day in Microsoft software that allowed the worm to jump from the infected thumb drive onto Natanz's computers. That exploit cleverly disguised itself as a benign .LNK file—the files used to display the tiny icons of a thumb drive's contents, like music MP3s and Microsoft Word files. Anytime you plug a USB stick into a computer, a Microsoft tool will automatically search a USB for these .LNK files. Attacking that scanning process triggered the worm to jump into action and deposit itself onto the very first Natanz machine without so much as a click.

Once the worm was on that first Natanz computer, a second Microsoft Windows zero-exploit kicked in—though technically, this second exploit wasn't a zero-day at all. It had been detailed in the obscure Polish hacking magazine *Haking*, which TAO and Unit 8200 hackers apparently kept close tabs on, but nobody at Microsoft or in Iran had bothered to read.

The flaw detailed in *Haking* went like this: anytime someone hits print on a Windows computer, a file is created with the printout's contents and control codes that tell the printer whether the document needs to be printed front to back, in black and white, or in color. By attacking that printer function, a person can crawl into any computer in the local network that the printer accesses.

Just like universities and labs that use open file-sharing networks to share spreadsheets, music files, and databases, Natanz had its own file-sharing network. The worm was able to spread from computer to computer in search of its final destination. Sometimes it did this using the printer-spooling exploit. Other times, it used another well-known remote code execution exploit that Iran's technicians just hadn't bothered to patch.

Once the worm was inside Natanz's local network, it used two additional Microsoft Windows zero-days to take control of each machine it infected, in its hunt for the computers that controlled Natanz's PLCs. To skirt Windows defenses, the worm used the digital equivalent of a stolen passport. It vouched for its components using security certificates from two different companies in Taiwan. This in itself is no small feat. Only a select few multinationals are deemed secure enough to issue the digital certificates that vouch that a website is secure or, in this case, that Windows' operating system could trust the driver the worm installed on each new machine. Companies keep the private keys needed to abuse their certificates in the digital equivalent of Fort Knox. The vaults that hold these keys are often protected with cameras, biometric sensors, and can only be accessed, in many cases, when two trusted employees present their piece of a two-part credential. The scheme is designed to prevent an insider from stealing keys that could, undoubtedly, fetch a pretty penny on the black market. (In this case, the fact that two Taiwanese certificate authorities shared the same office park led many to suspect the heists were, indeed, an inside job.)

Moving from computer to computer, the worm looked for anyone that had installed Siemens' Step 7 software for PLCs. That software would reveal how fast the centrifuges were spinning, and whether they were stopping or shutting down. Then the hackers used a tried-and-true hacking technique: the default password issued by the manufacturer (usually "admin" or "password"). With that, they were practically in.

The worm was able to break into the Step 7 database and embed malicious code in the data files. Then it simply waited for Natanz's employees to connect to the database, which would trigger an additional exploit to infect the employee's machine. And once they were on those machines, they had access to the PLCs and the centrifuge rotors they controlled. The worm was exceedingly careful, designed with lawyers' feedback in mind, so it only dropped its payload on PLCs that met exact requirements. The worm specifically looked for PLCs that controlled clusters of precisely 164 machines. That was hardly a random number: Natanz clustered its centrifuges in groups of 164.

When the worm found a match, it dropped its payload on the PLC. This one step was surreal in its own right. Until this point, there had never been a worm that worked on both PCs and PLCs. The two machines are completely different, with their own languages and microprocessors. The first thing the worm did on the PLCs was sit and wait. For thirteen days, it did nothing but measure the speed of the centrifuge rotors. It was checking to make sure the rotors ran at speeds between 800 and 1100 hertz, the exact frequency range used by Natanz's centrifuges. (Frequency converters that operate past 1000 hertz are actually bound by U.S. export controls because they are primarily used for uranium enrichment.) Once that thirteen-day waiting period was over, the payload got to work. The code was designed to speed up the rate at which the rotors spun to 1400 hertz for exactly fifteen minutes, before returning to normal for twenty-seven days. After those twenty-seven days were over, it would slow the speed of the rotors to just 2 hertz for fifty minutes, before returning to normal for another twenty-seven days, and repeating the whole process all over again.

To keep Natanz engineers from noticing anything untoward, Olympic Games architects pulled a *Mission Impossible*. Like bank thieves who replace security camera footage with prerecorded footage during a heist, the worm sent prerecorded data to the Step 7 computers that monitored the PLCs while it spun Natanz's centrifuges wildly out of control or stopped them from spinning entirely, so technicians would be none the wiser.

By late 2008 the joint operation known as Olympic Games had infiltrated Natanz's PLCs, and nobody appeared to suspect a cyberattack. As the worm started to spread, Bush and Alexander were pleased with the progress. And the Israelis were too. While they continued to press for airstrikes, it seemed that for the moment, the immediate threat had passed. But Olympic Games began to take on greater urgency as they approached the 2008 presidential campaign that November.

Increasingly, it appeared that Barack Obama, not John McCain, would be Bush's successor, and in Israeli perceptions, Obama was a wild card.

DAYS BEFORE BUSH handed the Oval Office over to Obama in early 2009, the president invited the president-elect to the White House for a one-on-one.

As my colleague David Sanger later reported, it was at that meeting that Bush urged Obama to preserve two classified programs. One was a U.S. drone program in Pakistan. The other was Olympic Games.

For a president with no particular technical background to speak of, Obama became deeply engaged with Olympic Games. Not one month after his inauguration, the worm achieved its first major success: a cascade of centrifuges at Natanz had sped out of control, and several had shattered. Obama phoned Bush to tell him that his "third option" was paying off.

Ahmadinejad had stated that Iran was determined to eventually install more than 50,000 centrifuges, but after steadily building up capacity between 2007 and 2009, International Atomic Energy Agency (IAEA) records show a gradual dropoff starting in June 2009 that continued over the next year.

The program was working just as Alexander hoped it would. By the start of 2010, the worm had destroyed 2,000 of Natanz's 8,700 centrifuges. With every new strike on the rotors, Obama would meet with his advisors in the White House Situation Room. Not only were the centrifuges breaking down, the Iranians were losing confidence in the whole program. Inspections were turning up nothing, and Natanz officials started turning on each other, suspecting subterfuge. Several technicians were fired. Those remaining were told to physically guard the centrifuges with their lives. Meanwhile, their computer screens continued to show everything functioning normally.

While pleased, Obama worried about the precedent these attacks would set. This was the world's first cyberweapon of mass destruction. If the worm got out, it would reshape armed conflict as we knew it. For the first time in history, a country could reach across borders and do with code what previously could have only been done with aircraft and bombs. If Iran, or any other adversary, learned of this new weapon, it would almost certainly embolden them to do the same.

American companies, towns, and cities were proving themselves to be massively vulnerable. Even a short list of recent cyberattacks—the Russian compromise of the Pentagon's classified and unclassified networks in 2008; a series of 2009 North Korean attacks that jammed the websites of the Treasury Department, the Secret Service, the Federal Trade Commission, the Department of Transportation, the Nasdaq, and the New York Stock Exchange; the nonstop Chinese raids on American military and trade secrets—illustrated the problem. Any adversary that wanted to do U.S. interests harm in the cyber realm was achieving its goals. America's vulnerabilities were manifold, its defenses inadequate, and the attack surface was only expanding with every new computer, phone, and PLC coming online. How long before our adversaries discovered the potential for serious harm? How long before they acquired these same capabilities? And how long before they tested them on American soil?

In spring 2009 Obama created a new cybersecurity position at the White House to help coordinate between the various government agencies handling our cyber defense and to advise the American companies most vulnerable to an attack. In a speech announcing the moves, Obama warned Americans that the mass migration to the web held "great promise but also great peril."

For the first time, Obama spoke of his own brush with cyberattacks: hackers had breached his campaign offices and those of his 2008 presidential rival, John McCain. "Hackers gained access to emails and a range of campaign files, from policy position papers to travel plans," he said. "It was a powerful reminder: In this information age, one of your greatest strengths . . . could also be one of your greatest vulnerabilities."

All the while, the worm was spinning the centrifuges away.

THE FIRST INKLING nuclear inspectors had that something was off came in January 2010. Security camera footage outside the Natanz centrifuge rooms showed frantic Iranian technicians in white lab coats and blue plastic shoe covers carting out centrifuge after centrifuge.

Officially, the International Atomic Energy Agency did not have permission to ask Iran's technicians why the centrifuges were being discarded. And the Iranians refused to admit anything was wrong.

From the American point of view, the worm seemed to be working spectacularly. Until it escaped.

Nobody knows exactly how the worm got out. But that June, then CIA Director Leon Panetta, his deputy Michael Morell, and the vice chairman of the Joint Chiefs of Staff, General James "Hoss" Cartwright, briefed Obama and Vice President Joe Biden that somehow the worm had left the building, and they walked them through the horror show that would soon play out.

Their leading theory was that the Israelis—impatient with the worm's progress—introduced a new spreading mechanism that led to its escape. But to this day, that theory has never been confirmed.

David Sanger would later report that Biden latched onto the Blame Israel theory. "Sonofabitch," the vice president is reported to have said. "It's got to be the Israelis. They went too far."

Another theory is that a Natanz technician or maintenance worker may have plugged an infected computer into his or her personal device, which allowed the worm to flee its coop. For all the care that had gone into designing the first version of the worm and fine-tuning the conditions under which it deployed the payload, Olympic Games' architects had never contemplated what would happen if it ever seeped through the air gap.

Obama asked Panetta, Morell, and Cartwright the question they all feared: "Should we shut this down?" From the first time Bush had briefed him on the program, this had always been the worst-case scenario. How quickly would the Iranians put two and two together, and realize that the leaked code had come from their machines? How far would the worm spread? And what collateral damage would come of it?

Obama's advisers had no good answers. What they did know was that it would take time for sleuths to catch up to the code, its origins, and its target. Until their cover was blown, Obama figured they should use whatever time they had left to do as much damage as possible. He ordered the generals to accelerate the program.

Over the next few weeks, TAO and Unit 8200 pushed out another aggressive round of strikes, destroying another round of centrifuges, while the worm began zigzagging all over the web, aimlessly searching for more PLCs. There was no telling how many systems would soon be infected. But it would not be long before someone detected it and ripped the code apart.

It was Obama's young advisor Benjamin Rhodes who warned, "This is going to show up in the *New York Times*." And he was right.

That summer, virtually all at once, security researchers in Belarus, Russian researchers at Kaspersky in Moscow, at Microsoft in Redmond, two researchers at Symantec in California, and Ralph Langner, the German industrial security expert in Germany, began picking up traces of the worm as it darted from Iran to Indonesia, India, Europe, the United States, and on to a hundred other countries, infecting tens of thousands of machines in its wake. At the same time, Microsoft issued an urgent advisory to its customers. Forming an anagram from the first few letters of the code, they called the worm Stuxnet.

INSIDE HIS SLEEK office in Hamburg, Langner stewed. For years he'd been a voice in the wilderness, warning his clients in Germany and all over the globe that the little gray PLC boxes they were plugging into their auto, chemical, and power plants, their dams, hospitals, and enrichment facilities, would one day be a target for sabotage or worse—explosions, digitally-triggered tsunamis, widespread power outages. But until now these concerns had been purely hypothetical. As Stuxnet's code and payload came into focus, Langner realized that the attack he had long dreaded was staring right back at him.

Inside the confines of his lab, Langner's team infected computers with Stuxnet to see what the worm would do. "And then some very funny things happened," he recalled. "Stuxnet behaved like a lab rat that didn't like our cheese. Sniffed but didn't want to eat."

They tested the worm on several different versions of PLCs. But the worm didn't bite. It was clear that it was looking for a very specific configuration of machines, and that its authors had designed the code with insider information on their target.

"They knew all the bits and bytes they needed to attack," Langner said. "They probably even know the shoe size of the operator." The spreading mechanisms were impressive, but it was the worm's payload—what Langner called its "warhead"—that blew him away. "The payload is rocket science," he said.

The mastery of the code suggested that this was not the work of some cybercriminal thug. This was the work of a well-resourced nation-state. And it had been designed, Langner concluded, to "drive the maintenance engineers crazy."

Langner was also struck by one number that kept popping up in the payload: 164. He told his assistant to get him a list of centrifuge experts, and see if this number held any resonance with them. It did: at Natanz's enrichment facility, operators bundled centrifuges into cascades in groups of 164. *Bingo!*

Back at the *New York Times*, my colleagues David Sanger, William Broad, and John Markoff were starting to piece together the mystery of the Stuxnet code as well. In January 2011 the three published a lengthy account of the worm in the *Times*, detailing Israeli involvement.

Two months later, in March 2011, Ralph Langner was in Long Beach. He'd been asked to deliver a ten-minute talk breaking down the Stuxnet code at the annual TED ideas conference. Langner had never even heard of TED Talks; the entire concept behind it is antithetical to everything Germans stand for. Germans don't do small talk, and they don't do bullshit. Feel-good messages and blatant self-promotion have no place in Germany. Doing your job well is not a good reason to deliver a long, self-aggrandizing speech. That March, Langner was in the middle of a bitter divorce, and he figured a paid trip to California, and a few walks on the beach, might offer some respite. But when he arrived that Sunday, he got the sense that this was no ordinary cybersecurity conference. The speaker list and audience included Bill Gates, Google cofounder Sergey Brin, Pepsi CEO Indra Nooyi, and Stanley McChrystal, formerly the top U.S. commander in Afghanistan. Among the first speakers was an astronaut who beamed into Long Beach from space to talk about life on the International Space Station.

Langner never would get his walk on the beach. He threw away his technical speech and spent the next few days holed up in his hotel room, trying to pull a slide deck together and come up with a layman's explanation for "programmable logic controllers." He emerged only to attend special invite-only events like the speaker dinner, where he put a buffet plate together and went to eat in peace at the nearest table, only to be approached by Sergey Brin, apparently a fan of his analysis of Stuxnet. Brin had questions. This only added to his stress.

When Langner took the stage three days later, he delivered what was arguably the most coherent description of the world's first digital cyberweapon of mass destruction. He ended with a warning. Stuxnet had been specifically designed for Natanz, but it was also generic, in the sense that there was nothing in the code to prevent others from firing the very same weapon at the very same Windows and Siemens computers—computers that control the world's water pumps, air-conditioning systems, chemical plants, power grids, and manufacturing plants. The world should be forewarned, Langner said, that the next worm might not be so narrowly confined.

"The biggest number of targets for such an attack are not in the Middle East," Langner said. "They are in Europe, Japan, and the United States. We will have to face the consequences, and we better prepare right now."

"Ralph, I have one question," TED founder Chris Anderson began as soon as Langner had wrapped. "It has been quite widely reported that Mossad is the main entity behind this. Is that your opinion?"

Until that moment, the Obama administration had reason to hope that, with reporters and researchers zeroing in on the Israelis, their own role in the attacks might never see the light of day—or that if researchers did have any inkling that another nation-state was involved, they wouldn't dare utter its name.

They would have no such luck with the German. "You really want to know that?" Langner asked Anderson.

The audience chuckled. Langner took a deep breath and straightened his suit. "My opinion is that Mossad is involved, but the leading force is not Israel," he said. "The leading force behind this is the lead cyber power. There is only one. And that is the United States.

"Fortunately, fortunately," he added, "because otherwise our problems would be even bigger."

THE IRANIANS NEVER did acknowledge the destruction that Stuxnet wrought on its enrichment programs. Ali Akbar Salehi, the head of Iran's Atomic Energy Organization, claimed that his team had "discovered the virus exactly at the same spot it wanted to penetrate because of our vigilance and prevented the virus from harming [our equipment]."

In truth, Iranians were already looking for ways to exact revenge; and the United States and Israel had shown them a terrific shortcut. The United States may have thwarted a conventional war, but in releasing Stuxnet on the world, it opened up an entirely new battlefront. The worm had crossed the Rubicon from defensive espionage to offensive cyberweapon, and in just a few years, it would come boomeranging back on us.

Perhaps Michael Hayden, the former NSA director, put it best when he said "This has a whiff of August 1945," the month and year the United States dropped the world's first ever atomic bomb on Hiroshima. "Somebody just used a new weapon, and this weapon will not be put back in the box."

CHAPTER 10

The Factory

Reston, Virginia

S TUXNET DID A couple turns around Asia before it came home.

The first American company to acknowledge that its computer systems had been infected was Chevron, the country's second biggest energy company. And while the cautionary elements of the code prevented the worm from damaging these computers, it was a wake-up call for every chief information officer in America; they were collateral damage in an escalating global cyberwar.

"I don't think the U.S. government even realized how far it spread," a senior official at Chevron told reporters. "I think the downside of what they did is going to be far worse than what they actually accomplished."

With the worm out of its box, Iran onto the U.S.-Israeli plot, and American infrastructure still so vulnerable, you might think that the agency charged with both offense and defense might shift its gaze inward, take stock of its own vulnerability, and begin the slow, grueling work of locking up its own embattled kingdom.

But this was the age of acceleration. Everything that was analog was being digitized. Everything that was digitized was being stored. And everything that was stored was being analyzed, opening up entirely new dimensions for surveillance and attack. Smartphones were now real-time trackers, digitizing a person's every movement, relationship, purchase, search, and noise. Smart homes could adjust thermostats, lightbulbs, and security cameras, play and capture music and sounds, and even heat the oven up on your evening commute. Train sensors could identify broken wheels and eliminate downtime. Smart traffic lights, equipped with radars, cameras, and sensors, could manage traffic flow, adjust signals for inclement weather, and catch someone running a red light. Retailers could now trace a customer's purchase to the "smart" billboard they drove by days earlier. Even cows were being outfitted with glorified pedometers and sensors to alert farmers when they were sick or in heat.

The cost to record, store, disseminate, and analyze this data was becoming virtually free thanks to leaps in the cloud, storage, fiber-optic connectivity, and computational power. In February 2011 IBM's Watson computer made its first public debut on *Jeopardy!*, besting the game's all-time human champions and proving that machines were now capable of understanding questions and answering them in natural languages. A short eight months later, Apple introduced the world to Siri, our new voice assistant, whose high-quality voice recognition and natural language processing let us send emails and texts and set reminders and playlists.

The resulting mix of mass mobility, connectivity, storage, processing, and computational power gave NSA unprecedented openings and capabilities to track every last person and sensor on earth. Over the next decade the NSA continued to probe every last pore of this new digital dimension for exploitation, surveillance, and future attack. Once Pandora's box had been opened, there was no going back.

With Stuxnet under way in June 2009, the Obama administration created a dedicated Cyber Command at the Pentagon for offensive cyberattacks. More hacking—not better defenses—was the Pentagon's response to the Russian attacks on its own classified networks. The success of Stuxnet, however short-lived, meant there was no going back. By 2012 the U.S.'s three-year-old Cyber Command's annual budget had tripled from $2.7 billion to $7 billion (plus another $7 billion for cyberactivities across the Pentagon), while its ranks swelled from nine hundred dedicated personnel to four thousand, and eventually fourteen thousand by 2020. Starting in 2012, Obama ordered his senior intelligence officials to produce a list of foreign targets—"systems, processes and infrastructures"—for cyberattack, according to classified directives leaked by Snowden the following year. It is not clear whether the command planned to attack those targets, or if the Pentagon was just "preparing the battlefield," but the directive made evident that attacking those targets could provide "unique and unconventional" ways "to advance national objectives around the world with little or no warning to the adversary or target and with potential effects ranging from subtle to severely damaging."

By then, the challenge at NSA wasn't so much how to penetrate these targets as who exactly would man the government's growing surveillance and attack apparatus. Despite the NSA's global reach, the agency only had the manpower to monitor one-eighth of the tens of thousands of digital implants it had placed in foreign computers around the world, even as the agency's leaders lobbied for budgets to push the number of NSA implants into the millions.

The NSA had started test-driving a game-changing new robot, code-named Turbine, to take over management of its vast implant apparatus. Described internally as "an intelligence command and control" that would enable "industrial-scale exploitation," Turbine was designed to operate "like the brain." The Turbine robot was part of a broader "Owning the Net" NSA initiative and if all went well, officials believed it could ultimately supplant humans in operating the NSA's vast digital spiderweb.

It would now be up to this automated robot to decide whether to use an implant to hoover up raw data or inject malware that, like a digital Swiss Army knife, could do almost any job the NSA needed to get done. The agency's diverse arsenal of malware tools—many described in leaked NSA documents but many more that were not—could steal phone conversations, text threads, emails, and industrial blueprints. Other malware could hot-mic an infected computer and capture any conversations in close proximity. Still other tools could steal screenshots, deny a target access to certain websites, shut down computers remotely, corrupt or delete all their data, and grab their keystrokes, search terms, browsing histories, passwords, and any keys necessary to unscramble encrypted data. Some NSA tools supercharged the agency's malware, enabling their code to spread automatically from vulnerable server to server in a fraction of a second, rather than relying on human operators to manually infect each server one by one. An entire trove of NSA hacking tools was developed for obfuscation alone.

Snowden's leaks of NSA PowerPoint slides and memos referred to these tools in vague, open-ended terms. They were just enough to send me on this crazed mission, but nowhere close to what we would learn in late 2013, when the German publication *Der Spiegel* published a fifty-page classified NSA catalog that detailed some of the agency's most clever exploitation techniques—so clever, in fact, that officials began to suspect that the leaker was not Snowden but a second NSA mole or a foreign spy who had found a way inside TAO's vault.

The equipment catalog read straight out of Bond's Q Factory. There was Monkeycalendar, an exploit that relayed a target's geolocation back to the agency via invisible text message; Picasso, a similar exploit that could do all that and hot-mic a phone's microphone to eavesdrop on any conversations nearby. Surlyspawn, the modern-day equivalent of the Russian typewriter exploit in Gunman, could grab keystrokes off computers that were not even connected to the internet. The *Der Spiegel* leak is how the world came to know about Dropoutjeep, the TAO exploit developed specifically for the iPhone, the one that could do all the usual text, phone call, and location monitoring, hot-miking and photo snapping, even when the iPhone was offline.

Among the more intriguing tools mentioned in the catalog was a device called Cottonmouth I that looked like any old USB stick, but contained a miniature radio transceiver that passed data to yet another NSA gadget—dubbed Nightstand—miles away. Once details of these tools were leaked, security researchers came to suspect that they might hold the key to how the Americans and Israelis got Stuxnet into Natanz in the first place.

For the most part, the NSA's zero-days were still being discovered and honed by humans, inside the agency and also by the growing number of private hackers around the Beltway and beyond.

But their deployment was increasingly left to supercomputers. By 2013, Turbine was fully operational and began buffering TAO analysts from their operations. The robot was designed, in the words of one internal NSA memo, to "relieve the user from needing to know/care about the details." By the end of the year, the NSA anticipated that Turbine would manage "millions of implants" for intelligence gathering and "active attack." So fixated was the NSA on its new offensive cyber tools that year that offense trumped defense at the agency by a factor of two. The agency's breaking-and-entering budget had swelled to $652 million, twice what it budgeted to defend the government's networks from foreign attack. Its critics began to argue that NSA had abandoned its defensive mission entirely.

THE WORLD HAD changed in the thirty-odd years since Gunman. It was no longer the case that Americans used one set of typewriter, while our adversaries used another. Thanks to globalization, we now all relied on the same technology. A zero-day exploit in the NSA's arsenal could not be tailored to affect only a Pakistani intelligence official or an al-Qaeda operative. American citizens, businesses, and critical infrastructure would also be vulnerable if that zero-day were to come into the hands of a foreign power, cybercriminal, or rogue hacker.

This paradox began to keep Pentagon officials up at night. America's cyberweapons could no longer exist in a vacuum. The United States was effectively bankrolling dangerous R&D that could come boomeranging back on us. American businesses, hospitals, electric utilities, nuclear plants, oil and gas pipelines, transportation systems—planes, trains, and automobiles—relied on the same applications and hardware the NSA's arsenal exploited.

And that is not going to change anytime soon. As the Trump administration would later learn first-hand—in its losing battle to blackball Huawei from next-generation mobile networks—no amount of government lobbying can halt globalization when it came to technology.

The NSA faced a quandary: its solution to dealing with the bad actors in the world was escalating an arms race that only made the United States more vulnerable to attack. The NSA's answer to this problem was a system called Nobody But Us (NOBUS). The premise behind NOBUS was that low-hanging fruit—vulnerabilities that could easily be discovered and abused by American adversaries—should be fixed and turned over to vendors for patching. But more advanced exploitation—the kind of advanced zero-days the agency believed only it had the power, resources, and skills to exploit—would remain in the agency's stockpile and be used to spy on American enemies or degrade their systems in the case of a cyberwar.

Michael Hayden, the former NSA chief who ran the agency until 2005, addressed NOBUS this way: "You look at a vulnerability through a different lens if even with the vulnerability it requires substantial computational power or substantial other attributes, and you have to make the judgment: Who else can do this? If there's a vulnerability here that weakens encryption, but you still need four acres of [supercomputers] in the basement in order to work it, you kind of think NOBUS, and that's a vulnerability we are not ethically or legally compelled to try to patch—it's one that ethically and legally we could try to exploit in order to keep Americans safe from others."

But starting in 2012, NOBUS was falling apart. "Hacking routers has been good business for us and our Five Eyes partners for some time," one NSA analyst noted in a leaked top-secret memo dated that year. "But it is becoming more apparent that other nation-states are honing their skillz and joining the scene."

The NSA was finding evidence that Russian hackers were tampering with the same routers and switches it had exploited for years. Chinese hackers were breaking into American telecoms and internet companies and stealing passwords, blueprints, source code, and trade secrets that could be used to exploit these systems for their own ends.

The agency still had a strong lead in signals intelligence, thanks to Gosler and others who came before and after him, but the advantage was fading. One might think this would force the agency to reckon with the failings of its NOBUS approach and the realities of a flatter internet age. Instead, to stay ahead of the game, NSA doubled down, ramping up its search for and stockpiling ever more zero-days, and contracting out the hunt and development of these tools to private companies around the Beltway.

In 2013 the NSA added a new $25.1 million line item to its classified black budget. That was how much it planned to start spending each year to procure "software vulnerabilities from private malware vendors." By one estimate, that would enable the agency to purchase as many as 625 zero-days a year, in addition to the vulnerabilities the agency was developing in-house.

All this appetite for vulnerabilities and exploits created a surge in the market for offensive cyberweaponry. It wasn't just NSA. After Stuxnet, the CIA, DEA, U.S. Air Force, U.S. Navy, and FBI started pouring more dollars into zero-day exploits and malware tools. At TAO, young hackers with the skills to develop these tools were learning that they could make far more money on the outside, developing and selling their surveillance and attack tools back to the government, than they could on the inside. In Russia and China anyone with cyber skills could be coerced, threatened, and blackmailed to conduct offensive hacking operations, but the U.S. government had no such luxury. Increasingly it was losing its best hackers and analysts to higher-paying jobs at private defense contractors like Booz Allen, Northrop Grumman, Raytheon, Lockheed, and Harris, and at boutique zero-day hunting and development firms around the Beltway.

The Snowden leaks in 2013 only exacerbated the brain drain. After the NSA's public flogging that year and the demoralization that followed as the leaks forced the agency to shutter one program after another, its analysts began leaving in droves. (The agency disputes this, but cops to a severe skills shortage.)

Increasingly, the only way to acquire the same capabilities the agency once developed in-house was to buy them from hackers and contractors. And once the intel agencies started allocating more of their budgets for zero-day exploits and attack tools off the private market, there was even less incentive to turn over the underlying zero-day flaws to vendors for patching. Instead, they started upping the classification levels and secrecy around these programs.

The irony is that such secrecy did little to make Americans more secure. Zero-days do not stay secret indefinitely. One study by RAND Corporation, the research corporation that concentrates on U.S. defense planning, found that while the average zero-day exploit can stay secret for nearly seven years, roughly a quarter of zero-day exploits will be discovered within a year and a half. Earlier studies determined that the average life-span of a zero-day is ten months. After Stuxnet opened the world's eyes to the power of a zero-day, American allies, adversaries, and authoritarian regimes started searching for and stockpiling their own. America's classification levels and nondisclosure agreements did nothing to stop them. It just kept journalists, like me, from exposing the government's dirty little secret.

ONE DAY IN 2008, almost simultaneously, five of the NSA's most elite hackers turned in their security badges and pulled out of the Fort's parking lot for the last time.

Inside the agency, these men had been revered as "the Maryland Five," and time and time again, they had proved indispensable. They were each members of a premier TAO access team that hacked into the systems nobody else could. If the target was a terrorist, an arms dealer, a Chinese mole, or a nuclear scientist, you wanted the Five on it. Rarely was there a system, or a target, they could not hack.

But the bureaucracy, the turf wars, the midlevel managers, the secrecy, and the red tape grated on them. Like so many hackers before them, money had never been much of a motivating factor. Still in their twenties, they didn't have mortgages or tuition to pay. It was autonomy they sought. But they also couldn't help notice how much their employer was increasingly willing to funnel to front companies to pay hackers, brokers, and defense contractors on the outside for some of the very same work they were doing inside the Fort.

And so they left NSA to start their own firm—a boutique zero-day exploit shop, about an hour's drive from Fort Meade. And for twelve long years, they managed to keep their entire operation a secret.

One cloudy day in March 2019, I got in a taxi near the Pentagon and wound through the Virginia suburbs to an unassuming six-story office building in Reston, Virginia, encased in mirrored glass. I ordinarily would never have given this building, squeezed between a day care center and a massage parlor, a second thought. It was hard to know if I was in the right place. There were no signs for the company I knew would be inside. They didn't advertise. The building looked no different from any other office cluster built in the 1990s.

I didn't have an appointment that day, nor did I think I would get very far. Over the years, I'd heard whispers of the men who worked inside. But when I reached out to executives or employees, I never heard back. As I stepped inside headquarters that day, I fully expected to see security cameras, turnstiles, or armed guards. I was eight months pregnant at the time, and I figured if anyone gave me a hard time, I could just tell them I was looking for a restroom. But when I waddled inside, there was no one to stop me. I made my way to the elevators and pushed the button to the third floor, Suite 300, to see what I could glimpse of Uncle Sam's Q's lab, though with a far less cinematic name: Vulnerability Research Labs.

Discretion, I knew, was among VRL's founding virtues. Its sparse website posed the question "Why haven't you heard of us?" and then answered it: "VRL does not advertise. We hold all business relationships in the strictest of confidence." The only hint that the company played any role in weaponizing the digital universe was its motto, borrowed from the ancient Chinese philosopher Sun Tzu: "Know thy enemy and know yourself; in a hundred battles you will never be in peril."

Even as VRL tried to spotlight the defensive elements of its business—and it did defense too—I knew from the visible panic on zero-day hunters' and brokers' faces anytime I mentioned its three little initials that VRL was among the premier buyers and sellers of zero-day exploits, surveillance tools, and cyber-weapons on the market. VRL had strict policies of saying as little as possible about its offense work. It fired anyone who so much as whispered its customers' names beyond these glass doors. If the three-letter agencies caught any whiff that VRL was discussing its business in public, contracts would vanish.

But LinkedIn always yielded interesting results. Former NSA SIGINT liaison officers were now VRL's "Managers of Offensive Tools and Technologies." A former U.S. counterterrorism specialist was now VRL's operations manager. Some VRL employees spelled out their job description as "redacted." On job search sites, I found VRL postings for engineers skilled at discovering "critical vulnerabilities in software and hardware" to "bolster the cyber hunting capabilities of our customers." The company was constantly on the lookout for kernel exploitation specialists and mobile developers who could string exploits together into push-button spy tools. Nowhere on VRL's website was its growing zero-day procurement business mentioned—and yet it somehow sought and employed the country's premier zero-day researchers, especially those who had cut their teeth at VRL's best customers: the NSA and CIA.

On one job posting site, the company made its value proposition known: "VRL's unique knowledge of the cyber armament stockpiles of foreign adversaries, as well as the distinguishing characteristics of their tradecraft, place VRL in a class by itself."

SECRECY DEFINED THIS business, but the names of VRL's early U.S. competition—companies like Endgame in Virginia, Netragard outside Boston, and Exodus Intelligence in Austin—were relatively well known. VRL was not. It took me years of muckraking and prying to understand why everyone in this business took such great pains to keep VRL under wraps.

Nobody inside the company would ever talk to me. Few of VRL's contracts are public. The ones I did find don't tell me very much. In the government's procurement database, I tracked down several VRL contracts for the army, the air force, and the navy for millions of dollars. One contract showed that the air force had paid VRL $2.9 million for vague "computer peripheral equipment." The paper trail largely disappeared in the years after VRL was acquired by Computer Sciences Corp., the defense contracting behemoth, in 2010. VRL didn't have much of an online presence for a company whose CEO bragged at the time of the CSC acquisition that VRL had "unparalleled capabilities" in the cybersecurity space. "We believe that a talent pool such as the one we've assembled has never before existed in a commercial entity," Jim Miller, VRL's president, said at the time.

But in the years after the sale, I started coming across former VRL employees at the big international hacking conferences in Buenos Aires, Vancouver, Singapore, and Las Vegas. Many were tight-lipped. But the more I shared what I already knew—that VRL was among the government's most secretive purveyors of offensive spy tools and cyberweaponry—the more they began to open up. Some only offered to fact-check the basics of what I had found. Others were more forthcoming, in part, they told me, because they had started to question the wisdom of poking holes in the world's computer systems, smartphones, and infrastructure, baking them into slick spy tools, and throwing them "over the fence" to government agencies, with no clue as to how they would be used.

This sort of second-guessing had taken on a new dimension with the election of Donald Trump. With Trump's strange affinity for dictators, his inability to hold Russia's feet to the fire for its 2016 election interference, his abandonment of the Kurds, one of America's closest allies, and his refusal to clearly condemn the Saudis' gruesome killing of *Washington Post* columnist Jamal Khashoggi, America was losing its moral authority. (Even after a CIA assessment concluded that Saudi crown prince Mohammed bin Salman personally ordered the hit on Khashoggi, Trump responded, "Maybe he did and maybe he didn't.") With these cases piling up, one former VRL employee told me in late 2019, "It's getting harder to know if you're selling these tools to the good guys or enabling the bad."

As it turned out, former VRL employees told me, my search for their contracts had been doomed from the start. VRL's tools were procured primarily through "special contract vehicles" set up by the CIA and NSA to hide their dealings with contractors. (By now I was starting to see why Panetta had told me I was going to "run into a lot of walls.")

But back in 2008, these men told me, the original premise behind the company was this: "It's a pain in the ass to find a zero-day. Weaponizing a zero-day is an even bigger pain in the ass. Testing it and making it reliable is the biggest pain in the ass of all."

VRL did all of the above—for a price. Starting in 2008, the Five hired away some of the NSA's best zero-day hunters, contracted with hackers in places like Argentina, Malaysia, Italy, Australia, France, and Singapore, and poured their money into giant "fuzz farms"—tens of thousands of computers in virtual server farms—that threw terabytes of junk code at VRL's tools to ensure that nothing they sold intelligence agencies would crash in the course of an operation, or tip off a target to the fact they'd just been hacked by Uncle Sam.

VRL's tools developed a reputation for being top-notch, leagues above the crap iDefense, Tipping Point, and other companies were buying back in the day.

"These were multimillion-dollar zero-days, one hundred percent reliable, and jealously guarded," a former employee told me. "You didn't casually throw these around. You only used them in critical precision attacks. Because once you deployed them, the risk they'd be discovered was too great. These were virtually never deployed 'in anger.' You had to wait for an emergency bad enough worth burning them for."

VRL provided tools for the government's most sensitive targeted operations, the kind of operation that might net Ayman al-Zawahiri or shut down North Korea's missile launch systems. They weren't just selling zero-day exploits; they sold the agencies turnkey spy tools and cyberweapons that, in one employee's words, "could make things go boom!"

"The difference between buying a bug and a reliable, weaponized exploit is night and day," one told me. "These were push-button systems."

A decade ago, if a skilled hacker discovered a zero-day in the morning, he might have a weaponized exploit ready to use that afternoon. But as software vendors like Microsoft began introducing stronger security and antiexploitation mitigations, it took more time and man-hours to develop a reliable exploit. "It went from a few hours, to a few weeks, to a few months," one former NSA analyst told me.

VRL had found a powerful niche, weaponizing zero-days and selling the agencies turnkey hacking tools that worked across a wide range of systems.

As a contractor, VRL also could play where government agencies couldn't: in the shadier elements of the market, buying zero-days, exploits, and attack techniques from foreign hackers. VRL would then turn those raw materials into click-and-shoot spy tools and cyberweapons, and their customers at the CIA and other U.S. government agencies would never need to know where the underlying exploits came from.

Occasionally VRL would accept "walk-ins," buying one-off exploits from hackers who approached them with their discoveries. But more often VRL would develop a concept for a spy gadget or cyberweapon and task foreign hackers with making it happen. I asked one employee if he had any qualms about working with foreign hackers, or if he ever wondered who else they might be selling to on the side.

"We didn't care," he told me. "We just did whatever it took to produce the goods."

EVEN AFTER VRL was acquired in 2010, it still was operating relatively autonomously by the time I camped outside its glass doors that March. I saw no fancy cafeteria, no rock-climbing gym, no other start-up perks. There were no hipsters in thick-framed glasses and skinny jeans.

If anything, this was the anti-Silicon Valley. Inside, VRL's employees were secretly making life hell for the security engineers at companies like Google and Apple, poking holes in their products and weaponizing their code. The tools VRL was developing were the kind that, if discovered, would send security engineers straight to war rooms in Mountain View and Cupertino, to figure out just how far their own government had crawled into their systems, and how to get them out.

"Everyone who works there has a monklike dedication to the job," one former employee told me. "They are more like the Knights of Templar than what you'd see at Google or Facebook."

These former employees used one example over and over again to demonstrate the loyal, Navy SEAL–like culture of the company: the story of an employee who discovered he had kidney failure. Nearly everyone at the company got tissue-typed until they found a match, and a colleague ended up giving him a kidney. This kind of thing was not happening at Google or Facebook.

I stood outside VRL's doors and watched twentysomething male programmers stream in and out. They were just how you might imagine them: unassuming, laser-focused on their iPhones, oblivious to my presence. There were no ironic T-shirts or oversize retro headphones like you saw in Silicon Valley. These were not the loud voices I saw screaming on Twitter every day. I doubted if they even used social media.

Nobody outside classified circles, and the underground zero-day market, even heard of these men, but inside these walls, VRL had more advanced cyber capabilities than the vast majority of countries around the globe. And yet, as far as I could tell, there was nobody holding them accountable. VRL employees who came from the big agencies had some concept of how their tools would be used. But it had been years since many of them had any operational role. "Once I threw things over the fence, I never knew what happened to them," one former VRL employee told me. Sometimes, he might catch something in the news and think, Was that me? "But the other side is so opaque and so classified, you really don't know."

VRL made a point of selling its hacking tools only to American agencies. And for most employees, that was all they needed to know.

"We might not have agreed with everything our government was doing, but if you're going to sell exploits, at least the U.S. government is one of the more ethically responsible ones. Most of us came from the NSA. This was a good way to transition out. We didn't have to deal with the bureaucracy. We could make good money, and still be aligned with the agency's mission."

If they ever had any hesitation, well, all they had to do was look at the work their former NSA colleagues were doing across the Beltway—with some of America's less than reputable allies—to feel better about themselves. "I'd look at what some of my friends were doing, and think, "Now those are the real bad guys. I guess if you want to justify something to yourself, you'll find a way. In exploits, and in life."

IN THEIR EAGERNESS to pay top dollar for more and better zero-day exploits and spy tools, U.S. spy agencies were helping drive a lucrative and unregulated cyberarms race, one that gradually stopped playing by American rules.

While companies like VRL only did business with U.S. agencies, and others like Azimuth and Linchpin Labs worked exclusively within Five Eyes, Stuxnet's darkest legacy is that it showed other countries what could be accomplished with a few zero-days strung together. After the worm was discovered in 2010, countries with abysmal human rights records began feverishly pulling together their own offensive cyber units. But without the talent of the coders at the NSA or Unit 8200, these countries started flooding the zero-day market, outbidding Western governments and front companies for zero-day exploits in pursuit of the kind of success, albeit temporary, that Stuxnet had achieved in Iran.

"I think it is fair to say that no one anticipated where this was going," one U.S. senior official told me. "And today, no one is sure where it is going to end up."

By 2013 Five Eyes was still the market's biggest sponsor but Russia, India, Brazil, and countries in the Asian Pacific, like Malaysia and Singapore, were buying too. North Korea and Iran were in on the market. And soon, Middle Eastern intelligence services would become the market's biggest spenders.

That year, American hackers started receiving urgent emails from foreign brokers that weren't exactly subtle: "Need code execution exploit urgent," read the subject line of an email one hacker shared with me. "Dear Friend," it began. "Do you have any code execution exploit for Windows, Mac, for applications like Browser, Office, Adobe?"

"If yes," the email added, "payment is no issue."

The year I stepped into the closet of Snowden's classified secrets, the zero-day market had become a full-fledged gold rush. But there was little incentive to regulate a market in which the United States government was still its biggest customer.

That year, having ironically spawned the zero-day market and launched the world into the era of cyberwar, Keith Alexander, Stuxnet's architect, was asked what kept him up at night. "My greatest worry," Alexander told a reporter, was the growing likelihood of zero-day exploits falling into the wrong hands.

PART IV

The Mercenaries

A man got to have a code.

—OMAR LITTLE, *THE WIRE*

CHAPTER 11

The Kurd

San Jose, California

REGULATING THE GLOBAL sale of zero-days has long been a messy and flawed endeavor. Most can agree that restricting the sale of hacking tools to authoritarian regimes is noble in theory. Proponents point out that the State Department regularly blocks weapon sales to dictatorships and argue the same logic should apply to digital tools that can be used to spy on entire populations or trigger a deadly explosion.

But critics argue that, in practice, the rules would backfire. Security researchers argue that restrictions on zero-days would actually handicap cybersecurity, in that it would keep researchers from sharing vulnerability research and malware code across borders. American companies that conduct business abroad argue it would open them to select prosecution, and ultimately benefit states like China and Russia that only enforce regulations when convenient. Others argue that zero-days are code and regulating the exchange of code would be akin to regulating math and ideas, infringements on free speech. As the two Italians, Luigi and Donato, put it: "We don't sell weapons, we sell information." The pushback has helped keep the zero-day market largely unregulated. And so long as the United States continues to be one of the zero-day market's biggest sponsors, that is unlikely to change.

The closest the United States has ever gotten to controlling the export of hacking tools and surveillance technology is the Wassenaar Arrangement.

Named for the Dutch town where the arrangement was originally signed in 1996, the Wassenaar Arrangement on Export Controls for Conventional Arms and Dual-Use Goods and Technologies was designed to replace the previous set of Cold War norms used by Western states to keep weapons and military technology from making their way to Russia, China, and their communist satellites. Wassenaar's goal was to control the sale of conventional weapons systems and dual-use technologies—sophisticated computers, centrifuges, drones—from reaching despots in Iran, Iraq, Libya, and North Korea. Its original signatories included the United States and forty-one other countries, including most of Europe, Argentina, Australia, Canada, India, Japan, Mexico, New Zealand, Russia, South Africa, South Korea, Switzerland, Turkey, Ukraine, and the United Kingdom, and though the arrangement is nonbinding, member-states agree to establish and enforce their own domestic laws to control the sale of items on Wassenaar's control list, which they update every December.

In 2012 and 2013, working with security researchers at the Citizen Lab at the Munk School of Global Affairs at the University of Toronto, I wrote a series of stories for the *New York Times* about the sale and spread of a British company's spyware to countries like Bahrain, Brunei, Ethiopia, and the United Arab Emirates, where authorities were caught using their code to monitor journalists, dissidents, and human rights activists. The coverage prompted Wassenaar's members to add surveillance technology to their control list. And while European countries started mandating that companies get licenses to export spyware and other surveillance and intrusion tools abroad, the United States never did.

The closest the United States ever came was in May 2015, when regulators pushed to incorporate Wassenaar's changes into law. The Department of Commerce tried to mandate that security researchers and technology companies obtain a license to export "cybersecurity items" like "intrusion software." But the proposal failed after everyone from the Electronic Frontier Foundation to Google threw a fit. The most vocal critics were companies in the $1.7 billion "penetration testing" industry, where White Hats get paid to hack their clients' systems with exploits and intrusion tools to harden their defenses. They were terrified that Wassenaar's overly broad wording would put them out of business and they pushed U.S. regulators to abandon the proposal and to lobby

Wassenaar's members to narrow the scope of the items added to their control list. Eventually their arguments won out: Wassenaar's language was narrowed to the systems that "command and control" intrusion software. And even then—even after Wassenaar's member states adopted the limited language into law—the United States, without explanation, never has.

As a result, the United States exploit market remains largely unregulated. The exception are tools that fall under older encryption export controls. Americans cannot sell intrusion tools to embargoed countries like North Korea, Iran, Sudan, Syria, and Cuba. But there is nothing stopping hackers from selling exploits and intrusion tools to "favorable" countries, which includes most Western allies, but also many countries with questionable human rights records like Turkey. To sell these tools to other foreign groups, cryptographic controls require sellers to obtain a license from Commerce's Bureau of Industry and Security, which often grants them for four years or more, and asks only that sellers report biannual sales in return. Pen-testers, exploit brokers, and spyware makers argue encryption controls are adequate; digital rights activists call that ludicrous.

Once it became clear the U.S. would not adopt Europe's stricter rules, several spyware sellers and zero-day brokers moved from Europe to the United States and set up shop near their best customers around the Beltway. Between 2013 and 2016, the number of companies selling surveillance technology inside the United States doubled in size. A firm that once catalogued spyware technologies in what it called the "Little Black Book of Electronic Surveillance" had to rename its catalog "The Big Black Book" in 2016. Its 2017 edition included 150 surveillance companies. These companies began to cross-pollinate with foreign law enforcement agencies, some in places with dubious human rights records, and soon a new kind of security outfit was born—one that did not just sell to U.S. government agencies or Five Eyes, but some of the most recognizable human rights offenders around the globe.

AMONG THE BETTER-KNOWN businesses run by a former NSA hacker is the Miami-based company called Immunity Inc. founded by Dave Aitel. A lean, tanned hacker with angular features, Aitel was known to push boundaries inside the agency, pissing off superiors. He had taken to parking his Toyota

Camry in spaces reserved for top NSA officials at the front of the parking lot. As an extra *F you*, Aitel stuck a FREE KEVIN sticker on the back, a nod to Kevin Mitnick, a FBI-hunted hacker who did time in prison for hacking. (After prison, Mitnick reinvented himself as a White Hat, but it wasn't long before he drifted into gray territory selling and brokering zero-day exploits to unnamed governments and corporations on the side). When parking officials called up Aitel's bosses to complain about his parking job, Aitel didn't budge. Instead he launched an insurrection on the NSA's internal mailing list. Riffing on the agency's unofficial slogan, "One Team, One Mission," Aitel wrote colleagues: "One Team, One Parking Lot" and encouraged NSA's rank-and-file to park wherever they damn well pleased.

But what really pissed off Aitel's superiors is what he did after he left the Fort. He co-authored a book with several well-known hackers called *The Shellcoder's Handbook: Discovering and Exploiting Security Holes*. It became a bible for aspiring hackers. In it, Aitel detailed specific exploits and attack methodologies that his former bosses felt went too far in disclosing NSA spycraft. At Fort Meade, they put up a dartboard with Aitel's face on it and encouraged his successors to aim right between his eyes.

In 2002, Aitel started his security company, Immunity, from his apartment in Harlem and began consulting with the big financial services companies. But soon he developed an automated exploitation tool called Canvas that allowed his customers to test genuine threats—some known, some zero-day exploits Aitel developed himself—on their systems to mimic the techniques of advanced nation-states and cybercriminals. It proved a hit with the banks and, soon, governments, which were less interested in defending their systems from attack than in learning how to use zero-days to hack their enemies, and in some cases their own people.

Aitel never would tell me about his exploit work for governments. Whenever I probed for specifics, he grew evasive.

"Have you ever sold a zero-exploit to government agencies, domestic or foreign?" I asked him point blank.

"I would never comment on my customers," he replied.

The fucking salmon.

*   *   *

TO UNDERSTAND THAT dimension of the exploit trade, I had to go to Aitel's first employee, a Kurdish hacker by the name of Sinan Eren.

Eren was Kurdish by heritage, and had grown up in Istanbul. He learned how to hack as a form of resistance. His father, a Kurdish activist, spent almost a year in prison after the 1980 Turkish military coup and still had a bullet lodged in his shoulder, courtesy of a Turkish policeman who shot him at a protest.

The younger Eren, however, stayed out of politics. He tuned it out by playing bass guitar in his indie band, Conflict. His mother was the heiress of a well-off steelmaking family, and Eren resembled her. Unlike most Kurds in Turkey, who have a distinctive accent, Eren could easily pass as an affluent Istanbulite. Growing up, the police left him alone.

It was only as Turkey's crackdown on the Kurds intensified that Eren became a regular target. Turks were required to carry identification by law, and anytime a police officer stopped Eren and saw that he was a Kurd, it was game over.

"Anything could happen," he told me late one summer afternoon. Most times they detained him just for being a Kurd, rounding him and his friends up in a bus, where the police would make them stand for hours, "just to see if you would piss yourself."

Eren and his friends grew to fear one particular Turkish cop, who swatted Kurds with a two-foot-long whip to remind them who was in control. Those who were beaten and let go came to believe they were the lucky ones. In the 1990s, murders of Kurdish activists became common. The Turks had a legal term for these killings—"*faili meçhul*" which roughly translates as "unknown perpetrator"—and in the mid-1990s Kurdish Turks started disappearing by the thousands.

"You would see people picked up by the same make and model of Renault that we knew belonged to the secret police, and you'd never see them again," Eren told me. "People would scream their names, their relatives' names, and their telephone numbers out the back of the car, so at least their families would know their fates. I came of age in this era."

Eren needed me to understand why he had gotten into hacking and exploit development, and why—as more unsavory governments entered the exploit market—he had to get out.

He turned to hacking in college as a form of political resistance. For as long as Kurds could remember, college campuses had been considered safe zones from police brutality. But that began to change during Eren's time at Istanbul Technical University. With alarming frequency, administrators started inviting the secret police onto campus for "any little thing." Members of the Kurdish Student Union started to set up internet channels—early versions of Slack—to warn others that the police were on campus. That became the first of many small acts of resistance. Eren and his friends also stole a page from the script kiddie hackers who defaced websites in the United States. On the anniversary of Turkey's military coup, Eren and his friends vandalized Turkish government sites.

Eren dropped his band and began to spend all his time on hacking forums, communicating with hackers in Argentina and the United States, learning their tradecraft and the art of the zero-day. He could see that hacking wasn't just a resistance tool; it could also be a powerful intelligence tool. Using the tools he gleaned from hacking forums, he began hacking into university officials' emails, and he could see that school officials were complicit in the crackdowns. "We couldn't get access to everything, but we could collect the breadcrumbs—meeting minutes, appointments, schedules—and leak that to the press."

Eren was becoming one of the web's early "hacktivists." His family didn't understand his contributions. "They would say, 'We're risking life and limb, and you're playing with crayons.'"

But he was transfixed. Eren could see the direct impact his defacements and leaks were having in the press, and the power of this new virtual movement. He and his friends posted tear sheets around the neighborhood, offering free internet dial-up numbers, and hacked usernames and passwords to give people free internet access so they could join in this new digital resistance. Hacking and hacktivism became a compulsion.

As graduation neared, the military started knocking on Eren's door. Turkish students, even Kurds, were being forced to enlist.

"I knew that with my background, they would try to turn me against my own people."

Eren looked for work abroad. An Israeli security company was hiring engineers in San Jose, the industrial city at the heart of Silicon Valley. This was the post–9/11 cybersecurity hiring boom. Eren became one of the lucky few foreigners to acquire an H-1B visa to work in the United States. This was his out.

San Jose and Istanbul might as well have been on distant planets. Eren spent most of his waking hours at work; weekends on hacking forums. The job was a far cry from the hacktivism he was doing back in Istanbul. He missed his family. When McAfee, the security giant, acquired his company and moved him to its headquarters up the road, he knew he had to get out. The culture was "dry and boring," and professionally it felt like a dead end.

At night Eren would settle into his old familiar hacking forums, seeing what hackers published on BugTraq and ripping their tools apart. It was there that Eren met Aitel, the former NSA hacker. Aitel dumped a new intrusion detection tool on the list, and Eren immediately set to work attacking it and publishing ways to fool it. The two went back and forth a few rounds, and the cocky American came to appreciate the Kurd's tenacity. Aitel asked Eren if he wanted to be Immunity's second employee.

The two set to work developing zero-day exploits to bake into Immunity's Canvas framework. Compared with his previous gig, Eren found the work exhilarating. Before long, Immunity started attracting attention from the big security companies; McAfee, Symantec, Qualys, and others were looking to license their platform and their tradecraft. Canvas and consulting paid the bills, but the real profit, Eren and Aitel discovered, was in getting paid to train security contractors in zero-day exploitation techniques.

Suddenly Booz Allen, the defense contractor, was knocking. Then Boeing, Raytheon, and Lockheed Martin.

Then the French police. The Norwegian government.

Before long, Immunity's biggest customers were foreign-born.

Perhaps it was inevitable—as with all things on the web—that "less savory actors," as Eren put it, would show up.

In the middle of a training session one day, Eren came face-to-face with his worst enemy: a Turkish general. The general did not recognize Eren as a Kurd—why would he? He only recognized Eren's Istanbulite inflection.

"I didn't know you had a Turk on staff!" the general exclaimed to Aitel. "Why didn't you tell me?"

The general demanded that Eren train him directly. With the Turk slowly circling him, Eren found it hard to breathe. He thought of his father and uncles back home. He thought of all the ways the Turkish military would use Immunity's exploitation techniques to make his people's lives hell. He flashed back to the bullet in his father's shoulder, the roundups, the Turkish cop with the whip, the busload of Kurds urinating on themselves. He could almost smell the exhaust of the Renault Toros—the cars in which his friends and fellow Kurds disappeared forever. And he began to shake.

Even as Eren relayed this interaction to me years later, I could hear the anger in his voice. I had heard that voice before. I heard it every time my own grandparents—Jews who fled the Holocaust but whose siblings and parents were all murdered at Auschwitz—mentioned Nazi Germany.

The Turk was asking him to betray his own blood. It took everything in his being not to grab the general by the throat. He was in fight-or-flight mode, and he chose flight. Eren politely excused himself and conferred with Aitel. He told him he would rather be dragged to prison than teach the Turkish military his tradecraft.

I asked how Aitel reacted. "He's very American," was Eren's response. "This was business. He was willing to work with everybody."

Aitel said he did not recall the episode but did not dispute it.

It was not the last time Eren would be approached by someone from his homeland. Turkish front companies would regularly show up to Immunity's workshops looking for trainings or zero-day exploits. Eren learned to sniff them out. "I don't want to say we hacked anyone," he told me. "But we developed our own methods to find out who was who. And I turned down a lot of Turkish companies that smelled like complete fronts."

It wasn't just the Turks. Other Immunity customers started to give Eren pause. Even "friendlies," customers like French government agencies, kept him up at night. He thought of the Algerians. Even Spain had its reasons for using exploits for shadier purposes. "I'd think, 'What about the Basques? The Catalans? What if there's an uprising?' This business gets complicated quickly. There weren't always good answers."

He became weary. "The people I was working with started to keep me up at night and haunted my dreams."

In 2009 Eren quit. With another ex-Immunity employee, they founded their own security company and vowed to be pickier about their customers. He knew he couldn't avoid governments completely—that's where the good money was, especially as authorities in Israel, Britain, Russia, India, Brazil, Malaysia, and Singapore started creating their own mandates and quotas for zero-day exploits and tools. Stuxnet had cracked Pandora's box wide open. Suddenly governments who would never match the United States when it came to conventional warfare saw what could be done with code. And even if they didn't have the cyberwarriors to conduct these operations themselves, they had the cash to acquire them.

The NSA was expressly forbidden from using these tools on Americans. The governments approaching Immunity, Eren's new venture, and their competitors were far less constrained. Sure, there were many who planned to use these exploits to spy on foreign adversaries and terrorists, but increasingly they were looking for tools to spy on their own people. "Considering my background, it was an eerie dilemma to play in this market," Eren said.

Eren tried to find a middle ground. He and his business partner scoured Amnesty International reports and mapped out governments with proven records of democratic norms, respect for civil liberties, press rights. They vowed to only work with governments in the upper right quadrants. It was a familiar story, one I would hear time and time again from hackers who believed that by coming up with their own ethical code, they could keep the darker forces of the internet—authoritarianism, repression, the police states—at bay a bit longer.

"Israel called me once," Eren told me. "I didn't even pick up. I was turning down ex-Soviet nations. I'd work with agencies in North America and Canada, but not Mexico, and only some Europeans."

The ethical wrangling only got more complex as time went on. And then the adrenaline just stopped. In 2015 Eren sold his company to Avast, the Czech antivirus giant, and after the golden handcuffs were off, he got out.

When I caught up with Eren in 2019, he'd gone fully in the other direction, working on a mobile app that could detect government surveillance on smartphones. It was an ironic twist, he acknowledged. He was now protecting everyday citizens against his former customers.

By 2019 dozens of foreign players had entered the market. But what really shocked the industry, or at least the reporters covering it, was the discovery that year that some of the very best NSA hackers were moving overseas, many to the Gulf. Their cover story: they were helping American allies defend against cyber threats and terrorists. The reality was grimmer and more sordid.

IN JUNE 2019 I received a cryptic message from a former NSA hacker, David Evenden, who was ready to come clean. Evenden had read a story I had written on the underground sale of hacking tools. "Interested in learning more?" he wrote me on Twitter.

He had been recruited to a boutique security contractor called CyberPoint, a competitor of VRL. The difference between the two: VRL worked strictly for American government agencies, while CyberPoint had *diversified*. CyberPoint lured former NSA hackers, including Evenden and some of his closest friends, with offers to double—in some cases quadruple—their salaries. There were also promises of a lavish contractor lifestyle in Abu Dhabi. The work would be exactly the same as at the NSA, he was told, only on behalf of a close American ally.

When Evenden and his wife landed in Abu Dhabi in 2014, the red flags were everywhere. CyberPoint's headquarters weren't in one of the many corporate skyscrapers downtown, but a secret, fortified mansion known as the Villa on the outskirts of town. This in itself was not too strange, Evenden told me; he'd heard several American start-ups worked from similar villas.

But then there were the two folders.

On his first day at work that August, Evenden's new bosses opened up one folder and read carefully from his new job description. He was here to help the UAE defend its networks from cyber threats. "Got it? Good," his bosses told him. Because just as soon as they closed that first folder, they pulled out the second one. Everything they had just told him was a lie. The first was just his cover story—the "purple briefing," they called it. He was to memorize it word for word and recite it over and over again, like an actor rehearsing his lines. If anyone asked him what he was doing in Abu Dhabi, Evenden was to give them the cover story. The second folder contained his real role, what CyberPoint called the "black briefing." He'd be hacking into terror cells and foreign networks on behalf of CyberPoint's Emirati clients. Even this, Evenden told me, was not so strange. Every former-NSA-hacker-turned-contractor he knew, around the Beltway or overseas, was now getting the equivalent of a purple and black briefing. Everyone was told to talk up the defensive elements of their work to nosy reporters like me, and never, ever, to speak of the offensive work they were doing for their government clients.

Still, two folders in Abu Dhabi were different from two folders in Virginia; despite the murky laws governing the offensive hacking trade, the NSA has its own rules for what employees can and can't do after they leave the Fort. The first rule: former NSA employees are expressly prohibited—for life—from revealing classified information and tradecraft to anyone, without specific agency approvals.

That is why Zero-Day Charlie had to seek permission from his bosses to publish his paper about selling his exploit.

Evenden didn't heed this red flag either, however. His bosses assured him that everything was kosher, that his assignments for the UAE had been cleared at the highest levels—the State Department, the Commerce Department, the NSA. The project even had a code name: Project Raven. It was all part of a much larger U.S. defense contract with the UAE—initiated in 2008 by Richard Clarke, the former U.S. counterterrorism czar under Bill Clinton and George W. Bush—to help the Gulf's monarchies develop their own terrorism tracking technologies. The overarching contract had an ominous name— Project DREAD—short for Development Research Exploitation and Analysis Department. And it relied heavily on subcontractors like CyberPoint and dozens of talented former NSA hackers like Evenden.

Among Evenden's first jobs was to track ISIS terror cells being groomed in the Gulf. This was not so easy. Islamic terrorists were "consistently inconsistent" when it came to their technology. The allegedly unsophisticated enemy was constantly adapting. They knew they could never expect to beat the West in a cyber game of cat and mouse, so they tried to stay off the web and blend into the crowds. They abandoned phones for burners and continually switched from one technology platform to the next. Evenden's colleagues at CyberPoint were constantly approaching zero-day brokers for exploits in the obscure platforms ISIS was picking up and using on the fly. It was the only way to keep up. The terror group was proving as brilliant as it was revolting.

But in a matter of just a few months, Evenden's bosses pivoted. "They told us, 'We have these reports that the country of Qatar is funding the Muslim Brotherhood. Can you guys prove that?'"

"Not without getting access to the country of Qatar," Evenden told his bosses. Not unless he hacked Qatar, in other words.

Go for it, was their response.

Others may have asked more questions. Evenden concedes he didn't ask enough.

Once he was inside Qatar's systems, his bosses appeared to have zero interest in ever getting out. Typically, the goal in such operations is to get in and stay only as long as you need. But his bosses made it clear that they wanted him to spread as deep and as far into Qatar's networks as he possibly could. The UAE, and its close allies the Saudis, had long had it out for neighboring Qatar. The brewing battle for Gulf Supremacy—"Game of Thobes" as some call the clash between dueling Arab monarchs draped in flowing white robes known as thobes—is among the least understood outside the region, and certainly by the NSA hackers being lured to the Gulf. The gist is that Qatar, once a backwater for pearl divers and fisherman, started pissing off its Gulf neighbors forty years ago, when Qatar struck gas off its coast. The tiny nation has since become the world's largest exporter of liquefied natural gas, at precisely the point its oil-rich neighbors started facing the biggest slump the oil market had seen in years. Qatar's freewheeling, influential news network, Al Jazeera, regularly blasted its Gulf neighbors. And the Qataris supported the Arab Spring in 2011, even as the UAE and Saudis grew especially terrified of, but narrowly avoided, uprisings of their own."In retrospect, we had no idea what we were doing there," Evenden told me. "Officially we were there to track terrorists. Unofficially, the Emiratis just used reports that Qatar backed the Muslim Brotherhood to get NSA hackers to get them access to Qatar's systems."

Evenden's team never found proof that Qatar was funneling money to the Muslim Brotherhood; nor did they find anything to show that Qatar had bribed FIFA soccer officials to secure the 2022 World Cup, something the Emiratis also wanted to know about. Still, the requests just kept coming. Soon Evenden's team was hacking FIFA officials in Europe, South America, Africa. The Emiratis were particularly fixated on Qatari air travel. His clients wanted to know where every single Qatari royal was flying, who they were meeting with, who they were talking to. This too, his team was told, was all part of their mission. In the War on Terror and the offensive cyber trade, you could rationalize just about anything.

And they did. Before too long, Evenden's team was tailoring spearphishing emails to human rights activists in the UAE and journalists in England. They never clicked send, Evenden told me. Emirati officials just wanted to know—hypothetically, of course—how one might find out if their critics had terrorist connections. Perhaps he could write a template, they suggested.

He wrote a booby-trapped email addressed to Rori Donaghy, a London-based journalist who was calling out human rights abuses in the Emirates at the time. The email invited Donaghy to a fabricated panel on human rights. It was never supposed to be sent.

But sent it was, along with spyware that could trace Donaghy's every keystroke, password, contact, email, text, and GPS location. By the time researchers uncovered the spyware on Donaghy's computer, CyberPoint's digital fingerprints were all over the hacks of some four hundred people around the world, including several Emiratis, who were picked up, jailed, and thrown into solitary confinement for daring to insult the state on social media or even so much as questioning the ruling monarch in their most intimate personal correspondence.

Perhaps it was inevitable, in retrospect, that CyberPoint's dragnet would scrape up data on American officials too. But few predicted that one day, former NSA hackers would ensnare Americans at the very top.

ALL THE RATIONALIZATIONS were stripped away the day Evenden's team hacked the First Lady of the United States.

In late 2015 Michelle Obama's team was putting the finishing touches on her big week-long trip to the Middle East. Qatar's Royal Highness Sheikha Moza bint Nasser, the second wife of Qatar's former ruler Sheikh Hamad bin Khalifa al-Thani and mother to his successor, Emir Tamim bin Hamad al-Thani, had personally invited Michelle Obama to speak at her annual education summit in Doha. And Michelle Obama saw it as the ideal venue to discuss her "Let Girls Learn" education initiative. It was also the perfect opportunity to meet with U.S. troops stationed in the Qatari desert at the Al Udeid Air Base.

The First Lady's team was coordinating with comedian Conan O'Brien to provide some necessary relief for the estimated two thousand soldiers on the ground. Obama would also make a short stop in Jordan to visit a U.S.-funded school for Syrian refugees. There was $700,000 in logistics to figure out. Obama's team was in constant communication with Sheikha Moza.

And every last email between Her Royal Highness, the First Lady, and their staffs—every personal note, every hotel reservation, flight itinerary, security detail, and itinerary change—was beaming back to CyberPoint's servers. Evenden's team were not just hacking Qataris anymore, or Emirati activists, or Western bloggers. They had just effectively hacked Americans. And not just any American.

If Evenden was looking for a sign he was on the wrong side of the law, or just adrift without a moral compass, seeing the First Lady's emails on his computer screen was a cold, hard slap in the face.

"That was the moment I said, 'We shouldn't be doing this. This is not normal. We should not be capturing these emails. We should not be targeting these people,'" he said.

Evenden went to his bosses and demanded to see the State Department letter that had green-lit the program. The first few times he asked, it was clear that his bosses hoped he would drop it. But he kept asking until the day they allowed him to view it in person. The document was indeed real; it bore a State Department emblem and a signature—but it was dated 2011. It was unimaginable that the State Department approved of what his team was doing now. It was then that Evenden realized that almost everything he had been told up to that point was a lie.

When he and his colleagues confronted CyberPoint's executives, they were told that this was all a terrible mistake. If they came across Americans' data, all they needed to do was flag it, and their managers would ensure it got destroyed. And so they did. But two, then three, then four weeks later, Evenden would query CyberPoint's databases and find that the data was still there.

Beyond the Villa, Evenden started to see Abu Dhabi in a harsher light. Its man-made islands and museums were just distractions from a state that imprisoned anyone who offered even the tamest criticism. Evenden started to read local news stories about American expats being thrown in "debtors' prison" because they were unable to pay their credit card debt. In traffic one day, he witnessed a bad accident in which an Emirati ran through a stoplight and collided with an expat. Though the accident was clearly the Emirati's fault, the local police let the Emirati get away and dragged the expat into custody. "The local government started to terrify us more than any terrorist organization," Evenden said.

Evenden's bosses seemed determined to look the other way. His immediate boss was making upward of half a million dollars a year. When Evenden and his colleagues raised their concerns, they got: "You guys are just being overly cautious."

CyberPoint's solution to all this ethical wrangling was not to stop targeting dissidents, journalists, or Americans; just the opposite. Evenden and his team were told that their contracts would be shifted from CyberPoint to an Emirati LLC called Dark Matter. They would no longer be on loan from the State Department. They would be working directly for the Emiratis, without constraints. Evenden's bosses gave CyberPoint's employees a choice: they could join Dark Matter, or CyberPoint would pay for their relocation back to the United States. No questions asked.

Half switched over to Dark Matter. Evenden warned his former colleagues to think this through. "You are going to be targeting Americans," he told them.

Either they were in denial or blinded by their fat Emirati paychecks. Some told Evenden they could never expect to make this kind of money back in the States. "They basically said, 'If I do this for another couple years, I'll be fine.'"

"A thick line was drawn," Evenden said. Those who opted not to join Dark Matter were ousted from friend circles. "People we regularly had drinks with, who we'd had over to our house, stopped communicating with us." And then they were kicked out of the shop. CyberPoint took away key cards and shut down personnel accounts. Evenden and the others who'd chosen to leave had to wait until the company could schedule their moves back to the United States.

And once they got back to the States, once they had reckoned with what they had left behind, Evenden sent a tip to the FBI.

By the time Evenden reached out to me in mid-2019, Dark Matter was under FBI investigation. One of Evenden's former colleagues who had opted to stay in the UAE—another former NSA analyst by the name of Lori Stroud—was picked up by FBI agents at Dulles Airport on her return to Dubai. Three years later, she'd given her story to reporters at Reuters. "That was her way of coming clean," Evenden told me. "It was her way of trying to say, 'Hey I'm one of the good guys here.' But the truth is, we told her very clearly, 'You're going to be targeting Americans if you stay.' She absolutely knew what she was doing."

All this begged the question of why Evenden had reached out to me in the first place. Was this his way of covering his ass? The FBI investigation was still active. Surely they didn't want him talking to journalists. But he was starting to get more calls from NSA operators, still in the service, who had been contacted by Dark Matter. "They said, 'Hey, this sounds really cool, what should I do?'"

His answers were unequivocal. As the number of people who reached out to him grew, he felt compelled to put out a public service announcement. "My mindset is, 'Hey former NSA Operators, here's what not to do when taking a job overseas,'" he began. "If the people sending you over there won't tell you what you're going to be doing before you get there, don't go. If once you get over there, you're given two folders, that's a red flag. If you're considering taking a contract for a lot of money overseas, you're probably not taking the job you think you're taking."

It struck me as a uniquely American naïveté, Evenden's initial trust in CyberPoint. It was like the story of the slowly boiling frog who fails to realize his peril until it's too late; I knew the damage had already been done. Evenden was too late. The frog had already boiled.

CHAPTER 12

Dirty Business

Boston, Massachusetts

I ALWAYS SAID WHEN this business got dirty, I'd get out," Adriel Desautels told me late one summer evening in 2019.

Desautels was a cyberweapon merchant who looked like a milkman. He had an unruly head of curls, frameless glasses, a gap between his front teeth, and a penchant for quoting the astrophysicist Carl Sagan. His original hacker alias, Cyanide, never sat well. He'd eventually change it to the more sensible "Simon Smith." But in a faceless business, looks meant little. Everyone who was anyone in the game knew Desautels was one of the country's preeminent zero-day brokers.

When I first started digging into the zero-day trade, Desautels's name was everywhere. But not in the ignominious sense. He seemed to be the one man with a moral compass in an industry that had none. I wanted to understand the nuts and bolts of the trade, but I also wanted to know how someone with such an interest in truth and transparency could play in a world shrouded in so much darkness. While other zero-day brokers seemed to relish their inscrutable Darth Vader reps, Desautels had stepped into the light. He seemed to understand something about his reputation that newer players on the market did not: it was his true currency. As a result, his clients—three-letter U.S. government agencies, Beltway contractors, and fixers who did not tolerate ostentation, double dealers, or stealth—trusted him.

Like Eren, Evenden, and so many others in this game, he never sought the zero-day market. It found him. He'd discovered a zero-day in Hewlett-Packard software back in 2002, and in what was now a familiar story, HP threatened to sue him for it under computer crime and copyright laws. Instead of rolling over, Desautels fought back, hired a lawyer at the Electronic Frontier Foundation, and together they forced the company to retract its threats, apologize, and set a new precedent for how companies should approach vulnerability research. It never occurred to him that the case would put him on the map. The year was 2002. iDefense's bug program was still sputtering up. He didn't even know a market for vulnerabilities existed until he took a call from an unknown number.

"What do you have to sell?" the man asked him. The question was a confounding one.

"I'm not sure what you mean," he replied. "Like security services?"

"No, I'm looking to buy exploits," the man told him.

To Desautels, the notion of buying an exploit seemed ridiculous. Why would anyone *buy* one, when they could just download it off BugTraq or Full Disclosure or any other number of hacker mailing lists? But the man persisted. "Just tell me what you're working on."

It just so happened that Desautels had been working on a clever zero-day MP3 exploit for fun. If he sent anyone a digital MP3 song file, and they played it, the zero-day would give him full access to their machine. Before he could even finish explaining how the exploit worked, the man interrupted. "I'll buy it. How much?"

Desautels still couldn't tell whether he was being serious.

"Sixteen thousand!" he replied, as a joke.

"Done."

One week later, his check arrived in the mail. He stared at it for a good long while and then quickly came to the conclusion Sabien and so many others around the Beltway were coming to: This could be a big business.

At the time, his penetration-testing company, Netragard, was just getting established. The company was more involved, let's say, than the competition. "Everything else on the market was crap," he told me. Netragard did the kind of in-depth hacking tests that made sure clients wouldn't be hacked by people like him. That was the motto: "We Protect You From People Like Us." Most other pen-testers did a basic scan of a company's network and handed back a report with a list of things to upgrade and fix. That's all most businesses wanted anyhow. They simply wanted to check off boxes on a compliance checklist. But in terms of keeping actual hackers out, the tests were useless. Desautels compared competitors' practices to "testing a bulletproof vest with a squirt gun." In his book, they were con artists, bilking clients for tens, sometimes hundreds, of thousands of dollars, and failing to keep hackers out. When Netragard performed a penetration test, they actually hacked you. They forged documents, hacked security keypads and work badges. When tried-and-true digital methods didn't work, they sent hackers up their client's freight elevators to grab a badge off a secretary's desk, bribed cleaning ladies, and broke into chief executives' offices. It was all covered in their contract. They called it their "get out of jail free card" and Netragard soon made a name for itself breaking into Las Vegas casinos, pharma companies, banks, and the big national labs.

He figured he could fund Netragard's business by selling zero-day exploits on the side and keep the venture capitalists at bay. The next time he got a call from the broker who paid $16,000 for his MP3 zero-day, he doubled his asking price. The next time, he doubled it again to $60,000. He just kept upping the asking price until he met resistance. Soon, he was selling zero-days for more than $90,000 a pop. By then iDefense had emerged with its measly $100 price lists. Desautels didn't know why anyone would sell to iDefense when they could go through what Desautels calls the "invisible, legitimate black market" and live off a single paycheck for years.

I asked Desautels if he had any qualms about how his exploits were used. He never told me the names of his buyers, only that he sold exclusively to U.S. entities in the "public and private sector"—in other words, three-letter U.S. agencies, defense contractors and occasionally security companies looking to test his zero-days against their own software. September 11 still felt fresh, and he told himself that his exploits were being used for good—to track terrorists and child predators. He started telling friends that if *they* had a zero-day exploit, he could help them sell it. iDefense and the bug bounty programs paid crap, he told them, compared to the five- and six-figure payments his buyers were offering. Pretty soon he was brokering more exploits than he was developing.

Desautels could sell exploits for ten times iDefense's bargain-basement prices and iDefense was getting his sloppy seconds. The zero-days Desautels brokered had to be in an "ideal state"—that is, they required zero interaction on the target's end: no spammy text messages or phishing emails like the ones Chinese hackers were known to send. The exploits Desautels developed and brokered had to work 98.9 percent of the time. And if they failed, then they had to "clean fail"—meaning they couldn't trigger a security alert or crash a target's computer. No one could know they were being hacked. The operations were simply too sensitive. If the target caught even the faintest whiff they were being targeted, it was game over.

Desautels seemed confident that his buyers played it straight, that they weren't using his exploits to track dissidents, journalists, or ex-lovers.

But what about your sellers? I asked. Who were they? What if they were double-dipping? What if they were selling the same exploits they sold Desautels to authoritarian governments like the UAE or China, who then used them to spy on their own people?

This wasn't a hypothetical. By 2013, the year Desautels began trading exploits, Chinese clients were flooding the market. A notorious young hacker, George Hotz, alias Geohot, who made a name for himself unlocking the first iPhone and hacking Sony's PlayStation game console, was recorded trying to sell an Apple exploit for $350,000 to a broker who suggested that China was the end client. (Hotz, who later did a stint at Google, denied the deal was ever consummated and insisted he has worked only with U.S. buyers, but added that he doesn't particularly dwell on the morality of who he does and does not sell to: "I'm not big on ethics.")

Desautels had a plan for that too. He paid hackers three times more for their zero-days if they agreed to sell them to him exclusively. Nondisclosure agreements crystallized these agreements. But how could he be so sure that hackers weren't selling the same zero-days on the side? These weren't conventional arms. They were code. Everything relied on Bushido, the samurai code of honor, he told me. He had to trust that hackers wouldn't blab about the epic zero-day they had just sold, or sell it to another party on the side, or burn him and his brokerage business in the process. Considering who these hackers were, *where they were*, I couldn't help but think this seemed like a shitload of trust.

The vast majority of sellers, Desautels told me, were in the United States, Europe, and Romania. He had a guy in Romania who could hack just about anything, and never tell anyone about it. *Romania?* I had to catch myself— Romania was the fraud capital of the world. But Desautels name-dropped the country like he was talking about Iowa on the Fourth of July.

"There were hackers in Iran who had really incredible capabilities, and we saw some really interesting attacks out of North Korea, but those guys never approached us," he told me.

He bought one zero-day off a Russian hacker early on for $50,000. But the second time the guy approached him, something didn't feel right, and Desautels shut it down. He had failed to pass what Desautels called his "sniff test."

I asked him where he developed this sniff test. Had he read interrogation manuals? Behavioral science? The psychology of manipulation? How did he read people? "I have an uncanny gift," he told me with an affable arrogance. "Even as a little kid, it only took me two minutes of talking to someone to know exactly what kind of person they were. I just pick up on their demeanor, micro-expressions, their tells when they speak." I shifted uncomfortably in my chair. There were a lot of lives riding on one guy's sniff test.

Fifteen percent. That was the percent of exploits he accepted. The rest he turned down either because they weren't good enough, or because he just got that shifty feeling. And he made a point of regularly reminding his sellers that if they ever violated their NDAs, there would be consequences, many out of his immediate control. "Basically, I told them that if their zero-day ever popped up elsewhere, our buyers had the means to understand where it came from. They would examine the pattern of exploitation and trace it back. And when they did, we wouldn't hesitate to give them the author's information, and they would deal with it. I basically said, I'll broker this for you, but if you take this somewhere else, you're going to get gutted."

There was a casual absurdity to his words. It sounded hyperbolic and he'd said it all matter-of-factly, as if it happened, it was just business. Nothing personal.

His approach won customers' trust. Soon Desautels was on retainer with several clients, earning a monthly, or quarterly fee, to negotiate with hackers on their behalf. The price of a zero-day fluctuated based on the software it exploited. At the bottom of the heap, a zero-day that could take over routers or USB sticks that an operator had physical access to, would net low five figures. Just above those were zero-days that could remotely exploit Adobe PDF software, Safari and Firefox browsers, or Microsoft applications like Word and Excel. Higher up the chain were $100,000 to $250,000 exploits that could remotely hack Microsoft's email products and Windows software. Time constraints also played a factor. When his buyers needed to exploit something *now*—say a terrorist's mobile phone, an Iranian nuclear scientist's computer, or Russia's embassy in Kyiv—his buyers would shell out $500,000 to $1 million for a zero-day they might otherwise pay $250,000 for. Often Desautels would take a small cut—sometimes as little as 3 percent—off the sale price but his fee could run as high as 60 percent if he had to source, vet, and test an exploit for a client on the fly.

As a business, it was turning out just fine.

NEWER PLAYERS—BUYERS AND sellers—were flooding the market. And they were a different breed. There was the Grugq, the South African living large in Thailand who happily allowed himself to be photographed for *Forbes* with a giant duffel bag of cash. There was the German spyware entrepreneur who went by the initials MJM and scrubbed any trace of himself off the internet. There were Luigi and Donato in Malta, readily advertising zero-days in industrial control systems, the bulk of them used in the United States. In Singapore, an animated businessman by the name of Thomas Lim was selling cyber munitions to countries and brokers with no serious exploitation skills of their own, but the deep pockets to buy them. A French Algerian named Chaouki Bekrar regularly taunted Google and Apple, bragging about the flaws in the products he was finding, buying and selling to governments, God knows where. Bekrar's Twitter handle readily embraced his villainy. It featured a picture of Darth Vader, and he liked to remind everyone that his critics called him the Wolf of Vuln Street.

These men didn't seem to care about public perceptions. They didn't care for Bushido. And they felt no more responsibility for weaponizing the internet than the tech companies who left gaping holes in their products. They behaved more like mercenaries than patriots. Desautels remembers the first time he saw new buyers and brokers lurking around hacker conferences in 2010. There was no point to a sniff test. His olfactory senses were overloaded with bullshit. "They pissed me off to no end. I knew there were enough rogue nations playing in the black market to make them fat and happy. Things got dirty really fast," he said, with a twinge of shame.

Desautels was intimately aware that there was a lucrative market for men like these. Middlemen from Israel to South Korea started aggressively approaching him at hacking conferences, pressing him to open his business up to foreigners. He stopped advertising his travels for fear these middlemen would track him down and pressure him into business. They found him anyway. In Vegas one morning, he woke up to his hotel landline phone ringing in his room at Caesars Palace. Nobody knew he was here. "Come downstairs," a strange voice told him. "We need to meet." It was an Asian middleman—an ally, but Desautels wouldn't say who—offering him a first-class visit to their country and a grand tour if he would open up a line of business. He always said no.

Other interested parties didn't even bother to ask. On a trip to Moscow, he made a point of renting an Airbnb apartment with heavy metal doors and huge locks, with steel reinforcements. Before he ventured out, he painted over the screws on his laptop with his wife's nail polish. It seemed paranoid, but by now he knew he had legitimate reasons to worry. If shadier players were coming into the industry, the shadiest were in Russia. Sure enough, when he returned, the polish was cracked. Someone had tampered with his laptop. If foreigners were willing to go to such lengths with him, they were almost certainly approaching newer players who made it clear they were open for foreign business.

BY 2013, THE year I was shoved in a closet, the market was spreading far beyond Desautels and the reliability of his sniff test. That year, the founder of an annual surveillance trade show estimated that the market had surpassed $5 billion from "nothing 10 years ago." It was also the year the NSA added its $25 million line budget for zero-days and CyberPoint was buying zero-days to hack the UAE's enemies and our allies. Desautel's zero-day procurement business had doubled, but so had that of his competitors. Vupen, the French company owned by the Wolf of Vuln Street, was seeing its sales to governments double each year. Israel, Britain, Russia, India, and Brazil were matching U.S. government prices. Malaysia and Singapore were buying too. In fact, it was rare to find a country—outside Antarctica—that wasn't in the zero-day trade.

Desautels's grip on the market was slipping away, and with so much cash from so many sources coming in, nobody cared about his sniff test. Sellers now had plenty of other options. Hackers started wiggling out of his exclusivity clauses. Desautels tried to stick to his guns, but his top exploit developers gave him little choice. They threatened to stop selling him zero-days unless he made exceptions. "We'd go to the buyers and tell them the seller wasn't interested in selling if it meant the sale was exclusive." His buyers started to cave, too. "They would say, okay, get us exclusivity for two or three months, and then we'll give them nonexclusive terms."

His longtime clients started capitulating in other ways, too. One of Desautels' trusted buyers approached him about potentially selling his zero-days to new customers outside the United States. They would never sell his wares to oppressive governments, they told him, but they had recently developed a relationship with a new buyer in Europe—in Italy, actually.

"Like an idiot," Desautels admitted, he assumed the Italians only worked with "friendlies"—American and European government agencies and their closest allies.

The Italian deal was with a newer Milan-based player named Hacking Team, and it would cost him the market.

DESAUTELS OPENED UP his computer and tried not to vomit.

At precisely 3:15 A.M. local Italian time that day—July 5, 2015—Hacking Team's typically quiet Twitter account posted an ominous message: "Since we have nothing to hide, we're publishing our emails, files, and source code."

As it turned out, Hacking Team, the Milan-based purveyor of hacking tools, had itself been hacked, by an ideological hacker who went by the alias Phineas Fisher. Over the next few hours and days, Fisher dumped 420 gigabytes worth of Hacking Team contracts, payroll documents, invoices, legal memos, customer support records, and five years' worth of email correspondence belonging to the chief executive on down. Despite Desautels's buyers' reassurances, Hacking Team was not just selling to "friendlies," it was baking his zero-day exploit into the spyware it sold to some of the worst human rights offenders on earth.

By now, it was hard to tell if Desautels was the milk man or the mercenary. Buried in the leaks was an email he sent directly to Hacking Team: "We've been quietly changing our internal customer polices and have been working with international buyers . . . We do understand who your customers are both afar and in the U.S. and we are comfortable working with you directly."

When I asked Desautels about this, he replied, "Like an idiot, I had not done enough due diligence." Something of an understatement. Had he been paying even the slightest bit of attention, he would have caught a series of disturbing stories about Hacking Team. For three years I had worked with researchers at Citizen Lab, a cybersecurity watchdog lab at the University of Toronto's Munk School of Global Affairs, uncovering Hacking Team's spyware in emails sent to dissidents in Bahrain, journalists in Morocco, and Ethiopian journalists in the United States. Though Hacking Team billed its tools as "untraceable," Citizen Lab's researchers were able to reverse-engineer and trace its spyware back to servers located in dictatorships around the world.

This wasn't data buried on some website. I'd covered the discoveries on the front pages of the *Times*. And I'd always made a point to reach out to Hacking Team's Italian executives for comment. The company's chief executive, David Vincenzetti, was emphatic that Hacking Team went to "great lengths" to ensure that its spyware was used only for criminal and terrorism investigations and never sold to governments "blacklisted by European countries, the U.S., NATO or any repressive regimes for that matter." Vincenzetti said his firm had even set up a board—comprised of engineers and human rights lawyers—with veto power over any sale. But as I sorted through Vincenzetti's hacked emails, it was clear the company had been lying to me all along.

What "Phineas Fisher"—his or her real identity is still unknown—unleashed in July 2015 confirmed my worst suspicions. For twelve years, Hacking Team had been selling its spyware to a growing list of government agencies around the globe, some with human rights records that were not just questionable but grotesque. Its customers included the Pentagon, the FBI, and the Drug Enforcement Agency, which used it to spy on cartels from the

American embassy in Bogota. Hacking Team had loaned out samples of its spyware to the CIA through a front company in Annapolis, Maryland. Leaks showed it had contracts with agencies across Europe, in Italy, Hungary, Luxembourg, Cyprus, the Czech Republic, Spain, Poland, and Switzerland. But it was a slippery slope. Hacking Team had also sold its "Remote Control System" to CyberPoint in the UAE, to the Saudis, the Egyptians, the Russian security services, authorities in Morocco, Bahrain, Ethiopia, Nigeria and the 'Stans of Central Asia—Azerbaijan, Uzbekistan, Kazakhstan—where authorities turned their spy tools on innocent civilians. Leaked emails showed executives trying to close a deal with Belarus—the country commonly known as "Europe's last dictatorship"—a Bangladeshi "death squad," and worse. Hacking Team had sold $1 million worth of spyware to Sudan, to the very intelligence service that had been engaged in a decades-long effort to savagely displace, kill, rape, mutilate, abduct, and pillage large portions of the Sudanese population. Sudan had, U.S. aid workers said, "one of the most horrendous human rights situations in the world" and Desautels had just armed the people responsible.

"I'd never been so repulsed in my life," Desautels told me. "I was disgusted."

Journalists all over the world began digging through the leaks. South Korean journalists found emails showing that Hacking Team's spyware may have helped South Korean's intelligence operatives rig an election. (One South Korean agent who used Hacking Team's spyware committed suicide after his emails were made public.) In Ecuador, journalists discovered that the ruling party had used its spyware to track the opposition party. The leaks showed that despite its repeated denials, Hacking Team was also selling to governments waging vicious crackdowns on critics and dissidents.

After the first series of Citizen Labs reports and articles came out in 2012, Hacking Team had apparently paused to take stock of some of its customers. The firm had ended its support for Russia in 2014 because, a spokesman said, "the Putin government evolved from one considered friendly to the West to a more hostile regime." Apparently, Putin's raid on Crimea pushed Russia into a different customer category—never mind that for years, Russian journalists and activists had disappeared under Putin's watch. As for Sudan, Hacking Team cut its contract in 2014 "because of concerns about the country's ability to use the system in accordance with

Hacking Team's contract." This after hundreds of thousands of Sudanese were already dead and millions more displaced.

It also offered a more mundane view into the inefficiencies of the zero-day trade: Warranty provisions should zero-days get patched; the wild pricing discrepancies in a market with zero transparency; months of frustrating back and forth with zero-day brokers at COSEINC in Singapore and at Vupen, only to get exploits containing bugs that had already been patched; and one almost comical exchange in which Hacking Team purchased a fake Microsoft exploit from a shady dealer in India.

The leaked emails also made clear just how little serious thought was given to the potential abuse of Hacking Team's products. In one email in which Vincenzetti seemed to predict the future, he joked "Imagine this: a leak on WikiLeaks showing YOU explaining the evilest technology on earth! ☺."

ANYONE WITH AN internet connection could soon discover that Desautels had brokered the Italians a zero-day exploit in Adobe Flash software that formed the raw material for Hacking Team's spyware. Desautels' Adobe zero-day made it possible for customers like CyberPoint to hack their targets using booby-trapped PDF documents disguised as legitimate documents. Desautels' only consolation was that with his zero-day out there for the world to see, it would be patched, and in time, Hacking Team's spyware would be neutered. But he felt sick thinking of how it might have already been used.

Desautels had believed he could control the market with morals and scruples. *Bushido*, I thought. *More like Bullshit*. The Flash zero-day Desautels had brokered to Hacking Team was the first he ever sold to a non-U.S. buyer. There was, we both knew, no way to verify this with 100 percent accuracy, especially given the secrecy and increasing complexities of the market, where murky middlemen often acted as front companies for various clients. But it was also the last zero-day Desautels ever sold. "I always said that once this business got dirty, I'd get out," he'd told me. And after the first leaks, he abruptly announced he was closing his zero-day business:

The Hacking Team breach proved that we could not sufficiently vet the ethics and intentions of new buyers. Hacking Team, unbeknownst to us until after their breach, was clearly selling their technology to questionable parties, including but not limited to parties known for human rights violations. While it is not a vendor's responsibility to control what a buyer does with the acquired product, Hacking Team's exposed customer list is unacceptable to us. The ethics of that are appalling and we want nothing to do with it.

In an industry not known for its scruples, Desautels's position was notable. But it was also too late. (I was learning that nobody in this trade ever seemed to take a stance until it was too late.) The Hacking Team leaks offered an incredible window into how zero-day exploits were being priced, traded, and incorporated into ever-more-powerful off-the-shelf spyware and sold to governments with the most abysmal of human rights. By then, none of this came as a shock. But there was one element of the leak I did not expect. I had always assumed that the stories I wrote about Hacking Team had helped shed a brighter light on a seedy industry. Occasionally the coverage would be cited by European regulators and human rights lawyers who vowed to investigate, to change export rules, to take a harder look at the cyberweapons trade.

But as I sifted through the leaks, I could see that the coverage had had the opposite effect: it had functioned as advertisement, showing other governments that did not possess these capabilities what they were missing.

By late 2015, no intelligence agency on the planet was going to miss out.

CHAPTER 13

Guns for Hire

Mexico, United Arab Emirates, Finland, Israel

J OURNALISTS WERE STILL reckoning with the fallout from the Hacking Team leaks when a source showed up at my house in the summer of 2016, made small talk for an hour, and then, without warning, popped open his laptop. "Take pictures of my screen, print them out, wipe any trace of them from your phone, computer, and printer and never tell anyone where you got these. Understood?"

It was an abrupt turn in conversation, but my source had proven reliable before, so I did as I was told. I started snapping what appeared to be emails, PowerPoints, proposals, and contracts. I printed them out and wiped any trace of them from my phone, my printer, and the cloud. By then my source was already pulling out of my driveway. It was left to me to make sense of the pile of documents laid out on my kitchen counter. For the next few hours, days, and weeks, I dug through detailed customer records, product descriptions, price lists, even photos secretly captured off phones.

They belonged to a highly secretive Israeli spyware company called NSO Group that I had only heard of in passing whispers. NSO did not have a corporate website. I could find only a passing mention of it in a single entry on Israel's Ministry of Defense website, in which the company claimed to have developed cutting-edge spyware. I found a few press releases and deal stories dated 2014, the year NSO had sold a controlling stake to Francisco Partners, a San Francisco private equity fund, for some $120 million. But that is where the digital crumbs trailed off. As I read through NSO's files, I felt compelled to call every journalist, dissident, white hat, and internet freedom fighter in my contact list, before tossing my phone in the toilet.

While the world consumed itself with Hacking Team, the leaks made clear that sophisticated surveillance states, intelligence, and law enforcement agencies had already moved on. The Israelis didn't bother with PCs; they could get everything their government clients might ever want or need simply by hacking phones. And judging by their pitch deck, NSO had found a way to invisibly and remotely hack into every smartphone on the market: BlackBerries, the Nokia Symbian phones still used by many in the third world, Android phones, and, of course, iPhones.

NSO's surveillance technology was originally developed by graduates of Israel's Intelligence Unit 8200. In 2008 two Israeli high school buddies— Shalev Hulio and Omri Lavie—marketed the technology to cell-phone companies as a way to troubleshoot customers' IT issues remotely. Their timing was fortuitous. The iPhone was less than a year old, and smartphones—more so than PCs—offered cops and spies an intimate, real-time window into their target's location, photos, contacts, sounds, sights, and communications. What more did a spy need, really? Word of NSO's capabilities traveled to Western spy agencies. And soon everyone wanted in.

Not only could NSO's technology transform a smartphone into a spy phone, but it gave governments a way to skirt encryption. Big tech giants like Apple, Google, and Facebook had started encrypting customers' data as it passed from server to server, device to device. For years law enforcement criticized the moves, warning that encryption made it far more difficult to monitor child predators, terrorists, drug kingpins, and other criminals. They called the phenomenon "going dark," and by 2011 the FBI, in particular, feared the worst was yet to come. For years the bureau had accessed data relatively easily through wiretaps. But with the move to decentralized communications— mobile phones, instant messages, email, and web-enabled phone calls—and the addition of encryption, even when agents obtained a warrant to access someone's communications, in many cases they only got a bunch of gobbledygook.

"We call this capabilities gap the 'Going Dark problem,'" then FBI general counsel Valerie Caproni testified to Congress in 2011. "The government is increasingly unable to collect valuable evidence in cases ranging from child exploitation to pornography to organized crime and drug trafficking to terrorism and espionage—evidence that a court has authorized the government to collect."

Over the next decade, the FBI continued to press for highly flawed solutions: a requirement that tech companies make wiretap-friendly backdoors for law enforcement agents. In theory, it sounded doable. In practice, it fell apart. A backdoor for one party, national security officials knew better than anyone, would become a target for all. The idea these companies would leave Americans more vulnerable to cybercriminals and nation-states to satisfy law enforcement's needs was off the table. Then there were the logistical challenges. Not every tech company was based in the United States. Skype, for example, was originally hosted in Luxembourg. How many backdoors, for how many government agencies, would these tech companies be required to produce?

What NSO offered law enforcement was a powerful workaround, a tool to keep from going blind. By hacking the "end points" of the communication—the phones themselves—NSO's technology gave authorities access to data before and after it was encrypted on their target's device. Not long after Caproni testified to Congress, Hulio and Lavie pivoted and began pitching their remote access technology as a surveillance tool. They called the tool Pegasus, and like the mythological winged horse it was named for, it could do the seemingly impossible: capture vast amounts of previously inaccessible data—phone calls, text messages, email, contacts, calendar appointments, GPS location data, Facebook, WhatsApp and Skype conversations—from the air without leaving a trace. Pegasus could even do what NSO called "room tap": gather sounds and snapshots in and around the room using the phone's microphone and video camera. It could deny targets access to certain websites and applications, grab screenshots off their phones, and record their every search and browsing activity. One of the biggest selling points was that the spyware was "battery conscious." One of the few tells that you may have spyware on your device is a constantly draining battery. All that spying and siphoning can be a battery hog. But Pegasus came with a nifty trick. It could sense when it was killing battery life, shut itself down, and wait for a target to connect to WiFi before extracting any more data. This was, as far as I could tell, the most sophisticated spyware on the commercial market.

The leaked contracts showed that NSO had already sold tens of millions of dollars' worth of hardware, software, and interception capabilities to two eager customers in Mexico and the UAE, and were now marketing Pegasus to other customers in Europe and the Middle East. I did a quick search for NSO in WikiLeaks' Hacking Team database, and sure enough, Vincenzetti's emails offered a portrait of a competitor in full panic mode. The Italians were scurrying to protect as many client relationships from NSO as possible, as longtime clients in Mexico and the Gulf threatened to take their business to the Israelis. NSO's deal with Francisco Partners had sent Hacking Team's management into a tailspin, searching for an equity partner of their own. But what terrified the Italians most was one particular NSO feature long considered the unicorn in the cyberarms trade.

In some cases Pegasus still required a target to click on a malicious link, image, or message to download onto the phone, but increasingly it required no interaction at all. Digging through NSO's pitch decks and proposals, the company marketed a new zero-click infection method that executives called "over the air stealth installation." NSO did not detail how exactly it had accomplished this. In some cases they alluded to rigging public WiFi hot spots, but it appeared they could also hijack a target's phone from long distances. However they did it, NSO's zero-click infection method was clearly their secret sauce. And the Italians were petrified that it would put them out of business.

"We are working on the NSO issue day and night," Vincenzetti wrote his team in early 2014. "We have zero tolerance on a feature we allegedly blatantly miss."

One year later, Hacking Team was still unable to match NSO's zero-click feature, and it was hemorrhaging customers left and right. I called other sources to learn anything I could about NSO. But it appeared that NSO was the one company in this space that still managed to keep a low profile, even as its spyware was clearly some of the best on the market. NSO's prices alone were a good sign that the Israelis' spyware was top-shelf; the company was now charging double Hacking Team's asking price. They charged a flat $500,000 installation fee, then another $650,000 to hack just ten iPhones or ten Android phones. Their clients could hack an additional hundred targets for $800,000; fifty extra targets cost $500,000; twenty, $250,000; and ten extra cost $150,000. But what this got you, NSO told customers, was priceless: you could "remotely and covertly collect information about your target's relationships, location, phone calls, plans and activities—whenever and wherever they are." And, their brochures promised, Pegasus was a "ghost" that "leaves no traces whatsoever."

NSO had already installed Pegasus at three Mexican agencies: the country's Center for Investigation and National Security, its attorney general's office, and its department of defense. All told, the firm had sold the Mexicans $15 million worth of hardware and software, and they were now paying NSO some $77 million to track a wide array of targets. The UAE was locked in, too. And custom NSO proposals, brochures, and pitch decks made clear there was a long and growing waiting list of interested parties.

WHEN FINLAND IS in the market, you know every other European country is too. Among the NSO pitch decks were custom proposals for Finland, and emails between the Finns and NSO's pitchmen suggested that the Finns were eager to sign on the dotted line. I had to do a double take. *Finland*—the land of saunas and reindeer—was in the spyware market?

Sure, Finland had no discernible terrorism problem to speak of. But the Finns do share an 830-mile-border with the world's savviest predator, Russia. And unlike Russia's smaller neighbors, Finland had opted not to join NATO for fear of pissing off Moscow. During the Cold War, the Finns served as buffer between the Soviet Union and the West. In exchange for sovereignty over their domestic affairs, Finland agreed not to pursue any foreign policy that would provoke Russia. But by 2014 the Russians had started nibbling, flying into Finnish airspace and sending a thousand Indian and Afghan migrants to the Finnish border in a move some compared to *Scarface* and the Mariel boatlift, when Fidel Castro emptied his jails and sent Cuba's rejects to Florida. In response, the Finns started modernizing their military, participating in joint military exercises with the United States and other NATO member-states. They were making it clear to Moscow that they would not roll over without a fight.

"We have to have a threshold high enough so that if someone wants to come here without invitation, he knows it will be expensive," Finnish president Sauli

Niinistö—careful not to mention the name Vladimir Putin—told a reporter in 2019. Niinistö had not allowed the reporter to bring a laptop into his Helsinki residence, and he boasted of his special window drapes, which blocked sensors that might record their conversation. (It turns out there was a good reason we were forced to hunker down in Sulzberger's closet, after all.) "The walls have ears," Niinisto joked, neglecting to mention that the Finns were investing in their own ears, too.

The Finland proposal and the list of inbound NSO requests from countries like Croatia and Saudi Arabia made it clear that every country with disposable cash and enemies, real or imagined, would soon become a customer. What NSO, Hacking Team, and other cyberarms dealers had done almost overnight was democratize the surveillance capabilities once reserved for the United States, its closest allies in Five Eyes, Israel, and its most sophisticated adversaries in China and Russia. Now any country with a million dollars could buy its way into this market, many with little to no regard for due process, a free press, or human rights.

I pored over the leaks for weeks, struggling to know what to do with them. The last thing I wanted to do was advertise NSO's services to the few government agencies and authoritarian regimes who weren't already on their waiting list.

And so I searched high and low for evidence of how Pegasus was being used, or misused, by NSO's customers. At the first sign of abuse, I told myself, I would go public with everything I had.

As it turned out, I didn't have to wait long.

JUST A FEW weeks after my source dumped NSO's innards at my door, I got a call from some White Hats who believed they had just discovered the first sign of NSO "in the wild."

An Emirati activist by the name of Ahmed Mansoor—a man I had come to know all too well—had passed along a series of strange text messages he'd received that purported to contain information about the torture of other Emirati citizens. Suspecting foul play, Mansoor had reported them to Bill Marczak, a Berkeley graduate student and longtime contact of mine, who served as a fellow at the Citizen Lab. Mansoor, a vocal critic of UAE oppression in the wake of the Arab Spring, had reason to be suspicious. I'd interviewed him just months earlier, after Marczak confirmed he had been the target of not one but two commercial spyware products—one by Hacking Team, the other sold by a commercial spyware maker out of England called Gamma Group. Since both companies claimed to sell their spyware only to governments, it was fairly obvious that the UAE was to blame. I'd written about the UAE campaign to spy on Mansoor for the *Times*. Surely the state wouldn't be so reckless as to hit him with a third spyware.

And yet that is exactly what this was. As Marczak unspooled Mansoor's text messages, he uncovered a class of spyware he had never encountered before. The code was wrapped in layers of encryption and jumbled in such a way that it was nearly impossible to understand. He infected his own phone with the spyware and found what appeared to be the motherlode: a zero-day in Apple's Safari mobile browser. A zero-day like that could easily be worth a high six figures, possibly seven, on the underground zero-day market. This was above his pay grade. A colleague suggested he reach out to Lookout, the mobile security firm just across the Bay, to help examine the code.

Sure enough, as Marczak and security researchers at Lookout unwound the messages, they discovered a chain of three Apple zero-day exploits designed to implant Pegasus on Mansoor's iPhone. The spyware was coming from a web domain in the UAE and carried a payload. Baked inside were files that contained several hundred mentions of "Pegasus" and "NSO." It was the first time NSO's "no trace" spyware had been caught trying to infect a target. Together the researchers had taken to calling Mansoor the "million-dollar dissident"; clearly the UAE security apparatus deemed him worthy of seven figures worth of spyware.

By then Mansoor's life was already a living hell. A mild-mannered poet, he had earned a bachelor's in electrical engineering and a master's in telecom-munications in the United States at the University of Colorado Boulder, which gave him his first true taste of a free society. In 2011, as the Emiratis began clamping down on even the mildest form of dissent, Mansoor could not let it go. Together with a group of Emirati academics and intellectuals, he petitioned for universal suffrage and began calling out the state for arbitrary detentions and arrests. He won international acclaim and awards for being one of the few credible, independent voices on human rights abuses in the otherwise sanitized state-owned media, and the UAE monarchy wasn't having it.

In 2011 Mansoor and four other men—the UAE Five, as they came to be known—were arrested and charged with insulting Emirate rulers. Under international pressure—and fearing the monarchy was only making martyrs out of the men—authorities released and quickly pardoned them. But that is when Mansoor's real troubles began. By the time I was able to get him on the phone in late 2015 and early 2016, he was a frequent target for a state-run media smear campaign. Depending on the day, he was either a terrorist or an Iranian agent. He had been fired from his job. His pension had been terminated, his passport confiscated, his bank account robbed of his life savings. When authorities "investigated," they discovered a forged check for $140,000, written in his name and bearing his forged signature, made out to a ghost. When he went to court, a judge sentenced the ghost to one year in prison, but he never got his money back. Authorities barely bothered to conceal their harassment. At one point, the police called him in for three hours of questioning, while Mansoor's car vanished from the police parking lot. He received death threats regularly. His wife's tires were slashed. His email was hacked, his location tracked. He only knew this because thugs had appeared out of nowhere to beat him up, twice, in the same week. The first time he successfully fought them off and made off with scratches and bruises. The second time, Mansoor's assailant punched him repeatedly in the back of his head and "really tried to give me a permanent disability.

"I've faced everything you can think of," Mansoor told me.

The day we spoke, Mansoor had not left home in weeks. Friends, relatives, and associates had stopped calling and visiting out of fear of retaliation. Some had had their passports revoked. Others were harassed. By then a British journalist sympathetic to Mansoor's cause had unknowingly been hacked by the team at CyberPoint. Mansoor's wife, a Swiss citizen, had pleaded with Mansoor to take their four kids and leave the country. Over the course of several conversations, I found myself pleading with him to leave, too. "This is no way to live," I told him. But without a passport, there was nowhere for him to go.

Besides, Mansoor told me, "I would like to be able to fight for my rights, and the rights of others, to gain my freedom from the inside. It's not easy, but I am doing this because I believe this is the most difficult way for anyone to express their patriotism to their country."

Trapped at home with no work, no money, and a hopeless future, Mansoor told me he'd taken to reading and writing poetry again. It was all he could do to keep from feeling so isolated. Other times, he knew better than to assume he was alone. His watchers had already found their way inside his laptops. There was a good chance they were listening to us now. "It is as bad as someone encroaching in your living room," he said, "a total invasion of privacy, and you begin to learn that maybe you shouldn't trust anyone anymore."

Years later, I would learn that it was worse than that. Evenden's colleagues at CyberPoint had not only installed spyware on Mansoor's devices but hacked his wife's devices too. They even had a code name for Mansoor, Egret—his wife was Purple Egret. And they had implanted themselves inside Mansoor's baby monitor, watching and listening as his child slept. And, yes, I confirmed, they had listened to our call too.

"You'll wake up one day and find yourself labeled a terrorist," Mansoor told me in 2016. "Despite the fact you don't even know how to put a bullet in a gun."

That would be the last time we spoke. Two years later, the state decided it was time to shut him up for good. In a secret trial in May 2018, Mansoor was convicted of damaging the country's "social harmony and unity," sentenced to ten years in jail, and has spent much of the past two years in solitary confinement. He has no bed, no mattress, no sunshine, and, in what must be particularly painful for him, no books. Last I heard, his health was declining. Due to prolonged isolation in a small cell, he can no longer walk. And yet, somehow, he is still fighting. After a particularly gruesome beating, he went on hunger strike. He was now six months in to a liquid diet. He has become a cautionary tale not just for Emiratis, but for any human rights activist, dissident, and journalist around the globe. There is not a day that goes by that I do not think of Ahmed Mansoor, the Mansoors whose names I do not yet know, the creeping surveillance state, and do not want to scream.

IN THE FALL of 2016 NSO finally agreed to speak with me—with a few caveats, of course. By then, the company had been dragged out of hiding. I had published everything I knew about their crown jewel, Pegasus, in the *Times*. Apple had alerted a billion iPhone users to NSO's tricks when it released an urgent patch for the three zero-day flaws NSO's spyware relied on. By then, researchers had been able to trace Pegasus back to some sixty-seven different servers, and found that it had lured more than four hundred people into loading spyware onto their phones. Unsurprisingly, the vast majority of targets were located in the UAE and Mexico, but Marczak was able to trace the infections back to operators in forty-five other countries, including several human-rights offenders: Algeria, Bahrain, Bangladesh, Brazil, Canada, Cote d'Ivoire, Egypt, France, Greece, India, Iraq, Israel, Jordan, Kazakhstan, Kenya, Kuwait, Kyrgyzstan, Latvia, Lebanon, Libya, Morocco, the Netherlands, Oman, Pakistan, Palestine, Poland, Qatar, Rwanda, Saudi Arabia, Singapore, South Africa, Switzerland, Tajikistan, Thailand, Togo, Tunisia, Turkey, Uganda, the UK, the USA, Uzbekistan, Yemen, and Zambia.

Of course the Israelis denied all of this. In one of the stranger conference calls I have ever had, ten NSO executives, who refused to give me their names or titles, insisted they were not cold-blooded mercenaries. They only sold Pegasus to democratic governments, they said, for the express use in criminal and terrorism investigations. Like Hacking Team before them, they told me that NSO had a strict internal vetting process to determine which governments it would and would not sell to. An NSO ethics committee comprised of employees and external counsel vetted its customers based on human rights rankings set by the World Bank and other global bodies, they claimed. And each sale required approval by Israel's Ministry of Defense. To date, they told me, NSO had yet to be denied a single export license. They would not confirm the names of their customers. *The fucking salmon.* And my pointed questions were often followed by long, silent pauses, as I was put on mute while they deliberated their responses. "Turkey?" I asked. By now, Turkey had become my test case. Ankara jailed more journalists than any other country on record that year. "Would you sell to Turkey?" I asked again. Long pause. "Please hold." Another long five-minute pause. "No," finally came the reply.

Clearly there was a lot the Israelis were still figuring out. What NSO was eager to tell me, however, was that their spyware had helped thwart a terror plot in Europe. It had also helped Mexican authorities track down and arrest Joaquín Guzmán—"El Chapo," Mexico's most powerful drug trafficker—not once but twice. The company had played an integral role in both cases, and executives seemed miffed this was not the headline.

But when it came time to answer for Ahmed Mansoor, and the dozens of other journalists and dissidents who I soon learned had also been snagged in Mexico's Pegasus dragnet, the Israelis demurred.

IN THE MONTHS after I published everything I knew about NSO—including the few details I had been able to gather about the company's contracts in Mexico—my phone started buzzing with calls from an array of improbable targets: Mexican nutritionists, antiobesity activists, health policymakers, even Mexican government employees—all of whom reported receiving a series of strange, increasingly menacing text messages with links they feared might be NSO's spyware. I convened with Mexican digital rights activists and Citizen Lab, which examined the messages and confirmed that each was an attempt to install Pegasus spyware.

Other than being from Mexico, I struggled to make sense of what the callers had in common. Eventually, after some digging, I came to this: each had been a vocal proponent of Mexico's soda tax, the first national soda tax of its kind. On its face, the soda tax made a lot of sense. Mexico is Coca-Cola's biggest consumer market; it is also a country where diabetes and obesity kill more people than violent crime. But the tax had opponents in the soda industry, and clearly somebody working in government didn't want their kickbacks getting cut off. Now it appeared that they were going to extraordinary lengths to monitor the doctors, nutritionists, policymakers, and activists who wanted to see the soda tax through.

The messages themselves were a study in desperation. They always started innocuously enough: "Hey, check out this news article." When that didn't work, the messages grew more personal: "My father died at dawn. We are devastated. I'm sending you the details of the wake." And when those didn't work, they went for the jugular. "Your daughter was in a serious accident and is in the hospital," or "Your wife is having an affair. Here's the photo evidence." Every message enticed them to click on a link. Some were so strange, the recipient never clicked. Those that did were ominously redirected to Gayosso, Mexico's largest funeral home, while Pegasus downloaded itself in the background. The hacking campaign was clearly a corrupt use of NSO's spyware. When I contacted a lobbyist for Mexico's soda industry, they told me, "This is the first we're hearing of it, and frankly, it scares us too."

NSO told me it would investigate. But rather than cut Mexico off, its spyware only continued to pop up in more disturbing cases still. Almost as soon as I hit publish on the article, my phone started buzzing anew with calls from highly respected Mexican anticorruption activists. Lawyers looking into the mass disappearance of forty-three Mexican students, two of Mexico's most influential journalists, and an American representing victims of sexual abuse by the Mexican police had all received similar text messages. The spying had even swept up family members, including the teenage son of one of Mexico's most prominent journalists. My colleague Azam Ahmed, our Mexico bureau chief, and I tried to speak with as many of the targets as we could. He began to recognize their texts. He had gotten one just like them six months earlier. After his phone sputtered for months, he'd ditched it and gotten a new one. Now he knew why.

Together, Azam and I spent the next several months tracking down other targets. NSO executives told me that if you rounded up every Pegasus target in the world, you would only fill a small auditorium. But in Mexico NSO targets were coming out of the woodwork, many of them outspoken critics of then Mexican president Enrique Peña Nieto, or journalists who had reported stories critical of him. A frequent target for the menacing hacking attempts was Carmen Aristegui, the Mexican journalist who broke the scandal of the so-called Casa Blanca, a real estate intrigue that involved Peña Nieto's wife getting a cheap deal on a mansion from a major government contractor. Not long after her story forced Peña Nieto's wife to give up the house, Aristegui started receiving messages pleading for her help in locating a missing child. Another alerted her to a sudden charge on her credit card. One reportedly came from the American embassy about a problem with her visa. And when that clickbait failed, the messages grew more strident. One warned she would be imprisoned. Her sixteen-year-old son, who was living in the United States at the time, started getting the texts too. Thugs started breaking into her office, threatening her safety, and following her. "It's been about getting revenge for the piece," Aristegui said. "There's really no other way to see it." Other victims also had tie-ins to Peña Nieto. The targets included lawyers for the women of Atenco—eleven students, activists, and market vendors who were arrested by police more than ten years ago during protests in the town of San Salvador Atenco, and brutally sexually assaulted on their way to prison. Aside from the grave abuse of power, the case was especially sensitive: the man who ordered the crackdown on the protesters was then governor, now president, Peña Nieto.

In Mexico, only a federal judge can authorize the surveillance of private communications, and only when officials demonstrate a sound basis for the request. But it was highly unlikely a judge had approved the cases we were uncovering. Illegal surveillance in Mexico had become the norm, and now authorities could conduct click-and-shoot spying with NSO's spyware. It didn't matter what NSO's contract did and did not allow. Once the company learns that its spyware has been abused—and by now, the most frequent tip-off its products were being abused was a call from yours truly—there is only so much it can do. NSO executives argued that they could not simply march into intelligence agencies, remove its hardware, and take back its tools.

"When you're selling AK-47s, you can't control how they'll be used once they leave the loading docks," is how Kevin Mahaffey, a security executive, put it.

WITHIN HOURS OF publishing our story, people had taken over the streets in Mexico City to call for Peña Nieto's resignation. The hashtag #GobiernoEspía—the government spies—started trending worldwide on Twitter. All of Mexico appeared to be up in arms. Our reporting had forced Peña Nieto to acknowledge that Mexico was using NSO's spyware—a first for any government leader. But Peña Nieto denied ordering the government to spy on his critics and journalists. And then Mexico's president strayed from his script: His administration, he warned, would "apply the law against those who have levelled false accusations against the government." Peña Nieto's underlings later backtracked. The president had misspoken and had not intended to threaten Azam and I, or the *New York Times*.

But for months afterward, I knew better than to click on the dozens of strange text messages pinging my phone, beckoning me to click.

PART V

The Resistance

You can't stop the gears of capitalism. But you can always be a pain in the ass.

—JARETT KOBAK, *I HATE THE INTERNET*

CHAPTER 14

Aurora

Mountain View, California

F OR SEVERAL HOURS one early Monday afternoon in mid-December 2009, a Google intern teased apart the equivalent of a sonar blip on his screen. Someone had tripped an alarm.

He sighed. "Probably another intern."

Google had just introduced new tripwires across its network, and alarms were going off incessantly. The company's security engineers were now spending all their time trying to decipher which blips marked an imminent attack, an engineer accessing a spammy poker site, or simply an intern stumbling down the wrong digital hallway. Almost always, it was the latter.

"There's a Fog of War, but there's also a Fog of Peace," Eric Grosse, Google's affable vice president of security engineering, told me. "There are so many signals triggering, it's hard to know which ones to go after."

Some inside the company likened it to Pearl Harbor. That Sunday morning in December 1941 on the Hawaiian island of Honolulu had started peacefully enough. Lieutenants were still familiarizing themselves with the naval base's new radar system when a radar operator on the far end of the island informed the on-duty lieutenant of an unusually large blip on his radar screen—signs of a fast-approaching aircraft fleet over a hundred miles away. The lieutenant's first reaction was, "Don't worry about it." He assumed the blip was a squadron of B-17 bombers due in from San Francisco, not the first wave of Japanese bombers.

With so many new blips popping up on Google's screens that December, it was simply human nature to prefer the simple, benevolent explanation—a disoriented intern—to the reality, an imminent nation-state attack.

"WE WEREN'T TRAINED to think about spies," Heather Adkins, the freckled, thirtysomething director of Google's information security team, would later recall. That Monday afternoon, Adkins was just wrapping up another Google meeting about China. The company had tiptoed into the Chinese market three years earlier and was still struggling to navigate Beijing's draconian censorship rules. Adkins was something of an anomaly among the mostly male, testosterone-fueled coders she managed. Most had a deep distaste for authority. They buried their heads in code by day and lived vicariously through virtual role-playing games by night. Adkins was more of a history buff, who spent her off hours reading up on the Middle Ages. She saw her security gig at Google as the digital equivalent of stopping medieval invaders in the ancient world. Her job was simple: "Hunt down evil."

As her meeting came to a close, Adkins glanced at the clock. It was 4:00 P.M. She might just be able to beat the rush-hour traffic if she left work early. But as she headed for the door, her intern beckoned, "Hey, Heather, check this out."

The blip on his screen had metastasized and was now moving at dizzying speeds in and out of employees' computers, across Google's network. Whoever was on the other side of the screen was no intern. "It was the fastest cyberattack we had ever seen," Adkins recalled. "Whoever they were, they were clearly practiced. This was not their first rodeo."

As late afternoon turned to evening, the blip grew more animated. It was now bouncing from computer to computer, winding its way through Google's systems in unpredictable patterns, in search of something. The intern stayed glued to his screen through dinner, when he broke to join the rest of the team in Google's café. There, over burritos, he relayed the strange trail of the blip that was taking on a life of its own. Seated at the table that evening was Grosse, Adkins's boss, and several other security engineers.

With his glasses and graying hair, Grosse had a Socratic, professorial quality. He was one of the few Google directors to forgo an office so he could sit with his engineers. It was not uncommon to find him sloped on a couch, computer on lap, or staying late to dine with twentysomething engineers. That night Grosse listened intently to the intern's account, asking questions, trading notes with others at the table. A consensus emerged: Whoever this was, they appeared to be in the beginning stages of reconnaissance. An insider? What were they after? Salary records? As the men wrapped up their meals and walked out to the volleyball court that evening, nobody had so much as guessed a foreign nation.

AS MOUNTAIN VIEW retired for the night, the sun was just peeking up over the Swiss Alps in Zurich when Morgan Marquis-Boire, the dreadlocked then thirty-year-old hacker, logged in. Google's engineers in Zurich—or "Zooglers," as they called themselves—referred to their offices as the "real Mountain View" for its Alpine backdrop. But Marquis-Boire always felt that Google's Swiss headquarters, with its oversize rainbow-colored Google logo, stood out like a leering, oversize clown in Zurich's old Hürlimannplatz.

For years, Hürlimannplatz had been home to an old Swiss brewery. But once the brewers discovered a spring bubbling up inside the building's brick walls, they started producing their own mineral water. Europeans from across the continent made weekend pilgrimages to the square's mineral-fed fountain to taste the purest water Europe had on offer. These days, the well had been converted into a thermal bath and spa. It was an oddly Zen setting in which to be triaging the very beginnings of a cyberwar.

THAT MORNING, MARQUIS-BOIRE picked up where the Mountain View intern had left off, following the blip as it ping-ponged across Google's network, looking more and more ominous. He barely noticed the snow silently blanketing Zurich's rooftops and steeples.

This was no intern. "Google isn't a nuclear enrichment facility," he told me, "but in terms of security, it comes pretty damn close."

Whoever this was, they had managed to bypass the toughest security measures he had yet to see. And now they were riding roughshod across Google's networks, indiscriminately accessing systems that did not fit the typical employee's digital path. The list of possible explanations for the anomalous blip was getting shorter and shorter, until there was no other explanation: Google was under attack.

"We've caught a live one!" Marquis-Boire shouted. It was hard not to jump on top of his desk, pound his chest, and yell, "Shit is on!"

For years, he'd been chasing imaginary ghosts and pointing out the dangers of weak security. Now he finally faced something real. It felt like vindication.

By the time he relayed his analysis back to Mountain View and left the office for the night, it was 11:00 P.M. The streets were muffled with snow. Usually he biked to his apartment in Langstrasse, Zurich's equivalent of Amsterdam's red-light district. But that night he decided it would be better to walk. He needed time to process. As his combat boots crunched through the streets, his mind flashed back to a presentation in Las Vegas two years earlier, where he'd boldly declared to a large audience of hackers that "threats from Chinese hackers are overrated." Recalling his words now, Marquis-Boire could only smile: "History sure has a way of coming to bite you in the ass."

COME MORNING IN Mountain View, it was clear that this was no fire drill.

By 10:00 A.M., Google's entire security team had been briefed on the attack. As morning turned to afternoon, however, the activity flat-lined. Whoever was behind their screens had retired for a few hours. But that evening the blip returned with a fervor. Several engineers elected to pull an all-nighter, tracing the attackers' movements into the wee hours of the morning.

The intruder was clearly a night owl—or operating in a different time zone. By the time bleary-eyed engineers briefed their fresh-faced cohorts the following day, there was no question they were facing the most sophisticated cyberattack Google had ever seen.

It was time to call in the specialists. Google's first call was to a cybersecurity shop in Virginia called Mandiant. In the messy world of security breaches, Mandiant had carved out a niche for itself responding to cyberattacks, and was now on the speed dial of nearly every chief information officer in the Fortune 500.

Kevin Mandia, Mandiant's founder, was like Harvey Keitel's meticulous, fast-talking character, the Wolf, in *Pulp Fiction*, called upon by corporate

America to clean up the aftermath of the bloodiest digital breaches, extortion attacks, and cyberespionage campaigns. Google asked Mandiant to get to Mountain View as soon as possible. "Just one thing," Google's executives said. "Don't wear suits."

The following day Mandiant's forensics team arrived at the Googleplex. They'd foolishly ignored their clients' advice and showed up in dark suits and glasses. Googlers in hoodies took one look at the men and concluded they had to be federal agents.

Grosse and Adkins ushered the men into their improvised war room, a small, nondescript conference room overlooking Moffett Field, the former naval air base. In the distance, Mandiant's team could just make out the glimmer of the San Francisco Bay before someone drew the shades and slapped a sign on the door: THIS CONFERENCE ROOM OFFLINE UNTIL FURTHER NOTICE.

The next hour played out in what Kevin Mandia affectionately calls "Upchuck Hour." Mandiant's team insisted that Google fork over everything: Firewall logs. Web logs. Emails. Chats. They grilled Grosse and Adkins's team about everything they knew so far, blasting them with a series of questions that could be best summed up as "Who the hell do you think might have done this?"

Time was of the essence. With every second that passed, the blip was gathering more data, more code. There was a good chance that the attackers had already planted backdoors in Google's systems for quick return access. Google's employees were practically vomiting information on the table—anything and everything that might offer Mandiant's investigators a digital crumb or fingerprint to trace the attackers' identity and motive.

At Google offices across the globe, internal investigators started summoning employees in for interrogation. Why had their device accessed that file, that system, that one piece of data? What were they after? But by the end of the day, it was clear that this was no insider. An attacker had infiltrated their machines from the outside. Mandiant's investigators honed in on the logs, looking for any malicious links or attachments that employees might have clicked on, inadvertently granting attackers entry to their systems.

They had seen this a thousand times. Mandiant's clients could spend millions of dollars on the latest and greatest in newfangled firewalls and anti-virus software, but security was only as good as the weakest link. And usually the weakest link was a human who clicked on a simple phishing email or message containing something nasty. The messages could be quite persuasive. The attacker might mimic a FedEx tracking notice or an HR manager. Somebody, somewhere in the organization, almost inevitably fell for it and clicked. As Mandiant's investigators wound their way from infected machine to machine, they picked up on a common thread: several Google employees in the company's Beijing office were trading messages with colleagues, partners, and clients using an external Microsoft chat service. As investigators sifted through their chats, they found one blaring red flag. Each had clicked on a link attached to the same menacing three-word message: "Go Kill Yourself."

FOR THOSE NEXT few days in December 2009, Google's war room became a tangle of data and minds as Grosse and Adkins began pulling engineers from every corner of the company, telling them to ask any friends with any security experience to come work at Google. They started poaching digital spies from Fort Meade and the Australian outback and security engineers from Google's competitors up and down Highway 101, offering immediate, no-questions-asked $100,000 signing bonuses.

The war room quickly became something of a curiosity to other Googlers, particularly as Sergey Brin, the company's energetic cofounder, became a regular presence on their floor. Brin, who spent his spare time performing as a trapeze artist, was hard to miss. Often he'd come speeding into the office on Rollerblades or clownish outdoor elliptical bikes, wearing full-body luge racing suits or, at a minimum, neon slippers.

Brin, a Russian-Jewish émigré, had taken a personal interest in the attack. He was a picklock. As a student at Stanford, he'd experimented with various lock-picking techniques. He also happened to be one of the world's foremost experts in data mining, extracting meaningful patterns from mountains of data. This kind of forensic pursuit was in many ways what Brin did best. But he also began to take the assault personally. Brin's identity—and, one could argue, Google's corporate identity too—was inextricably linked with his family's escape from the Soviet Union in the late 1970s. He saw the attack as a direct assault on the founding principles of Google itself, summed up by its three-word motto: "Don't be evil."

With each visit to the war room, Brin was further convinced that this was not the work of some basement dweller; this was a well-resourced attack. Baked inside that three-word clickbait, "Go Kill Yourself," was a link to a website hosted in Taiwan that executed a script containing a zero-day exploit in Microsoft's Internet Explorer browser. Once the employees in Google's China offices clicked on the link, they inadvertently downloaded encrypted malware that gave Google's attackers a foothold and allowed them to plug in and out of Google's network. No kid—Brin didn't care how good they were—was burning a Microsoft zero-day exploit on Google and encrypting their attack code out of curiosity. This attacker was after something bigger. And they'd taken unusual care in hiding their tracks. The level of obfuscation alone suggested that this was the work of a highly trained, well-funded adversary. Brin made it his personal mission to find out who.

As more engineers joined their effort, the investigation moved to a second, larger conference room, then a third, and finally to an empty building across campus, where some 250 employees were now tasked with finding who had broken into Google's network, what they were after, and why. Their quest had taken on such purpose that engineers were now refusing to go home. Several took to sleeping on campus.

"When the building is on fire, it's hard to keep the firefighters away," Adkins recalled.

As the holidays neared, Adkins encouraged her team to go home to sleep and shower, even as she was becoming a permanent fixture on campus herself—and a slightly ridiculous sight at that. She'd run out of clean clothes just as Googlers were raiding the on-campus merchandise store for last-minute holiday gifts. At five-foot-three, Adkins found herself managing the digital investigation of a lifetime in an extra-large neon-green Google sweatshirt.

Holiday trips were canceled. Employees were not permitted to tell their loved ones what was keeping them away. Adkins managed to make it to Vegas to see her mother for Christmas, but she spent the entire time chained to a computer. Grosse managed only a brief cameo on Christmas Day.

"I just had to tell my mom, 'Something big is going on. Trust me. It's important,'" Adkins said.

The obsession with the attack started giving way to paranoia. On the way to campus one morning, Adkins spotted a utility worker emerging from a manhole. "I thought, 'Oh my God, that person is trying to backdoor our fiber on campus.' I started to wonder if someone was listening to our phone calls.'"

In Zurich, engineers began to worry about their personal safety. They wondered how personal Google's assailants were willing to make this. They were civilians effectively doing counterintelligence against what was clearly a well-funded adversary. Several took to watching their backs on their late-night commutes.

As the weeks went by, Google's security team was learning that they had good reason for concern. The attacker had begun to show the telltale signs of a sophisticated adversary: a Chinese government contract group Mandiant had come across before, a group the National Security Agency tracked by the classified pseudonym Legion Yankee.

LEGION YANKEE WAS among the murkiest—and most prolific—of the more than two dozen Chinese hacking groups that NSA hackers tracked, as they raided intellectual property, military secrets, and correspondence from American government agencies, think tanks, universities, and now the country's most vibrant technology companies.

Chinese cyber theft took two tacks. The majority of hacking crusades were conducted by the China's People's Liberation Army's Second and Third Departments. It was clear from their targets that various PLA units were assigned to hack foreign governments and ministries in specific geographic locales, or to steal intellectual property in distinct industries that benefited China's state-owned enterprises and economic plans.

The other approach was less direct and more episodic. Increasingly, high-ranking Chinese officials at China's Ministry of State Security started outsourcing attacks on high-profile targets—political dissidents like the Dalai Lama, Uighur and Tibetan ethnic minorities, and high-profile defense contractors in the United States—to freelance hackers at Chinese universities and internet companies.

The state identified these hackers for their skills, which often far exceeded those of their PLA counterparts. Plus, if anyone ever traced back the attacks to these individuals, Beijing could claim ignorance. "That way Beijing can say, 'It's not us. It's these hackers we can barely control ourselves.' They may not be responsible for the bulk of the activity, but their sheer existence gives the government an out," James A. Lewis, a cyberespionage expert at the Center for Strategic & International Studies in Washington, told me.

It was Putin's playbook through and through. The Kremlin had successfully outsourced cyberattacks to Russian cybercriminals for years. It was a strategy that was easily imported to China, where the state's embrace of liberties and free markets had its limits. Those with any notable hacking skills weren't so much recruited to the state's hacking apparatus as they were conscripted.

In one case I was able to track down the personal blog of a prolific PLA hacker who went by the alias UglyGorilla. The PLA hacker complained he'd been forced to enlist. He lamented the low pay, long hours, tight living quarters, and his instant ramen diet. How exactly China's Ministry of State Security recruited private Chinese hackers to moonlight in these attacks was unclear. Frequently, security researchers traced breaches back to students at Chinese universities, notably Jiaotong University, which received significant state funding. In other cases American security researchers traced attacks back to employees at China's leading internet company, Tencent. Often China routed attacks through some of its most popular websites, like 163.com— China's equivalent of Yahoo—and Sina, the company that runs Sina Weibo, China's Twitter equivalent. 163.com was officially owned and run by a Chinese gaming billionaire, but its mail servers were operated by a Chinese government domain, giving the Communist party's minders access to all the messages, and digital traffic, routed through it. And the PRC had started using 163.com's servers as staging grounds for its attacks.

Some cybersecurity experts speculated that China's employees and students made extra cash hacking for the state. Others speculated that they had been given no such choice. Whatever the relationship, the NSA had no better answers than I did.

"The exact affiliation with Chinese government entities is not known, but their activities indicate a probable intelligence requirement feed from China's Ministry of State Security," was the clearest answer I found in one leaked NSA memo.

The NSA had no such patina of plausible deniability. Uncle Sam didn't force American engineers to hack on its behalf. And it didn't deploy the best TAO hackers to hack foreign industries to glean trade secrets it could then hand to

American companies for their benefit. Even if the NSA got its hands on valuable chemical formulas, or the Tencent source code, which company would it hand them to? DuPont? Monsanto? Google? Facebook? In a truly free market economy, the sheer notion sounded absurd.

In recent years, it was these contract groups with loose ties to China's Ministry of Security that began to plague the NSA's analysts and private security researchers, as Chinese contractors raided a growing number of disturbing new targets. They had broken in to companies in the defense sector, where, according to one classified document, they zeroed in on "aerospace, missile, satellite and space technologies" and, perhaps most disturbing of all, "nuclear propulsion and weaponry."

Google's attacker, Legion Yankee, had come to U.S. intelligence analysts' attention six months before Google spotted the blip on their screen. They'd surfaced in a number of hacks on defense contractors. State Department officials would later connect the Google hacks back to Zhou Yongkang, China's top security official, and Li Changchun, a member of China's top ruling body, the Politburo Standing Committee, and the country's senior propaganda official. Li had apparently googled himself and didn't like what he saw, according to leaked diplomatic cables. As a result, Li sought to punish Google, first by ordering state-owned Chinese telecoms to stop doing business with the company, and later by coordinating the contracted hit on Google's networks—but those answers did not come until much later.

That January, Mandiant's investigators were hardly surprised to find the group on Google's networks. Chinese hackers were brazenly hacking anyone and everything they could. Nothing shocked them anymore. But Google engineers and executives were apoplectic.

"We had never thought we could be hacked by the Chinese military," Adkins said. "That seemed so outside the realm of what companies could be expected to handle."

"We didn't think militaries were allowed to hack civilians in peacetime," said Grosse. "We didn't think that could be true because you assume the backlash would be so severe. Now, that's the new international norm."

Google soon discovered it was not the only victim. As investigators traced the attack further back to the attackers' command-and-control server, they found trails leading to dozens of American companies, many in Silicon

Valley—Adobe, Intel, Juniper Networks—and others that were not. The body count included defense contractor Northrop Grumman, Dow Chemical, Morgan Stanley, and many more that—to this day—have refused to even acknowledge that they were breached.

Google's security team tried to alert their counterparts at these other companies, but it was exhausting. "It was so hard to get through," Adkins said. "We had to go through someone who knew a guy, who knew a guy at our competitors, and across so many different industries. We couldn't believe just how widespread this was. If we got through, we'd say 'Look, you have a problem. Look at this IP address and you'll see something scary.'"

"You could hear someone's face going white across the phone line," Grosse told me. "Then radio silence."

THEY WERE AFTER Google's source code.

Most laypeople assume hackers are after short-term payoffs: money, credit card information, or bribe-worthy medical information. But the most sophisticated attackers want the source code, the hieroglyphics created and admired by the engineering class. Source code is the raw matter for software and hardware. It is what tells your devices and apps how to behave, when to turn on, when to sleep, who to let in, who to keep out. Source code manipulation is the long game. Code can be stolen and manipulated today and, like an invisible hole in the wall of the Oval Office, bear fruit immediately or years into the future.

Code is often the most valuable asset technology companies have—their crown jewels—and yet when China's contracted hackers started popping up across thirty-four Silicon Valley companies in late 2009, nobody had ever thought to secure it. Customer and credit card data merited fierce protection, but the vast majority of tech companies had left their source code repositories wide open.

A subsequent investigation by researchers at McAfee—who dubbed the Chinese operation Aurora—found that it wasn't just Google. Everywhere the Chinese hackers had gone—high-tech companies, defense contractors— they were disturbingly successful at cracking source code repositories. With that access, they could surreptitiously change the code that made its way into commercial products and attack any customers who used the software.

Extracting Chinese backdoors from code was the ultimate exercise in finding a needle in a haystack. It meant comparing any software against backup versions, an extremely laborious process even for the world's marquee search company, especially when you are dealing with massive projects with millions of lines of code.

Google's Aurora attack elevated a fundamental question: Can any computer system be made totally secure? It recalled Gosler's exercise at Sandia, more than two decades earlier, when the nation's elite hackers couldn't even find his implant in a couple thousand lines of code—code they knew to be compromised. The software needed to run all of Google's services—from Google Search to Gmail to Google Maps—runs to an estimated two billion lines of code. By comparison, Microsoft's Windows operating system, one of the most complex software tools ever built for a single computer, is estimated to contain some fifty million lines of code. McAfee never found hard evidence that China's attackers altered the source code at any of their targets. But with so many of China's victims in denial that they had been hacked, the only certainty was uncertainty.

Mandiant and Google's investigators were determined to follow the Chinese trail to the bitter end. And the trail made clear that their attackers had a very specific goal in mind. They were after Chinese dissidents' Gmail accounts. The Chinese could have easily cracked those accounts by spraying them with possible passwords. But passwords can be changed. Hackers can be locked out after a series of wrong tries. The Chinese were looking for more permanent access. By stealing Google's source code, China's hackers could potentially implant backdoors into Gmail software, guaranteeing long-term access to any Gmail account of their choosing.

And it became clear they were after their usual targets. Along with prodemocracy activists, China considers Tibetans, Uighur Muslims, proindependence Taiwanese, and Falun Gong practitioners to be what the state calls the Five Poisons—groups the Chinese Communist party deems the biggest threat to their standing. China had reserved its best zero-day exploits and its top hackers to menace its own people.

IN RETROSPECT, GOOGLE might have seen it coming. Three years earlier, Google had entered the Chinese market as some kind of savior. At the time

Brin and his cofounder, Larry Page, told employees it was better to give the Chinese censored search results than nothing at all. Google would help educate Chinese citizens about AIDS, environmental issues, avian flu, and world markets. The alternative, they argued, was to leave a billion people in the dark.

Such rationalization was common in Silicon Valley, where tech leaders and founders have come to think of themselves as prophets, if not deities, delivering free speech and the tools of self-expression to the masses and thereby changing the world. Many a tech CEO had come to think of himself as the rightful heir to Steve Jobs, whose megalomania was excused as a byproduct of his ability to deliver. But Jobs was in a class of his own, and when other tech CEOs followed suit, they often invoked the same language of enlightenment to justify their own relentless expansion into the world's fastest-growing, albeit authoritarian, internet market.

Almost immediately after Google entered China in 2006, Brin found the compromise hard to stomach. Chinese officials demanded that Google sanitize search results for any mention of the Falun Gong, the Dalai Lama, and the bloody 1989 massacre at Tiananmen Square. That much Google had anticipated. But soon that list grew to include anything that offended the Chinese Communist Party's taste and "socialist values"—talk of time travel, reincarnation, and later even Winnie-the-Pooh made the blacklist. When Mountain View didn't move fast enough to block offensive content, Chinese officials took to calling Google "an illegal site."

Google's presence in China didn't play well in Washington either. Brin and Page were likened to Nazi collaborators. Members of the House International Relations Committee compared Google to a "functionary of the Chinese government" and called its actions "abhorrent."

"Google has seriously compromised its 'Don't Be Evil' policy," one Republican congressman said. "Indeed, it has become evil's accomplice."

Some executives began to feel that way. But they knew that anything less than full compliance with Chinese authorities was dangerous. They'd all heard the stories. Chinese agents frequently raided corporate offices to threaten local executives with jail time if they didn't move quickly to block "problematic" content.

Compromising on censorship was one thing; being an unwitting accomplice to Chinese government surveillance was another. When Google entered

China, Brin and Page had intentionally decided not to make its email or blog-ging platforms available to Chinese customers, out of fear that they would be forced to turn over a user's personal information to the secret police. Two years earlier Yahoo had handed over a Chinese journalist's personal information to the state, after he leaked details on Chinese press restrictions to a prodemoc-racy site run by Chinese exiles in New York. Yahoo's former customer was now serving out a ten-year prison sentence.

In Brin's mind, what the Chinese had done attacking Google was essen-tially the same thing China had done to Yahoo. The only difference was that they hadn't bothered to ask Google for access to its users' information. The hack reeked of the totalitarianism of Brin's Soviet upbringing, and he took it as a personal affront.

Brin was born in Moscow and grew up under Soviet oppression. As a matter of policy, the Soviet Union was not anti-Semitic. But in practice, Jews were banned from Russia's prestigious universities and upper professional ranks. They were forced to take university entrance exams in separate rooms—dubbed "gas chambers"—and graded on a steeper curve. Brin's father had to forfeit his dream of becoming an astronomer because Jews were expressly forbidden from enrolling in physics departments at Moscow's prestigious universities. The Soviets didn't trust them with their nuclear rocket research, and the state deemed astronomy a subset of physics. Brin's parents escaped to the United States in the late 1970s to spare young Sergey the same fate. Now Brin was among the world's most successful entrepreneurs, and richest men. He was not about to roll over to another authoritarian regime now.

Neither was every sleepless soul in Google's war room that January 2010. They'd come to work at Google for the free perks, free food, free classes, free gyms, and "Don't be evil" ethos. Its more recent recruits had come to join the fight. Nobody standing there that January was going to stay on if they thought their work was somehow abetting Chinese surveillance, jailings, and torture.

"Our entire attitude changed," Eric Schmidt, Google's CEO told me. "We weren't going to let it happen again. We couldn't. We had to take decisive action."

But now what? Google was a search business. Defending dissidents from highly trained nation-state hackers was not exactly in the job description. To truly do what was required to kick the Chinese out of their systems—and keep them out—would take astronomical amounts of money and labor. The company would have to build out its own intelligence agency, poach nation-state hackers and spies of its own, and make a massive cultural shift. Google's corporate culture is famously centered upon innovation and "employee happiness." Security is famously a pain in the ass. Nobody has ever said, "I love long passwords." But the company could not expect to reap the rewards of hundreds of millions of dollars of renewed security investment if their employees were still creating weak passwords and inadvertently clicking on malicious links. In the end, even if Google was able to accomplish all of the above, could they really be so bold to think they could keep the Chinese military at bay? Most executives would see—*do* see—this as the ultimate fool's errand.

"We had a series of very heartfelt, spirited discussions about the actual costs required. We had to ask ourselves, 'Are you up for what it would take?'" Grosse told me later.

"Defending against the Chinese military seemed so outside the realm of what companies could be expected to do," Adkins recalled "There was this question of 'Are we even going to try to keep them out? Do we just give up?' To do what you would have to do, most companies would conclude that it's just not worth it."

It was Brin who ultimately decided to respond in force. He pushed Google to pull out of China, giving up a stake in the world's most sought-after market. China had twice as many internet users as there were residents of the United States, and the Chinese rate of internet growth far surpassed that of any other country. Caving to Chinese censorship had been hard enough. The attack left Brin no choice. It was time to abandon China and do everything in Google's power to make sure nothing like Aurora ever happened again.

IN THE DARK of night one evening in January 2010—without warning—Google's security team swept the company's offices, confiscating every machine hackers touched. Hundreds of baffled employees arrived at their desks the following morning to find a web of wires and a note where their computers had once stood. "Security incident," it read. "Took your machine."

Google's security team simultaneously logged every employee out of every single system at the company, resetting their passwords. When irritated employees and executives demanded explanation, they were simply told: "We'll tell you later. Trust us."

Meanwhile, Google executives began plotting how to confront their attacker. They would no longer abide by Beijing's censorship, but they needed to find a legal approach that wouldn't put their employees in danger. If they stopped filtering Google.cn, they'd run afoul of Chinese law, and there was no doubt in their minds that Chinese employees and their families would be called to account. Google's legal team hatched a plan to shut down Google.cn and redirect Google's Chinese internet traffic to uncensored search engines in Hong Kong. The former British colony had been part of China since 1997, but operated under a "one country, two systems" policy. Mainland authorities didn't bother censoring internet content in Hong Kong. The redirection to Hong Kong would be a thumb in the eye of China's minders, but still within legal bounds. They weren't uncensoring Google.cn, after all. It would simply cease to exist. This would put the ball in Beijing's court. China would have to do its own filtering to and from Hong Kong. Google would no longer do their dirty work.

They knew that China would respond in kind. Most likely, the Communist Party would kick Google out of the market entirely. No American company had ever publicly called out Beijing for a cyberattack, even as Chinese hackers were pillaging American intellectual property in what Keith Alexander, the NSA director at the time, later called "the greatest transfer of wealth in history." A security researcher by the name of Dmitri Alperovitch coined a much-copied phrase for the phenomenon: "There are only two types of companies—those that know they've been compromised, and those that don't know." Later variations were more specific. Three years after Google's attack, James Comey, then head of the FBI, put it this way: "There are two kinds of big companies in the United States. There are those who've been hacked by the Chinese, and those who don't know they've been hacked by the Chinese."

Most victims refused to speak out, fearing what disclosure might mean for their reputations, or their stock price. But American officials traveling for business to China had started bringing burner phones and laptops, or forgoing any digital devices at all, knowing they would be infected with key-logging software by the time of their return. A Starbucks executive told me that on a trip to Shanghai, a storm had shut off power to his entire hotel, with the exception of the fifth floor, where, conveniently, he and other American executives from Ford Motor Co., Pepsi, and other companies were staying. "Our floor had redundant power, redundant internet, and clearly everything we were doing was being surveilled," he told me. "We had identified the best locations for Starbucks coffee shops. After our visit, our Chinese competitors started opening coffee shops in the exact same locations."

Most American companies rolled over in the face of Chinese cyberespionage. And based on Google's conversations with other victims of the Aurora attack—in defense, tech, finance, and manufacturing—none of the others planned to call China out any time soon. If Google didn't take aggressive, decisive action, then who would? There would be consequences. But the status quo was far worse.

ON TUESDAY, JANUARY 12, 2010, at 3:00 A.M. Beijing time, Google made its attack known to the world. Fearing for employees' safety, Google had already tipped off the State Department. Then Secretary of State Hillary Clinton was personally briefed. American diplomats at the Beijing embassy prepared for a possible mass evacuation of Google's Chinese employees and their families.

And then they hit publish. "We have taken the unusual step of sharing information about these attacks with a broad audience not just because of the security and human rights implications of what we have unearthed, but also because this information goes to the heart of a much bigger global debate about freedom of speech," then Google's head lawyer, David Drummond, wrote in a blog post. "These attacks and the surveillance they have uncovered— combined with the attempts over the past year to further limit free speech on the web—have led us to conclude that we should review the feasibility of our business operations in China. We have decided we are no longer willing to continue censoring our results on Google.cn."

Those words had been vetted up and down the corporate ladder—and yet their full weight had not hit Google executives. For the greater part of the past month, engineers had tracked the itinerary of a tiny blip through an insanely complex path that led back to the Chinese government. Very little about the past few weeks had felt real, until now.

"A new mindset got implanted at that moment," Adkins recalled. "Our users were in danger. In that moment, we knew we were absolutely the shepherds of their safety."

Within minutes, a headline about the attack popped up on CNN: "Google Reports China-Based Attack, Says Pullout Possible." Google's phones started buzzing. Reporters from Bloomberg, Reuters, the *Wall Street Journal*, the *New York Times*, the *Christian Science Monitor*, CNN, the BBC, and tech blogs all over the Valley struggled to capture in words what this moment meant, for Google, for cybersecurity, for the internet. For the first time, an American company was calling out China for a cyber theft. And Google wasn't pulling any punches. Until that moment, if someone inside China googled "Tiananmen Square," they would find photos of smiling Chinese couples and tourism scenes of the square lit up at night. On January 12, 2010, anyone who performed the same search was directed to death tolls of the student-led Tiananmen protests and the iconic photo of a Chinese man, holding nothing but a shopping bag, blocking a column of twenty-five Chinese tanks from mowing down protesters. "Tank Man" had been captured by photographers just before he was dragged off by the secret police. His fate, even his identity, remained a mystery. It would have been in China's interest to produce him, to silence the global outcry. But nobody ever did. Most assumed he had been executed. Many had been tortured and killed for far less. As Beijing woke that morning, many made their way to Google headquarters to leave flowers outside in a show of gratitude or mourning for what everyone knew was likely to be Google's imminent exit.

Chinese censors feverishly redirected their internet filters—"the Great Firewall"—toward Google.com.hk. Soon, anyone looking for the photo of Tank Man found their internet connection reset. And then Chinese officials let rip. On Xinhua, China's state media service, a senior Chinese official lambasted Google for violating the promise it had made to filter search results when it first entered the Chinese market. Officials denied any responsibility for hacking Google and expressed their "discontent and indignation" at the mere accusation.

Officials called up Google executives directly. Schmidt would later joke, referring to Hong Kong's lack of censorship: "We told the Chinese, 'You said it's "One country, two systems." We like the other system.' They didn't appreciate that either."

Over the next few weeks, Google's decision touched off a diplomatic brou-haha between Washington and Beijing. In state-run outlets, Chinese officials continued to vehemently deny their role in the Google attack and accused the White House of masterminding an anti-Chinese propaganda campaign. In Washington, President Obama demanded answers from Beijing. Clinton asked China to conduct a transparent investigation into the Google attack. In a half-hour address on free expression, Clinton addressed Chinese censorship head-on.

A "new information curtain is descending across much of the world," Secretary Clinton told an audience, before sounding the clearest warning shot yet on Chinese cyberattacks: "In an interconnected world, an attack on one nation's networks can be an attack on all."

At the very moment Clinton was speaking, China's hackers were furiously unplugging and abandoning their hacking tools and command-and-control servers. It would be months before Legion Yankee would hit American radars again. One year later, they would resurface in yet another sophisticated cyber-attack at RSA, the security company that sold authentication keys to some of the most high-profile U.S. defense contractors, before using RSA's source code to hack Lockheed Martin. They would eventually go on to compromise thou-sands of Western companies across diverse swaths of industries—banks, NGOs, auto makers, law firms, and chemical companies—siphoning billions of dollars' worth of sensitive military and trade secrets in the process.

In the months that followed Google's disclosure, Brin told the *New York Times* that he hoped Google's actions might lead to a "more open Internet in China."

"I think that in the long term, they are going to have to open," he said.

He could not have been more wrong.

China blocked Google permanently. And three years later, under its new president, Xi Jinping, China took a stranglehold over the web. It codified into law criminal punishments for anyone who "damaged national unity." It pioneered new forms of digital surveillance—facial recognition software, hacking tools, and novel spyware—aimed not only at its own people but also at the growing Chinese diaspora abroad. And it started exporting its censor-ship overseas. At one point it seized control of foreign traffic intended for Baidu, China's biggest internet company, injecting code that transformed

Baidu's traffic into a fire hose, then aimed it at U.S. websites that hosted mirror images of content banned in China. Some called China's move "the Great Cannon," and it was a shot across the bow to anyone who thought Beijing might eventually tolerate anything less than total internet control.

AS FOR GOOGLE, even the most righteous of corporations have short memories when it comes to the world's largest market. Within a year of Google's pullout from China in 2010, some executives began pushing for its reentry.

As the search company expanded over the next decade. Google became a sprawl of different companies—Android, Google Play, Chromebook, photo-sharing sites, Nest thermostats, cloud computing, drones, pharmaceuticals, venture capital, even satellites—each with its own reasons for wanting to break into the world's fastest-growing market.

In 2015 Brin and Page reorganized Google's various businesses under a new name—Alphabet—and separated its moneymaking businesses from its moon-shots. They began to disengage from the day-to-day. They promoted their longtime second in command, Sundar Pichai, to chief executive and poached a new CFO from Wall Street, who made beating quarterly earnings a top priority.

Google's reentry into China became a topic of furious debate within the company. With more than 750 million internet users, China's internet popula-tion had surpassed the combined populations of Europe and the United States. Google's archrival, Apple, was heavily investing in China. Baidu, Google's rival in China, had set up shop right next to Google's complex in Silicon Valley. Other Chinese tech companies—Alibaba, Tencent, and Huawei—started setting up their own Silicon Valley research and development centers and poaching Google employees with higher salaries.

Human rights concerns fell by the wayside as the company refocused on its bottom line. Executives hell-bent on capturing market share from Microsoft, Oracle, Apple, Amazon, and Chinese competitors like Baidu had no patience for those who continued to push for a principled debate on human rights. By 2016, it was clear who Google's new chief executive sided with. "I care about servicing users globally in every corner. Google is for everyone," Pichai told an audience that year. "We want to be in China serving Chinese users."

What he didn't say was that, by then, Google was already plotting its reapproach. A tight group of Google executives were already working on a top-secret censored search engine for China, code-named Dragonfly. The following year Google would establish a new artificial intelligence research center in Beijing. And six months after that, Google began releasing seemingly inconsequential products to Chinese users—first an app, then a mobile game—apparently in hopes that by the time Dragonfly was ready for launch, it might be overlooked as simply the next logical step in Google's regression.

It wasn't just China. In Saudi Arabia, Google was now hosting an app that allowed men to track and control the movements of their female family members. In the States, Google contracted with the Pentagon on a program—code-named Maven—to improve imaging for military drone strikes, prompting dozens of Google employees to quit in protest. Google's advertising had long been a sore subject for the company, but after the 2016 election, it became clear that the company had profited off advertising on sites that peddled blatant disinformation and conspiracy theories. Google's YouTube algorithms were radicalizing American youth, particularly angry young white men. Even YouTube Kids programming came under fire after journalists discovered that videos encouraging children to commit suicide were slipping through Google's filters.

And I would learn that Morgan Marquis-Boire, the hacker who had played such an integral role in Google's 2010 attack, a person I had spent countless hours alone with, even days, had a much darker past than he had let on. In 2017 several women accused him of drugging and raping them. After one accuser leaked an exchange in which he admitted as much, Marquis-Boire vanished without a trace. I never heard from him again.

But back in those first years after the Aurora attack, the men and women inside Google's security team worked with new resolve. Security at Google, across the Valley, would never be the same.

Adkin's team summed it up with their new unofficial two-word motto: "Never again."

CHAPTER 15

Bounty Hunters

Silicon Valley, California

A URORA WAS SILICON Valley's own Project Gunman. It had taken a Russian attack to push the NSA to step up its game in offense. Likewise, Aurora—and Snowden's revelations three years later—pushed Silicon Valley to rethink defense.

"The attack was proof that serious actors—nation-states—were doing these things, not just kids," Adkins told me.

Google knew China would be back. To buy time, it shifted as many systems as possible to platforms that would be alien to the Chinese. And then they began the slow, arduous journey of hardening Google—eventually the entire internet—from the inside out.

Grosse and Adkins started with long overdue security measures, before moving on to wilder pursuits. Before too long, Google's team had settled into an intensive regime of bit-by-bit security reinforcement, until finally it was spearheading a full-fledged surveillance resistance movement. By the time its checklist was complete, years later, Google's mission would include one radical addition: neuter the world's stockpiles of zero-day exploits and cyberweapons in the process.

Google introduced new protocols not just for employees but for its hundreds of millions of Gmail users. The company had been toiling with a two-factor authentication system for some time—the additional security step that requires users to enter a second temporary password, often texted to their phone, whenever they log in from a strange device. Two-factor authentication—2FA for short—is still the best way to neutralize a hacker with a stolen password. And by 2010, stolen passwords were everywhere. Hackers religiously scanned the internet for weaknesses, broke into password databases, and dumped them on the dark web. A Russian white hat cold-called me that year and told me he'd just uncovered a trove of a billion passwords.

"Your password," he began, "is a boy's name followed by your address."

Yep. I changed every password to every account I ever had to absurdly long song lyrics and movie quotes and switched on 2FA. I didn't trust password managers. Most had been hacked. Even companies that bothered to scramble, or "hash," users' passwords were no match for hackers' "rainbow tables"— databases of hash values for nearly every alphanumeric character combination, up to a certain length. *Dear reader, use long passwords.* Some dark-web sites published as many as fifty billion hash values, and cracked passwords were available for as little as a buck a pop. Unless you had 2FA switched on, a single stolen password was all a hacker needed to gain access to your email, bank, cloud photo account, or brokerage account. Google had been rolling out 2FA to employees for some time but after Aurora, "It was *bingo!*" Adkins recalled. "Time to make this available for all Gmail users."

Before Aurora, Google had thirty designated security engineers. After Aurora, they signed on a couple hundred. The Valley was in the midst of a ruthless war for talent that year. Google bumped salaries 10 percent to quash defections to Facebook and doled out tens of millions of dollars more to keep just two of its top product managers from defecting to Twitter. Around Silicon Valley, engineers were being enticed with large equity stakes, hefty bonuses, and freebies like iPads, organic meals, shuttle service, a year's supply of beer, and $10,000 cubicle decoration budgets.

But Google now had one advantage over its competitors. "Telling the world about the attack ended up being the best recruiting strategy in the world," Adkins told me.

After naming and shaming China, hundreds of security engineers who'd been itching for a fight—many of whom had written off Google because they took issue with its privacy practices—began knocking on its door. Hackers at the NSA, the CIA, and their Five Eyes counterparts started sending in their resumes. Over the next decade, Google's security team surpassed six hundred engineers, all determined to keep China and other oppressive regimes out. Google weaponized its greatest resource: data—mountains of it—to search its code for errors. It deployed giant "fuzz farms" of thousands of computers to throw massive amounts of junk code at Google's software for days on end, in search of code that broke under the load. Crashes were a sign of weakness, a sign that its software might contain exploitable flaws.

Google knew that its own hackers, and the most powerful fuzz farms in the world, were still no match for a country hell-bent on tracking its own people. And so Google came to the same epiphany iDefense had years earlier: it started tapping the world's hackers for good. Until 2010, Google had only paid hackers for bugs in street cred. Anyone who responsibly disclosed a Google bug was rewarded with a T-shirt and a mention on Google's website. After Aurora, Google decided it was time to start paying its volunteer army real loot.

THEY STARTED PAYING hackers minimum bounties of $500 and maximum payouts of $1,337. That seemingly random payout was a clever wink at their target audience—the number 1337 spells out *leet* in hacker code, an abbreviated form of *elite*. Leet hackers were skilled hackers, the opposite of script kiddies. It was Google's peace offering to hackers who, for years, had come to perceive tech companies as Satan reincarnated.

It was not the first time a tech company had paid hackers for bugs. Years before iDefense, in 1995, Netscape started paying tiny sums to those who turned over flaws in its Netscape Navigator browser. That inspired Mozilla to do the same, doling out a few hundred bucks to hackers who found serious holes in its Firefox browser in 2004. But Google's program upped the ante. Google started offering to pay hackers who found bugs in Chromium, the open-source code behind Google's Chrome web browser. Just like iDefense, the first bugs Google paid for were crap. But as word spread that Google was serious, the company started getting more critical bug submissions. Within a matter of months, Google expanded the program, paying hackers for any exploitable bug that compromised users' data on YouTube, Gmail, and the like. And it upped its maximum reward from $1,337 to $31,337—*eleet* in hacker code—and started matching offers for bounties that hackers donated to charity.

Google would never win over the Zero-Day Charlies of the world, who'd been badly burned before. And it could never expect to match the fees Desautels and others were paying. But it was enough to entice programmers in Algeria, Belarus, Romania, Poland, Russia, Kuala Lumpur, Egypt, Indonesia, rural France, and Italy—even stateless Kurds—to spend their off hours searching for Google bugs. Some used their bounties to pay rent; others for vacations to warm locales. Missoum Said, an eighteen-year-old Algerian kid living in a small commune in Oued Rhiou, hung up his soccer cleats and started hacking. Making it into Google's list of top ten bounty hackers became an obsession, and soon he'd netted enough cash to buy himself a nice car, remodel his family's house, travel to countries he'd never dreamed of, and send his parents to Mecca. Two hackers in Egypt bought themselves an apartment. One used his bounty to buy his fiancée an engagement ring. In India's slums, programmers started submitting bugs and using their bounties to fund new start-ups. Restaurant owners in Romania, laid-off programmers in Poland and Belarus, began funding entirely new lives with Google's bounties. In the rugged northernmost tip of Washington State, a hacker donated his bounty to the Special Olympics. In Germany one of Google's top bounty hunters, Nils Juenemann, doubled his bounty by donating it to a school in Ethiopia. He started sending bounties to kindergartens in Togo and helped set up a solar plant at a girl's school in Tanzania. For years a small cadre of hackers had secretly been making hundreds of thousands of dollars, in some cases millions, in the stealthy government market for vulnerabilities and exploits. Now Google was paying hundreds more programmers and defense-minded hackers to make their jobs more difficult.

But as Google's bounty program picked up, the rate at which hackers turned over critical bugs began to slow. In part, that was because Google's bounties had their intended affect: Google's software got harder to exploit. So Google upped the ante further. The company started adding thousand-dollar bonuses and sponsoring hacking competitions in Kuala Lumpur and Vancouver, offering to pay $60,000 for a single exploit in its Chrome browser. Some scoffed at its Chrome awards, noting that the same exploit could earn three times that much in the government market. Why should hackers tell Google about defects in its systems, when they could make far more by staying quiet?

* * *

NO ONE TAUNTED Google about its bounties more than the Wolf of Wuln Street. Chaouki Bekrar, the French Algerian, stalked the same hacking competitions and conferences Google sponsored. Every year, hackers from around the world descended on Vancouver to hack software and hardware for cash prizes and free devices in the Pwn2Own hacking contest at the CanSecWest conference—the top-paying hacking competition in the world.

In the contest's earliest days, hackers vied to break into Safari, Firefox, and Internet Explorer in the shortest time possible. As smartphones became ubiquitous, top prizes were awarded to those who could hack iPhones and BlackBerries. In 2012 the system to beat was Google's Chrome browser. Three teams of hackers broke into Chrome that year. But only two received Google's cash prize. Bekrar's team of Vupen hackers refused to play by Google's rules, which required the winner to disclose the details of their exploit to the company.

"We wouldn't share this with Google for even $1 million," Bekrar told a reporter. "We want to keep this for our customers."

Bekrar was as direct and crude as any broker on the market. "We don't work as hard as we do to help multibillion-dollar software companies make their code secure," he said. "If we wanted to volunteer, we'd help the homeless."

From Vupen's headquarters in Southern France, in-house hackers churned out zero-day exploits for government agencies around the world. Among its bigger customers that year was the NSA itself. The NSA, Germany's equivalent—the BSI—and other Vupen clients were willing to dole out $100,000 a year just to glimpse vague descriptions of Vupen's exploits. To get the actual exploit code, Vupen charged governments $50,000 or more on top of that for a single exploit. Bekrar considered Google's bounties chump change. He came to these conferences to liaise with clients. Though he claimed he only sold to NATO countries or "NATO partners"—like the non-NATO members of Five Eyes—he readily admitted that code had a way of getting into the wrong hands. "We do the best we can to ensure it won't go outside that agency," Bekrar told a reporter. "But if you sell weapons to someone, there's no way to ensure that they won't sell to another agency."

Bekrar called that being "transparent." His critics called it "shameless." Among them was Chris Soghoian, a die-hard privacy activist, who compared Bekrar to a "modern-day merchant of death" selling "the bullets for cyberwar."

"Vupen doesn't know how their exploits are used, and they probably don't want to know. As long as the check clears," Soghoian told a reporter.

Soghoian's point was validated by the Hacking Team leaks three years later, which showed Hacking Team had been baking Vupen's zero-day exploits into the spyware it sold to countries like Sudan and Ethiopia. The media attention from the leak put both companies under the microscope. European regulators recognized the hypocrisy of being the world's toughest voice on privacy, and at the same time headquarters for its biggest cyberarms dealers. In short order, regulators revoked Hacking Team's global export license, meaning the company could no longer sell its spyware to countries without Italy's express permission. Next on the chopping block was Vupen. After authorities revoked Vupen's global export license, Bekrar packed up and moved his Montpelier offices to the cyber-arms market's global headquarters: Washington, D.C. He took a page from the disgraced military contractor Blackwater and rebranded Vupen as Zerodium. He set up a slick new website and, in an unprecedented move, started advertising the prices he was paying for zero-day exploits, in exchange for hackers' silence.

"The first rule of [the] o-days biz is to never discuss prices publicly," Bekrar wrote in messages to reporters. "So guess what: We're going to publish our acquisition price list." He offered to pay $80,000 for exploits that could defeat Google's Chrome browser, $100,000 for Android exploits. The top prize, $500,000, was reserved for the iPhone. As the number of Zerodium customers went up, so did Bekrar's payouts. In 2015 Zerodium tweeted out a $1 million offer for the gold mine: a remote jailbreak of the iPhone, which entailed a chain of zero-day exploits that would enable his government clients to spy on an iPhone user remotely. By 2020 Bekrar was offering $1.5 million for exploits that could remotely access someone's WhatsApp messages and Apple's iMessages without so much as a click. He was paying out $2 million for remote iPhone jailbreaks and—in a notable shift—$2.5 million for an Android jailbreak. Apple exploits had long commanded top dollar. Some called the price change proof that Apple's security was weakening. Others made the more nuanced point that there isn't one Android; every device manufacturer customizes it a bit differently. A jail-break of one Android model might not work on another, which makes a remote attack that works across Android devices that much more valuable.

The companies hated Bekrar for it. He erased any doubts that hackers had far better options than Google's bounties. As critical bug submissions to Google dropped, the company had no choice but to up its price lists once again. By 2020 Google had upped its top bounty to $1.5 million for a full remote jailbreak of its Android phone.

It was a full-fledged arms race.

But Google had one big edge on the Zerodiums of the world. Brokers required *omertà*. Google's bounty hunters were free to discuss their work openly and avoid the shadier side of the business.

GOOGLE HAD ANOTHER edge on the offense markets, one the company may not have fully appreciated it at the time. The government's stable of freelance hackers had started souring on the cyberarms market.

"The contractors can be downright exploitative," a hacker told me late one night at a club in Vancouver. The CanSecWest hacking competition had come to a close that night and hackers, brokers, and contractors were letting loose. Linchpin Labs, an exploit broker for Five Eyes, was well represented. So was Arc4dia, a Canadian shop that sells exploits to an undisclosed lists of foreign intelligence services. A few former VRL hackers were there. So was Bekrar. The feds were doing their damndest to blend in. I'd been introduced to the hacker—a William Shatner lookalike in his late forties whom I will call Cyber Shatner—by a mutual acquaintance who vouched I could be trusted and wouldn't reveal his real name.

Shatner had been selling exploits to big defense contractors for decades. But these days, it was clear from the avalanche of grievances pouring from this mouth that he wanted out.

"I'd sell Raytheon an exploit for $30,000 and they'd turn around and sell it to an agency for $300,000," Shatner told me. For a time, he worked on retainer for Raytheon. "Until I learned just how screwed I was getting."

In any market, there is a fool. It had recently occurred to Shatner that he was the fool. There are no copyright laws for zero-days, no patents on exploits. He told me he spent months developing an exploit for a firewall, but when he submitted it, they rejected it.

"Raytheon told me, 'It didn't work.' Then a year later, I learned from a friend at the company that they'd been using my exploit for months. And I never got paid. It's an arms race," Shatner told me. "And the net outcome is we're all getting screwed."

Shatner's efforts to rectify the situation weren't welcome. At one point, he told me he was invited to present his work at an annual, invite-only summit for defense contractors and Five Eyes customers. He'd seized the opportunity to advocate for a better way: an escrow for exploits. He proposed that contractors consider using a trusted, technically-adept third-party to assess the value of each exploit and determine a fair price. It would keep hackers from getting screwed, eliminate distrust in the market, and still maintain discretion. To him, it all made perfect sense. The contractors saw things differently.

"I was never invited back," he rued.

Shatner was not just getting lowballed, he was getting replaced by foreigners who did the same work for cheap. Federal regulations mandate that only U.S. citizens with security clearances can work on classified systems, but that still leaves a lot of wiggle room when it comes to the raw material—the actual code. Back in 2011, a whistleblower tipped off the Pentagon that its security software was riddled with Russian backdoors. The Pentagon had paid Computer Sciences Corporation—the same megacontractor that now owns VRL—$613 million to secure its systems. CSC, in turn, subcontracted the actual coding to a Massachusetts outfit called NetCracker Technology, which farmed it out to programmers in Moscow. *Why?* Greed. The Russians were willing to work for a third of the cost that U.S. programmers had quoted. As a result, the Pentagon's security software was basically a Russian Trojan horse, inviting in the very adversary the Pentagon had paid hundreds of millions of dollars to keep out.

Things were even dicier on the offense side. Nobody ever bothered to ask defense contractors where the underlying exploits came from. Brokers like Desautels, Bekrar, and employees of VRL readily admitted that some of their best exploits came from hackers in Eastern Europe, South America, and Asia. There was no oversight, and nobody had to know about it. This only made it harder for American exploit artists like Shatner.

Google's bounty program offered the Shatners of the world one road out. Google would never match the gray market. But it paid for bugs. Hackers didn't have to spend months weaponizing those bugs into a reliable exploit, paranoid someone would discover the same bugs or screw them in the end. There was also more peace of mind—they didn't have to worry how their tools would be used or who they would be used against.

* * *

ONE YEAR INTO Google's bounty program, two baby-faced Dutch hackers in their early twenties put together a list of one hundred companies to hack. They called it the "Hack 100."

Michiel Prins and Jobert Abma grew up across the street from each other in the picturesque northern Holland, where they'd bonded over their mutual disdain for the frigid North Sea winds and affinity for hacking. They punked each other constantly. Michiel would find ways to hijack Jobert's computer screen from across the street, scrawling MICHIEL WAS HERE. Jobert would send discs flying out of Michiel's hard drive from two hundred yards away. By the time they turned sixteen, their parents started pushing them to get out of the house and use their skills for good. They started knocking on the doors of neighbors who left their WiFi networks open, offering to close them—for a fee. Soon they were marching into businesses and government buildings all over Holland, selling their services. Sometimes this required cake; the Dutch love cake. If executives gave them a half hour, they promised to find a gaping hole in their website. If they failed, the teenagers would give them a cake. Nobody ever got their cake. For five years, they made thousands of dollars selling their services to the biggest brand names in Holland, until tedium took hold.

"We were just telling different people how to fix the same vulnerabilities over and over again," Jobert said.

In 2011 the two met a thirtysomething Dutch entrepreneur by the name of Merijn Terheggen. Terheggen was in the Netherlands on business but lived in Silicon Valley, and he regaled the young men with stories of start-ups and the venture-capital dollars that seemed to magically appear out of thin air. The two imagined Silicon Valley to be a techie's paradise, nestled among redwoods and verdant mountains—like Switzerland, only with smiling engineers pedaling up and down Sand Hill Road in logoed hoodies. Terheggen invited them to visit.

"Great, we'll be there in two weeks," they said.

When the two arrived in San Francisco that summer, they only managed a brief stop at the redwoods. They spent the bulk of their time driving up and down Highway 101, visiting the campuses of Facebook, Google, and Apple. In 2011 a staggering amount of cash was sloshing around Silicon Valley. Facebook, not yet public, was marked at an unprecedented $50 billion private valuation.

Twitter, which still had no business model whatsoever, was valued at $10 billion. Groupon, the online discount shop, was turning down $6 billion acquisition offers. The Dutchmen felt the need to capitalize. Nobody at the Valley's unicorns seemed too concerned with security. (I asked Jack Dorsey that year whether he worried about the fact that hackers were continually pointing out holes in Twitter and in his new payment start-up, Square. "Those guys like to whine a lot," he replied.)

If Prins and Abma could just show the Dorseys how easily they could be hacked, perhaps they could also convince the Valley's money men that a security start-up had the potential to be the next unicorn. They drafted a list of one hundred successful companies around Silicon Valley, and one week later they'd hacked them all. On average, it took fifteen minutes each.

When they alerted executives, a third ignored them. Another third thanked them, but never fixed the flaws. The rest raced to solve the issues. Fortunately, nobody called the police.

SHERYL SANDBERG HAD never gotten an email like it. One morning in 2011, Sandberg, Facebook's chief operating officer, opened her inbox to find an email labeled "sensitive." It detailed a critical Facebook bug that allowed some Dutch twentysomethings to take over all of Facebook's accounts. Sandberg didn't hesitate. She printed the email, sprinted over to Facebook's head of product security, and told him to deal with it.

Alex Rice, a freckled engineer in his early thirties, took one look at the email and was impressed, both by the bug and by Sandberg's hustle. Facebook's competitor at the time, MySpace, had a history of aggressively prosecuting hackers who pointed out bugs in its site. Facebook founder Mark Zuckerberg took the opposite approach. Zuckerberg considered himself a hacker. He sponsored all-night hackathons and made a point of engaging with—and in many cases hiring—anyone who came forward with a serious bug. Facebook's 2012 IPO prospectus was part SEC regulatory filing, part dreamy love note to hackers the world over.

"The word 'hacker' has an unfairly negative connotation from being portrayed in the media as people who break into computers," Zuckerberg wrote. "In reality, hacking just means building something quickly or testing the boundaries of what can be done. Like most things, it can be used for good or bad, but the vast majority of hackers I've met tend to be idealistic people who want to have a positive impact on the world."

It read "like jacket copy for a would-be Deepak Chopra," the *New Yorker* wrote. But Zuckerberg was being genuine.

Rice invited the Dutchmen to barbecue, worked with them to fix the bug, and used the case to press management to start a Facebook bug bounty program of its own. Soon Facebook was paying minimum bounties of $500, with no upper limit. Two years later the company had paid out $1.5 million in rewards to some 330 researchers for 687 bugs, 41 of which could have been used to make Facebook a cybercriminal or spy's playground. By 2014 Rice was calling up his old Dutch buddies to see if there was an opportunity to neuter the cyberarms market for good.

IT TOOK MICROSOFT two more painful years to get there. In 2010, Aurora showed countries the surveillance potential of a single Microsoft zero-day exploit. A few months later, Stuxnet showed the destructive potential of a few Microsoft zero-day exploits chained together. Then came the 2011 and 2012 discovery of Stuxnet's predecessors, Duqu and Flame. Duqu had infected computers across the Middle East via a Microsoft Word exploit. Flame's infection mechanism was worse. The Americans, the Israelis, or both together had turned customers' trust in Microsoft into a weapon of warfare. They had spread Flame through Microsoft's Windows software update mechanism. What made that so terrifying was that 900 million Microsoft computers got patches and updates that way. Infecting Microsoft's updates was the Holy Grail for hackers, and a nightmare for Redmond. In any other hands, Flame could have taken down the global economy, critical infrastructure, hospitals, the grid.

The discovery of Flame by Russian researchers at Kaspersky was a disaster for Microsoft. It sent Redmond's hackers into the war room for weeks. Flame was a beast of a virus—at 20 megabytes, twenty times the size of most malware—and yet it had been hiding in plain sight. Nobody at Microsoft discovered it until four years later. Well-respected security researchers started floating conspiracy theories that Microsoft was complicit in cyberwar, or that there was a CIA or NSA mole inside Redmond.

By 2011 the number of bug reports Microsoft was getting directly from hackers—which had come in torrents of hundreds of thousands of messages a year—started to drop. Increasingly, hackers were hoarding the bugs for themselves, or choosing to sell them to defense contractors who were willing to pay six figures for the same work Microsoft used to get for free.

Katie Moussouris, Microsoft's head of hacker outreach, knew this didn't bode well for the company, or the internet. Even with "Microsoft" on her business card, she still considered herself a hacker, and she made it a point to remind everyone as much, with a T-shirt she'd emblazoned with "Don't hate the FINDER. Hate the VULN." With her jet-black hair that she sometimes dyed bright pink, she was easy to mistake for a twentysomething hacker, even though she was now well into her forties. "I'm really old but really well preserved because I never go outside," she told me.

Moussouris's mission, as she saw it, was to charm the world's hackers into turning over their bugs, and hopefully deplete the world's stockpiles of cyberweapons in the process. Microsoft, more than any other company, was increasingly being weaponized by nation-states and authoritarian regimes for espionage, surveillance, ransomware, and, in the case of Stuxnet, the most destructive attack the world had ever seen. Stuxnet and Aurora had been wake-up calls, but the breadth of potential harm and the lack of any constraints ensured that unless Microsoft locked its systems down, a bad actor would inevitably use those same capabilities for a cyberattack of mass destruction or as a tool in brute authoritarianism. The stakes were only getting higher.

Moussouris had her work cut out for her. When she arrived at Microsoft in 2007, the online link to the company's vulnerability disclosure policy had been dead for years. It was like calling 911 and getting a full voice mail. She found it mind boggling that there was no official way to tell the world's dominant technology company that there was a critical bug in its systems. She believed that hackers, inside and outside Microsoft, were key to neutralizing the kind of attacks that could have life-threatening consequences.

She started buying hackers beer. A lot of beer. She also began inviting them to late-night karaoke sessions at the big hacking cons. She personally updated Microsoft's Coordinated Vulnerability Disclosure policy, which gave her something to point to when hackers told her they would never turn a bug over to Microsoft for fear of getting sued. Her work started paying off. Hackers began to give Microsoft a couple weeks' notice before dropping a Windows zero-day onstage at Def Con. Before too long, Microsoft was getting 200,000 vulnerability reports a year. Those reports gave Microsoft a wealth of data about how its products could potentially be abused. It also gave her team valuable insights into the researchers themselves. Microsoft's responders could see, over time, which hackers were out to waste their time, and which needed to be handled with white gloves because they could drop a critical Microsoft zero-day at any moment. By the time I joined the cyber beat in 2011, I'd made a point of asking hackers which tech company they hated least. "Microsoft" was almost always their answer. "They turned their shit around." Much of that could be credited to Gates's Trustworthy Computing initiative, but a lot of it could be credited directly to Moussouris.

The minute the bug reports started to dry up in 2011, Moussouris knew Microsoft had a serious problem. The company was starting to get more bug reports from brokers—likely after the bugs had already been exploited—than from hackers themselves. That wasn't just bad news for dissidents, activists, and journalists, with the potential for a destructive cyberattack that it implied; it was bad for Microsoft. Silicon Valley's brutal talent war was still raging. Redmond had become fertile poaching ground for younger start-ups like Twitter and Facebook. The company wasn't just losing out on bug reports; it was also cutting off access to a large hiring pool. If Microsoft couldn't compete for the best security talent, it would enter dangerous territory. With Google and Facebook now paying out bounties, Moussouris knew it was high time for Microsoft to do the same.

Convincing Microsoft's top brass to start paying hackers bounties was a Herculean task. For one, Microsoft would never be able to compete with governments in the market for cyberarms. There was also a logical limit at which bounties could cross the line into perverse incentives: the more money there was to be made on offense, the more people might abandon defense. How many talented security engineers would apply for, or stay in, their jobs if they could get thousands of dollars for a single Microsoft bug?

Moussouris started spending her off hours studying game theory to understand various incentive models and their downsides. Microsoft might never compete with the government market, but she also knew that money wasn't hackers' main motivator. She grouped their motivations into three categories:

compensation, recognition, and the "pursuit of intellectual happiness." If Microsoft wasn't going to pay top dollar, it needed to create the conditions under which fixing a bug was more attractive than weaponizing it and selling it to governments. Not everyone would be convinced. Some hacked for the money. Others justified their sales to governments as their patriotic duty. But there were eighteen million software programmers in the world. If Microsoft could recognize those programmers in a more meaningful way, it could tap their brainpower for good, and invite the best to come to Redmond.

When Moussouris pitched Steve Ballmer's generals in 2011, they were receptive but not ready to pull the trigger. They needed more data. For the next two years she compared herself to Cassandra, "doomed to know the future but nobody believes her until she can show them the data." By 2013 she had two years' worth of data showing that Microsoft was now losing bug reports to third-party brokers and intermediaries. That June, correcting the trend became a matter of urgency. The *Guardian* dropped Snowden's first NSA leaks that month, detailing an NSA program called Prism. One NSA slide appeared to show that Microsoft, and the other tech companies, gave the NSA direct access to their servers. Some leaks described Prism as a "team sport" between the tech companies, the NSA, the FBI, and CIA.

Of all of Snowden's leaks, those slides would end up being the most damning—and misleading. The tech companies had never heard of Prism. Yes, they complied with narrow, court-ordered requests for specific customer accounts and metadata, but the notion that they were somehow NSA collaborators, handing the agency real-time access to customers' private communications, was flat-out wrong. Their denials, however, were complicated by the fact that, legally, they were forbidden from disclosing the full nature of their cooperation and resistance to the secret court orders.

The trust that Microsoft had spent years building was in danger of evaporating. It started bleeding customers, ranging from Germans, who likened Prism to the Stasi, to the entire government of Brazil. Foreigners demanded that they move their data centers overseas, where—the illusion went—their data would be safe from the prying eyes of the U.S. government. Analysts projected that U.S. tech companies could lose a quarter of their revenues over the next few years to foreign competitors in Europe and South America. The hackers were disgusted.

Moussouris knew that unless Microsoft acted fast—not just on the public relations front but with meaningful action—the company would lose its best allies in securing the internet. She knew from tracking the data that the biggest dropoff in direct bug reports was for Internet Explorer bugs. Clearly, there was an offensive market for IE bugs, given that it was still one of the most widely used browsers on the market. A single IE exploit could produce a wealth of intelligence about a target: usernames, passwords, online banking transactions, keystrokes, search histories, travel plans—essentially a spy's wish list.

Moussouris started there. What if Microsoft offered to pay hackers to scour beta (preview) versions of Internet Explorer and Windows upgrades—obviously another top government target—before they hit the market? Governments weren't interested in exploiting software that its adversaries, terrorists and dissidents weren't yet using. If Microsoft deployed hackers to probe its beta software, they wouldn't overlap with the underground market. The company could offer bounties as small as $500 up to $100,000 for truly novel exploit techniques. Ballmer's team signed off on a month pilot program. That June, Microsoft flipped the switch and started paying hackers to turn over IE bugs. In what would become something of a trend, the first bounty went to an engineer at Google. But by the end of the month Microsoft had gotten two months' worth of critical bugs, and Ballmer signed off on a permanent program that November. One year into the program, Microsoft had paid out $250,000 to researchers— roughly the annual salary of a talented security engineer—and worked out the security of its software before it hit the market. In the process, it jump-started a process that would drain hundreds of bugs from government stockpiles and, in time, Moussouris hoped, just tilt the scales in favor of defense.

GOOGLE, FACEBOOK, AND Microsoft were fierce competitors for the best security engineers, but they all had a vested interest in securing the internet. They regularly traded threat intel, met up at the big hacking conferences, and shared their bounty war stories. By 2014 Rice, Moussouris, and the three Dutchmen— Terheggen, Abma, and Prins—wondered if there wasn't something bigger that could be done. Initial conversations were casual, but eventually they started sketching out the outlines of a company that would manage bounty programs for as many companies as they could sign up. Bounties entailed psychological hurdles for executives. If they could manage the back-end logistics and the payments and offer a trusted platform through which hackers might engage with companies across industries, they could make a far greater dent in the exploit stockpiles than they could working from their respective siloes.

In April 2014 Rice and the Dutchmen walked down San Francisco's Market Avenue, took an elevator up to the gilded open-space offices of Benchmark Capital, a renovated building above the city's historic theater, the Warfield, and pitched five of the most competitive venture capitalists in the Valley. Unlike buzzy firms like Andreessen Horowitz and Accel Partners, which larded up their payrolls with marketers, PR handlers, and in-house designers, Benchmark was known for its focus. The firm had gotten rich off early bets in eBay and, later, Dropbox, Instagram, Uber, Yelp, Twitter, and Zillow. Its funds had paid out more than $22 billion to investors, a 1,000 percent gain, over the previous decade. And it had done so by sticking to a simple formula: five equal partners, each with equal stakes in the firm's funds, invested in early financing rounds, for the biggest equity stakes and a board seat. Where other storied firms like Sequoia Capital and Accel started expanding into China and India, Benchmark stayed its course. The partners despised firms in the Valley, like Andreessen Horowitz, that engaged in relentless self-promotion. They coined a name for it—"parade jumping"—because they believed the real credit should go to entrepreneurs running the day-to-day operations. And they were notoriously tough on their entrepreneurs, and tough to pitch. Every investment had to be unanimous. On rare occasions the partners would exchange a look at these pitch meetings. The look meant *Sold. Where do we sign?* The day the HackerOne team pitched Benchmark, they got the look, and $9 million.

"Every company is going to do this," Bill Gurley, a former college basketball player and one of the most competitive VCs in the Valley, told me. "To not try this is brain dead."

Gurley was usually right. Within a year, HackerOne had persuaded some of the biggest names in tech—including Yahoo and even Dorsey's two companies, Square and Twitter—as well as companies you might never expect, banks and oil companies, to start paying hackers bounties for bugs in their platform. And within another couple years, they would add automakers like General Motors and Toyota, telecoms like Verizon and Qualcomm, even airlines like Lufthansa, to engage hackers to neuter the bugs that threatened to turn cell towers, banks, cars, and airplanes into weapons for surveillance and cyberwar. By 2016 the company had even managed to sign on the most unlikely player of all: the Pentagon.

FRANKLY, IT WAS a lot to take in. When secretary of defense Ash Carter first announced the Hack the Pentagon bounty program at the RSA security conference that year, I heard audible groans in the audience. I could have sworn the guy sitting a few seats away snarled. Every hacker has seen *WarGames*, in which teenage Matthew Broderick unwittingly hacks the Pentagon's computers, nearly sets off World War III, and gets picked up by the FBI. Getting picked up by the FBI seemed like the only logical end to hacking the Pentagon. It didn't seem to matter that the Pentagon was *actually inviting them* to hack its systems. Nobody wanted to play. I imagined Snarl Guy felt a lot like the hackers who bitched about the program over cocktails that night, convinced it was just another way for the government to track them. That seemed overly paranoid, but I had to admit they had a point. The program only paid bounties to hackers who submitted to background checks—not ideal for people who savor anonymity.

But by then the government knew it had to do *something*. The previous year, the U.S. Office of Personnel Management—the very agency that stores the most sensitive data for the one million or so federal employees and contractors, including detailed personal, financial, and medical histories, Social Security numbers, even fingerprints—revealed that it had been hacked by Chinese hackers on a scale the government had never seen before. The Chinese had been inside OPM's systems for more than a year by the time they were discovered in 2015. When I dug in to see which other government agencies stored sensitive data and might also be vulnerable, I found the cyber equivalent of a pandemic. At the Nuclear Regulatory Commission, which regulates nuclear facilities, information about crucial nuclear components was left on unsecured network drives, and the agency had lost track of laptops with critical data. Computers at the IRS allowed employees to use weak passwords like "password." One report detailed 7,329 vulnerabilities because the agency hadn't even bothered to install software patches. At the Department of Education, which stores data from millions of student loan applicants, auditors were able to connect rogue computers to the network without being noticed. The SEC hadn't installed a firewall or intrusion protection on a critical part of its network in months.

But a year and a half in, to my surprise and many others, the Pentagon's bounty program was actually taking off. More than 1,400 hackers had signed on to the bounty program—triple what officials anticipated—and the Pentagon had paid out $75,000 in bounties ranging from $100 to $15,000. That was nothing compared to what the NSA and other agencies were paying hackers, but it was something. And it wasn't just HackerOne anymore. DOD had also signed on a HackerOne competitor, Bugcrowd, and another company called Synack, started by a former NSA hacker, that crowdsourced pen tests to vetted hackers around the world. Synack's cofounder, Jay Kaplan, assured me that the Pentagon's program was the real deal, not some bureaucratic head fake, and convinced me to fly out to Arlington to see it for myself.

I arrived in April 2018, eight months pregnant. I hadn't seen my toes in weeks. I also didn't realize the Pentagon was so . . . big. I waddled back and forth, what felt like miles, between the Department of the Air Force and the tiny office that housed the Pentagon's nascent Digital Defense Service. Besides TAO, DDS was one of the few DOD offices sans dress code. Carter had set it up to bring hackers and Silicon Valley talent into the Pentagon for one-year tours of duty, to shake things up and oversee the bounty initiative.

"We have actively *dissuaded* people from telling us about vulnerabilities," Chris Lynch, DDS's hoodied chief, told me. "It's time to 'Make America Safe Again.'"

Lynch, a serial tech entrepreneur, was frank and not afraid to use colorful language. Often found at Carter's side, he summed up his mission at DDS as "Get shit done." With his urging, the Pentagon had expanded its bounty program from a tiny set of unclassified websites to far more sensitive systems like the F-15's Trusted Aircraft Information Program Download Station (TADS), the system that collects data from video cameras and sensors in flight. Synack hackers had found several severe zero-days in TADS that, if exploited, could have been used to take full control of the system. They'd also discovered a critical zero-day in a Pentagon file-transfer mechanism used by warfighters to transfer mission-critical information between the Pentagon's networks, even classified ones. The system had been pen-tested by government contractors before, but one of Synack's hackers had found a way into the Pentagon's classified networks in under four hours.

"This game of zero-day exploits is a big deal," General Bradford J. "B. J." Shwedo told me in what I now knew to be a gross understatement. "If you wait to find out about a zero-day on game day, you're already mauled. Sealing yourself off from the world is not the future of warfighting. In cyber, it's spy-versus-spy all the time."

In the vast bureaucracy that was the Department of Defense, one agency was now paying hackers to patch its holes, while others were paying them far more to keep the world's holes wide open.

I recalled something Chris Inglis, the retired NSA deputy director, once said: "If I were to score cyber the way we score soccer, the tally would be 462–452 twenty minutes into the game. In other words, it's all offense and no defense."

Six months later the Pentagon decided to put real money behind its bounty program—$34 million—chump change compared to what it spent on offense, but perhaps it would finally lower the score.

CHAPTER 16

Going Dark

Silicon Valley, California

I F ONLY THE NSA had left out the smiley face, they could have made everything a hell of a lot easier on themselves.

Throughout the summer of 2013, Silicon Valley was still reckoning with the fallout from Prism. The Snowden leaks had introduced a thicket of innuendo, and companies were fielding call after call from reporters and angry consumers accusing them of being in cahoots with the NSA. But that October, the *Washington Post* published the most damning Snowden leaks of them all. Top-secret NSA slides showed that, in addition to all the data the agencies were getting from companies on the front end—the legal demands for their customer data—they were taking truckloads more out the back.

Without the companies' knowledge or cooperation, the Snowden revelations that fall showed that the NSA, and its British counterpart, GCHQ, were sucking up companies' data from the internet's undersea fiber-optic cables and switches. In agency jargon, this was called "upstream" collection, as opposed to "downstream" methods like Prism, in which agencies demand customers' data from companies through secret court orders. In a single day, top-secret NSA slides showed that—unbeknownst to Yahoo, Microsoft, Facebook, and Google—the agency had collected "444,743 Yahoo email address books, 105,068 from Hotmail, 82,857 from Facebook, 33,697 from Gmail, and 22,881 from unspecified providers."

That wasn't the worst of it. The slides appeared to show that the NSA and GCHQ were directly hacking Google and Yahoo's internal data centers to intercept customer data before it was encrypted and passed over the open web—essentially a man-in-the-middle attack. The NSA-GCHQ code name for these attacks was Muscular. On one level, it was helpful in explaining that the companies were not willing accomplices.

"It provided us a key to finally understand what was going on," Brad Smith, Microsoft's president, told *Wired* magazine. "We had been reading about the NSA reportedly having a massive amount of data. We felt that we and others in the industry had been providing a small amount of data. It was hard to reconcile, and this was a very logical explanation."

On another level, it touched off a full-fledged cryptowar between tech companies and their own government. Included in the October leaks was a hand-scribbled NSA analyst's drawing on a large yellow Post-it note. The diagram pointed to the sweet spot where the NSA and GCHQ were tapping Google's data, in unencrypted form, before it was scrambled and moved online. Between two drawings of clouds—one labeled "Google Cloud," the other marked "Public Internet"—an NSA analyst had scribbled a little smiley face, a triumphant *gotcha!* emoji that drove the companies into full battle mode.

Without the top-secret smiley face, Silicon Valley might have dismissed the slides as some kind of explainer on how data moved from Google's data centers to the open web. But the analyst's triumphant emoji indicated that the agencies were already there. The effect on foreign customers, activists, and anyone who cared deeply about privacy was a kind of hopelessness. It did not matter if Silicon Valley's lawyers pushed back on the government's secret orders for data; as the doodle made clear, the NSA was getting everything anyway.

Google distributed its customer data between custom Google front-end servers—see the "GFE" in the scribble—around the globe, partly for speed, partly for security. A Gmail user in Bangladesh didn't have to wait for data to travel halfway around the world from Silicon Valley to open a Google Doc. And users' data wouldn't be held hostage by regional natural disasters or local outages. The front-end servers were also a security mechanism that could detect, and stop, denial-of-service attacks. Google encrypted users' data as it moved from these front-end servers to the open internet, but it didn't bother encrypting data internally between its data centers. Encrypting the links between data centers was in its long-term plans, Google said, but until Snowden, encrypting data as it flowed between its own data centers had always seemed like an unnecessarily expensive endeavor.

The agencies' hackers had used this holding pattern to their advantage. By hacking into Google's servers, the NSA could access all the Gmail inboxes and messages, Google Map searches and locations, calendars, and contacts they could ever want in plain text. It was a feat of digital spycraft that made Prism—and everything else the NSA was doing to devour the world's data— look superfluous.

Officially, Google said it was "outraged." Eric Schmidt told the *Wall Street Journal* that the NSA was "violating the privacy of every single citizen of America" in order to find a few evil people. Behind the scenes, Google security engineers were more forthright. "Fuck these guys," a Google security engineer named Brandon Downey wrote in a post to his personal Google Plus page. Downey and hundreds of other Google engineers had just dedicated the previous three years to keeping China from hacking its customers, only to find out they'd been had by their own government. In a window to how Valley's engineers think, Downey made the obligatory *Lord of the Rings* reference: "It's just a little like coming home from War with Sauron, destroying the One Ring, only to discover the NSA is on the front porch of the Shire chopping down the Party Tree and outsourcing all the hobbit farmers with half-orcs and whips." He added, "The U.S. has to be better than this."

In Zurich a British Google engineer named Mike Hearn echoed Downey in "issuing a giant 'Fuck You' to the people who made these slides."

"Bypassing that system is illegal for a good reason," Hearn wrote. "Nobody at GCHQ or the NSA will ever stand before a judge and answer for this industrial-scale subversion of the judicial process." In the absence of that, he added, "We therefore do what internet engineers have always done—build more secure software."

"NO HARD FEELINGS, but my job is to make their job hard," Eric Grosse told me six months later. We were seated inside the Googleplex, and I couldn't help but notice the giant stick behind him. It was a staff, Grosse told me. His engineers had given it to him not long after the smiley face slide dropped. It was a nod to the staff raised by Gandalf, the wizard of *Lord of the Rings*, when he confronts an evil Balrog and intones: "You shall not pass!" Grosse was now Silicon Valley's Gandalf, standing on the stony bridge that was Google's front-end servers, where he would rather die than let the world's intelligence agencies cross into Google's data centers.

Over the past six months Grosse and his team had sealed up every crack the NSA had brilliantly exploited. Grosse called the unencrypted links between Google's data centers "the last chink in our armor," and he was now encrypting Google data internally. Other companies were following suit and migrating to a stronger form of encryption called Perfect Forward Secrecy, which made it far more labor-intensive for NSA to decode their data. Google was also now laying its own fiber-optic cable beneath the world's oceans and rigging it with sensors that would alert the company to undersea taps.

"At first we were in an arms race with sophisticated criminals," Grosse told David Sanger and me when we visited him in the spring of 2014. "Then we found ourselves in an arms race with the Chinese. And now we're in an arms race with our own government. I'm willing to help on the purely defensive side of things, but signals intercept is totally off the table."

Once Google had checked everything off its list, it rolled out a new user-friendly email encryption tool to customers. Buried in the code was a winking smiley face ;-)

STILL, ENCRYPTION DIDN'T do much to protect users from a nation-state with a zero-day. That was the beauty of a zero-day. A good zero-day could pierce the world's encryption and get you onto a target's device where everything was in plain text. It was a lot more time-consuming to hack these end points, and much harder to do at scale. But this was, Snowden said, his goal in leaking from the start. He hoped his leaks would steer governments away from mass surveillance toward a more targeted, constitutional form of intel gathering.

The bug bounties eliminated the lowest-hanging fruit and made it harder for nation-states to hack the platforms, but only marginally, according to several government analysts I spoke with. And so, in 2014, Google's engineers convened and decided to take things to another level. Chris Evans, a British security engineer with serious eyes and a square jaw, knew locking down

Google would not be enough. The security of Google's big products, like its Chrome browser, still relied on the security of third-party code like Adobe Flash, antivirus software, and elements of Windows, Mac, and Linux operating systems. Attackers always targeted the weakest point. If Google didn't do something to address the flaws in other systems, everything else Google was doing would be for naught. Evans took security personally. He was a gentle soul, but he got downright angry when he found an Adobe Flash zero-day was being used to track Syrian citizens and freedom fighters. Since Aurora, he'd kept a spreadsheet of every newly discovered Flash zero-day, noting when it was used against Syrian citizens, dissidents, the aerospace industry. He had no tolerance for companies that dragged their feet to patch the zero-days that put people at real risk. "It's unacceptable," he'd told me.

Inside Google, Evans quietly began enlisting elite hackers to take on the world's cyberarms market. He summoned the company's best hackers to an offsite at a cabin in Lake Tahoe in August 2014 to ask a simple question: "What can we do to make zero-days hard?" Not all vulnerabilities were equal; some caused disproportionate harm. The bounties were a step in the right direction, but the program was a free-for-all. What if Google mobilized a team to focus on strategic targets—the Adobe Flash software that allowed oppressive regimes to hack their own people; antivirus software; the Java code that touched cell phones and laptops, data centers and supercomputers, the internet? The iPhone and Android jailbreaks that Zerodium and NSO Group and others now paid top dollar for required chains of zero-day exploits to work. If they could neuter just one flaw in the chain, they could starve spies of their intrusion tools, or at least set them back months, possibly years. With zero-day prices going up, attack research was going underground at a phenomenal rate. What if they could blow it open? Where there was one bug, there were more. If they made their research public, maybe they could inspire other defenders to find the rest. It had happened before. By the end of the weekend, the mission had a name, Project Zero, and a goal: bring the number of critical bugs down to zero.

Among Project Zero's initial recruits was Ben Hawkes, a muscled rugby player from New Zealand who specialized in Adobe Flash and Microsoft zero-days; a British researcher named Tavis Ormandy, among the most prolific bug hunters in the world; Geohot, the talented hacker who had broken into Sony and developed the first iPhone jailbreak, the one that law enforcement agencies now salivated over; and Ian Beer, a British researcher who'd go on to neuter dozens of Apple iOS exploits, including those used by Chinese hackers to monitor the Uighurs, exploits that would have otherwise been worth tens of millions on the underground market. Almost immediately Project Zero's researchers found critical zero-days in Apple's Safari browser, design flaws in some of the most reputable security products, and a Microsoft zero-day that would have given spies full control over Windows machines. Their work triggered equal parts admiration and disdain, particularly from Microsoft, which accused them of disclosing the bug before it had a patch. Project Zero's team gave vendors ninety days. After that, they would dump their bugs online. Part of the goal was to light a big fat fire under vendors' asses.

The researchers also got criticized for potentially making spies' jobs easier; by publishing their work, vendors argued, especially before patches were rolled out, they were giving the NSOs of the world an opportunity to exploit the lag time in between. Data shows that flaws reach peak exploitation right after they are published—Day Zero—as companies race to roll out a patch, and customers to install it. In a dig, an exploit developer called NSO Group "the 'commercial arm' of Project Zero." But the alternative—keeping research silent—didn't do anything to improve security long-term.

There was another benefit to publicizing Project Zero's work. They were sending a message to skeptical customers and governments—especially those that had come to see Google as complicit in NSA surveillance—that Google took their security seriously. The exposure also helped lure the world's top exploit developers to defense. One of Hawkes's first recruits was a twenty-one-year-old South Korean hacker named Jung Hoon Lee, alias Lokihardt, who had caught exploit brokers'—and Google's—attention at the Pwn2Own contest in Vancouver that year. Surrounded by his bodyguard, a translator, and a team of hackers in hats and dark glasses, Lokihardt had gutted Chrome, Safari, and Microsoft's latest Windows software in a matter of minutes. I'd watched Zerodium's Bekrar drool. Months later, Lokihardt joined Project Zero to neutralize the bugs Bekrar and co. paid top dollar for. Bekrar told me at the conference that year that Lokihardt was the best hacker he'd seen in years. Now the two were working at cross-purposes.

Over the next few years, Project Zero identified more than sixteen hundred critical bugs, major flaws not just in the world's most targeted software and security tools but also in the Intel chips inside nearly every computer in the world. Its researchers eliminated entire classes of bugs, making spies' jobs a heck of a lot harder.

"Espionage groups don't reach out and say, 'You burned my zero-day!'" Grosse told me. But, anecdotally, "We're hearing that we've made their jobs a lot harder. I can live with that."

TIM COOK WAS getting flooded with personal letters from every corner of the globe—Brazil, China, his home state of Alabama. He received more personal notes from Germans throughout 2013 and 2014 than the sum total of notes he had received in his seventeen years at Apple. And the words in these notes were not merely dramatic and emotional. They were heartfelt. Germans had lived through Stasi surveillance, when every workplace, university, and public venue was monitored by soldiers, analysts, tiny cameras, and microphones to root out "subversive individuals." Sixty-five years later, the horrors of East Germany's past were all too real.

"This is our history," they wrote Cook. "This is what privacy means to us. Do you understand?"

Cook was famously private himself. He had grown up gay in conservative Alabama, a fact he kept private until 2014, the year after the Snowden revelations dropped. In Alabama, his lingering childhood memory was watching Klansmen burn a cross on the lawn of a black family in his neighborhood while chanting racial slurs. He'd screamed at the men to stop, and when one of the men lifted his white hood, Cook recognized him as the deacon of a local church. Civil liberties were a matter of urgency for him, and he took the Snowden revelations as a personal affront. As Cook saw it, there were few things more precious than privacy. He'd watched as new businesses and start-ups around the Valley eroded privacy, little by little, bit by bit, and worried about an Orwellian future. Apple was something of a redheaded stepchild in Silicon Valley. It sold things—phones, tablets, watches, and computers—not data. It didn't make its money tracking purchases, or searches, or from targeted advertising. It was, Cook told me over one of our meetings that year, what he valued most about Apple. And the letters were getting to him.

So when President Obama invited Cook, along with Randall Stephenson, the chief executive of AT&T, Vint Cerf, one of the internet's pioneers, and civil liberties activists to the White House to discuss the collateral damage from Snowden in August 2013, Cook carried the letters with him. By then the companies were accelerating long-held plans to encrypt customers' data, and Washington—the FBI in particular—worried that things were not just "going dark" but blind.

At the closed-door meeting, Obama made the case for a balanced approach to privacy and national security. Cook listened intently, and when it came time to speak, he shared what he'd heard from Apple's customers abroad. There was now a deep suspicion of America's technology companies, he told the president. America had lost its halo on civil liberties, and it might be decades before it ever earned it back. Leaving anything open to surveillance was, in his mind, a civil liberties nightmare, not to mention bad business. People had a basic right to privacy, and if American companies couldn't protect them, they would take their business overseas. Cook was putting the government on notice: Apple was going to encrypt everything.

ONE YEAR LATER, September 2014, Cook took the stage in Cupertino to debut the new iPhone 6, "the biggest advancement in iPhone history," the phone for the post-Snowden era. From now on, Apple automatically encrypted everything on the phone—messages, call logs, photos, contacts—using a complex mathematical algorithm that used the user's own unique passcode to unwrap a larger key on the device. Apple no longer held the spare keys to customer data. They'd given the only pair to the users. If governments wanted access to their data, they were going to have to ask the customers directly.

Up to that point, if a government needed Apple's help unlocking an iPhone, they had to physically fly to Cupertino and bring the phone into a secure sensitive compartmented information facility (SCIF), where a trusted Apple engineer unlocked it. The trips could be comical. In one case, a foreign government sent an iPhone, along with a government minder, by chartered jet to Cupertino, only to get into the SCIF and learn from Apple's engineer that the owner had never even bothered to set up a passcode. Now Apple was telling governments that there was no need to travel to Cupertino; Apple couldn't unlock an iPhone, even if they wanted to.

Guessing or "brute forcing" the passcode wouldn't help. Apple's new iOS had an extra security feature that wiped a phone's hard drive if someone entered an incorrect password ten times in a row.

I asked Apple engineers if they had anticipated what would happen next. "The only surprise," they told me, "was that the government was surprised we would do this. This had been our mission all along."

THE FBI WENT ballistic. The agency had far fewer intrusion tools at its disposal to hack phones than the NSA and CIA did to intercept foreign communications. That had always been the tension between the FBI and the intelligence agencies; that intelligence hoarded their best talent and tools. And Apple had timed the roll-out of its advanced encryption just as a new terrorist threat had emerged. ISIS was quickly outpacing its predecessor, al-Qaeda, as the world's preeminent jihadist group, in violence, cruelty, reach, and recruiting. Increasingly, ISIS was taking refuge in encrypted apps and using social media to coordinate its attacks and recruit sympathizers in Europe, the UK, and the United States.

Over the next few weeks, then FBI director James Comey embarked on a Going Dark tour.

"No one in this country is beyond the law," Comey told reporters at FBI headquarters a week after Apple's announcement. "The notion that someone would market a closet that could never be opened—even if it involves a case involving a child kidnapper and a court order—to me does not make sense."

He went on *60 Minutes* and, in an hour-long at the Brookings Institution, told an audience that the "post-Snowden pendulum" had "swung too far," noting that Apple's new encryption "threatens to lead us to a very dark place."

Comey was essentially making the same arguments the White House had two decades earlier after a programmer named Phil Zimmermann released end-to-end encryption software to the masses. Zimmermann's Pretty Good Privacy (PGP) software made it far easier for people to communicate over end-to-end encryption, which scrambles messages in such a way that they could only be deciphered by the sender and recipient. Fearing that PGP would make surveillance impossible, the Clinton administration proposed a "Clipper Chip," a backdoor for law enforcement and security agencies. But the Clipper Chip provoked a backlash from a coalition of unlikely bedfellows, including Missouri Republican senator John Ashcroft and Massachusetts Democrat senator John Kerry, the televangelist Pat Robertson, Silicon Valley executives, and the ACLU. All argued that the

Clipper would kill not only the Fourth Amendment but also America's global technology edge. In 1996, the White House backed down.

Each new novel encryption effort generated anxiety. When Zimmermann introduced a Zfone in 2011, NSA analysts circulated the announcement in an email titled "This can't be good." The Zfone stagnated, but Apple's new iPhone, and its iOS software, were essentially the Zfone for the post-Snowden era—only with exponentially more data for the government to lose. In late 2014 the FBI and Justice Department made preparations to confront Apple in court. It was now simply a matter of waiting for the right case.

One year later, two shooters, Syed Rizwan Farook and Tashfeen Malik, armed with assault rifles and semiautomatic pistols, shot out a holiday party for the city health department in San Bernardino, California, killing fourteen people and injuring another twenty-two before they fled the scene. By the time they were killed in a shootout four hours later, all that was left behind were three pipe bombs that had failed to detonate, a Facebook post in which Malik had professed her allegiance to ISIS, and Farook's locked iPhone. Comey had his case.

FOUR MONTHS AFTER the shooting, I attended a cyberarms bazaar in Miami. Nobody wanted me there. They tried to disinvite me. Twice. It was my book, they told me. Nobody here wanted anything to do with it. "Don't come," they said. "You're not welcome. Stay home." When I told them I'd already bought my plane ticket and would be showing up whether they wanted me there or not, they backed off. But when I arrived at the entrance to this grassy expanse overlooking Miami Beach, where a short invite-only list of the world's top hackers, digital weapons developers, and spies gathered, the conference organizers had something else up their sleeves: a glow stick.

"Here, put this around your neck," the large guy seated at the Fontainebleau registration desk told me. "This" was a beaming, fluorescent-green glow stick designed to alert every person here that I was the journalist, the pariah, a woman to be studiously avoided. I flashed the man a look that said I didn't find this particularly amusing. "Consider yourself lucky," he mused. "We thought about making you wear a giant helium balloon around your neck."

I was accustomed to this treatment, well-versed in the length that hackers, feds, and contractors will go to protect their little secret. And I knew that the only way they were going to let me into this cyberarms bazaar, the only hope I had of finding the men I had come here for, was to follow their rules—even if it meant putting this degrading rave gear around my neck.

So I played their game, snapped on the glow stick, stepped out into the Miami heat, walked east toward the horizon where thunderclouds gathered over the Atlantic Ocean, and joined the gathering of two hundred or so of the world's top hackers, cyberweapons brokers, feds, and spies. Out of the corner of my eye, I watched carefully as the conference organizers slurped down their gin and tonics and mint mojitos, waiting for the moment they would be sufficiently liquored up that I could rip this damn neon dog collar off. This was my final stop, I told myself. The FBI's iPhone hacker was surely here.

FOR MONTHS, THE Department of Justice had been pressing Apple in court to help it bypass its own airtight encryption so the FBI could access the iPhone used by Farook, the gunman in the San Bernardino terrorist attacks four months earlier. It was the FBI's last-ditch attempt to keep from going blind.

For the FBI, the San Bernardino attack made for the perfect precedent-setting case. Two ISIS sympathizers had pulled off an attack on American soil. Both had been careful to cover their digital trails. They had deleted all email correspondence, smashed their computer hard drives and personal phones, and used disposable burner phones. All they had left behind was Malik's Facebook post, in which she'd pledged her allegiance to ISIS, and Farook's work iPhone. The phone, if the FBI could just manage to unlock it, potentially stored critical evidence: Farook's GPS coordinates just before the attack, or any last contacts with associates or other home-grown ISIS terrorists planning their next attacks. Farook had not backed up his data to the iCloud, which would have otherwise given the FBI a window into the contents of his phone. The only way to know what was on that phone was to force Apple to unlock it. If Apple refused, the government could make the case that the company was providing a terrorist safe haven. The FBI believed it had other arguments in its favor: Farook was dead, and thus had no Fourth Amendment interest. Technically, the phone was not his. It was owned by his employer—the county—which consented to the search.

For six weeks, the Department of Justice tried to convince a judge to force Apple to dial back its security, while Apple stood firm. Comey had personally asked Apple to write new software that would allow the FBI to bypass its new encryption scheme. And though he promised the FBI would use the software judiciously, it was clear that the government was asking for a Clipper Chip anew. Hackers and privacy activists coined a name for what the FBI was asking Apple to do. They called it "Government OS." Not only would it make Apple users less secure, it had the potential to destroy Apple's market share in its biggest market—China—as well as in countless other countries like Germany and Brazil who were calling on Apple to do more to protect their data from American spies.

Capitulating would also set a perilous precedent, guaranteeing similar demands for less secure software and backdoors from Beijing, Moscow, Ankara, Riyadh, Cairo, and every other foreign country with which Apple did business. That list now included most of the world, with the exception of a short and shrinking list of nations—Iran, Syria, North Korea, Cuba—under U.S. trade sanctions. If Apple only complied with American officials but denied foreign officials the same access, the company stood to lose more than a quarter of future revenues to competitors abroad.

And then there was the reputational fallout. As Apple pushed back on the FBI, its brand had taken on the halo of America's wealthiest civil rights organization, the last front in the fight to preserve some measure of privacy in the digital age. If the richest company in the United States couldn't stand up to a democratic government, who stood a chance? In this emerging narrative, Cook had become Mr. Privacy, global human rights crusader. Rolling over was not an option. And Cook had made that perfectly clear to customers in a 1,100-word letter he posted to Apple's site.

"The implications of the government's demands are chilling," Cook wrote. "Ultimately, we fear that this demand would undermine the very freedoms and liberty our government is meant to protect."

Cook summoned my colleague David Sanger and me to a meeting at the Palace Hotel in San Francisco to personally make his case. "Their argument is to put in an exploit for the good guys—it's bizarre—and it's a flawed argument from the get-go," Cook told us. "This is a global market. You should go out and ask governments around the world: if Apple does some of the things the FBI is suggesting—a door, a key, a magic wand that gets them an in—how would you feel about that? The only way to get information—at least currently, the only way we know—would be to write a piece of software that we view as sort of the equivalent of cancer. It makes the world worse. And we don't want to participate in anything that makes the world worse."

Cook was almost emotional. "The most intimate details of your life are on this phone," he said, holding up his iPhone. "Your medical records. Your messages to your spouse. Your whereabouts every hour of the day. That information is yours. And my job is to make sure it remains yours. If you place any value on civil liberties, you don't do this."

Cook had another solid argument. Even if Apple did what the government was asking, even if it wrote a backdoor just for this one case, that backdoor would become a target for every hacker, cybercriminal, terrorist, and nation-state under the sun. How could the U.S. government ever guarantee it could keep Apple's backdoor safe, when it could not even manage to protect its own data? The breach at the Office of Personnel Management was still fresh in Cook's memory. The breach had exposed the very data you would think the government had the most personal incentive to protect: Social Security numbers, fingerprints, medical records, financial histories, home addresses, and sensitive details for every American given a background check for the last fifteen years—which included Comey, and the most senior officials at the Department of Justice and the White House. If they could not even manage to keep their own data safe, how could they ever be expected to safeguard Apple's backdoor?

Back and forth Apple and the Justice Department went, in a courtroom dispute that captivated the country. Everyone from the president to Edward Snowden to the comedian John Oliver weighed in. Polls showed Americans were split but tilting toward Apple, more so every day.

Even Carole Adams, the mother of one of the victims who had died in San Bernardino, offered public support for the company. Nobody deserved to know more about her son's killers, but even she didn't think the FBI's request was worth the risk to privacy.

"I think Apple is definitely within their rights to protect the privacy of all Americans," she told reporters. "This is what makes America great to begin with."

The FBI was losing in the court of public opinion, but the case was still up to a judge, and Apple made it clear it was willing to take its fight all the way to the Supreme Court. But the week before I arrived in Miami, there was a twist. Without warning, the Justice Department dropped its case, informing the judge it had found another way to access Farook's data. It no longer needed Apple's help. Unnamed hackers had approached the FBI with an alternative way in, a hacking method that allowed the government to bypass Apple's encryption to access Farook's iPhone—a zero-day exploit.

And perhaps most miraculously of all, Comey admitted that his agency had paid these cryptic hackers more than his salary for the seven-plus years left in his term. Journalists and hackers did the math: the FBI had just publicly copped to paying hackers $1.3 million for a way to bypass Apple's security. And the FBI claimed it did not know what the underlying flaw was and had no plans to help Apple fix it.

It was the first time in history that the government had openly copped to paying private hackers top dollar to turn over vulnerabilities in widely used technology.

"In an odd way, all the controversy around the litigation stimulated a bit of a marketplace around the world, which didn't exist before then for people trying to break into" the iPhone, Comey told an audience that April, in a twist on the dark truth of the government's marketplace. For two decades, the cyberarms market had operated in the shadows. Now the public had been given its first window into what hackers, intelligence agencies, and a growing Beltway army had known for two decades. The public was apoplectic, but to me it was old news. I'd been tracking the government's cyberarms market for years, from Sulzberger's closet to this Miami Beach soiree where I now stood, a neon-green glow stick around my neck. I still had unanswered questions. Among them: Who the hell did the FBI pay for its million-dollar hack?

There was a very good chance that person was standing right in front of me.

THE CYBERARMS BAZAAR had ground rules. No name tags. Only color-coded wristbands: black for speakers, red for audience, and, apparently, neon-green glow sticks for me. The list of 250 attendees—hackers, spies, and mercenaries—was kept secret. If you didn't know who you were talking to, the conference organizers reminded us, you should not be asking.

I immediately spotted the delegation of NSA hackers. They were easy to recognize: young men in their twenties and thirties with pale skin who stuck mostly to themselves. Getting sloshed on gin and tonics at the bar were their British GCHQ counterparts. The Beltway brigade was here too. A nice guy named Pete, last name unknown, who worked at "a small company outside D.C.," offered to grab me a drink. The big exploit developers were here from Trail of Bits, Exodus Intelligence, and Immunity, the brainchild of Dave Aitel, the former NSA hacker, who put on the conference and happily trained any government rep who would pay in the dark arts of cyber exploitation techniques. There were French, German, Italian, Malaysian, and Finnish "security specialists" from companies that, a former NSA analyst told me, were cutouts, looking to buy exploits for their governments back home. A group of Argentine exploit developers were here too.

I said hello to Nate Fick, a former U.S. Marine reconnaissance captain who served in Iraq and Afghanistan while the rest of his Dartmouth class made a beeline for Wall Street. He was a classics major, not the testosterone-fueled "boo-rah" soldier Hollywood had conditioned us to. His thoughtful bent made him an ideal protagonist for the HBO series *Generation Kill*. Venture capitalists thought it made him the ideal candidate to turn around Endgame, a controversial contractor known as the "Blackwater of hacking."

When Fick took over as CEO in 2012, Endgame was selling the government exploits and offensive tools with unfortunate names. Its leading product, "Bonesaw," mapped out which software U.S. adversaries relied on, dropped down a menu of all the ways to hack it, and exfiltrated it. An actual Civil War bonesaw hung on the wall, signed by the team who developed the software. "There are two ways to learn about your adversary: Read white papers about your adversary or *be* the adversary," Fick told me.

Endgame was helping the U.S. be the adversary, but the seedy elements of the exploit trade never sat well with Fick. Neither did the scalability. "Maybe it works for a Romanian teenager living in his parents' basement, but not for a company with venture backing," he told me.

And so in the years since Fick had taken over, he'd pulled the bonesaw off the wall, weaned Endgame off exploits, reoriented the business around defense and started agitating for cyber norms, which—considering the current audience—made him something of a fish out of water.

I spotted Thomas Lim, the jovial founder of a boutique exploit company in Singapore called COSEINC, that brokered exploits to the list of one hundred—and growing—countries that wanted in on the cyber exploit game, but did not yet have the coding skills or the exploit talent of Five Eyes, Russia, China, or the Israelis.

The Israelis were out in full force. One Israeli firm in particular, Cellebrite, which specialized in unlocking encrypted iPhones and Androids, was the leading suspect for the FBI's iPhone jailbreak. Cellebrite had timed a strangely public product announcement about its iPhone cracking software for the week the FBI revealed that someone had helped it crack Farook's iPhone. An Israeli newspaper surfaced a $15,278.02 FBI contract with Cellebrite, dated the same day the FBI informed the judge about their new access. The media flooded Cellebrite's phones and Twitter feeds to confirm reports. And Cellebrite's Israeli hackers were now tweeting out "no comment" but with winking emoji faces. As far as I could tell, the rumors that Cellebrite was the FBI's accomplice had started with Cellebrite itself, a convenient marketing ploy to market its latest phone-cracking service. On background, government reps emphatically denied that Cellebrite was their guy. It was hard to know what to make of all this, but $15,278.02 was a long way from $1.3 million.

And so for the next two days I loitered around hackers' jiujitsu tournaments and the Fontainebleau. I watched hackers "pwn" Apple and Java software by day, then explain over cocktails how to reverse-engineer Tinder to geolocate NSA spies from the Fort Meade parking lot. I asked anyone who didn't shoo me away if they had any intel they could share on the FBI's iPhone hacker. I got nowhere. Even if they were standing right next to me, nobody was going to violate their $1.3 million NDA. That *fucking salmon.*

The cyberarms market was an incoherent mess. Hackers all over the world were casually selling the tools of digital espionage and war to countries who were using them on their own people—and soon, if not already, on us. Encryption had only added another hurdle and goosed the competition. The market was spreading to far corners of the globe. The prices for zero-days were only going up. And so were the stakes. Nobody would talk about it. Or consider what it meant for our defense. There were no norms—none that anyone could articulate, anyway. And in that void we were establishing our own norms, ones I knew we would not want to live by when eventually—inevitably—our adversaries turned them back on us.

I never did find the FBI's iPhone hacker over the course of those three days in Miami.

Sometimes the most intriguing bits of string take a more serendipitous path. Two months after I left Miami, I found myself at a packed bar at the Waverly Inn in New York. I was there to celebrate a friend who had just wrapped a book about the dark web. Packed in next to me were literary agents, big-name media types, and friends of one of the FBI agents featured in the book. I introduced myself to the feds, and over Negronis, we got to talking. I told them I was still on the hunt for their iPhone hacker.

"Oh, he's long gone," one agent told me. "He quit his job and has been hiking the Appalachian Trail."

The Fed wouldn't give me his name. But he did tell me this: he was not Israeli. And he'd never worked for Cellebrite. He was just another American hacker moonlighting as a mercenary. The whole time I'd been hunting for him in Miami, he was off the grid, walking a lonely narrow dirt path somewhere between Georgia and Maine.

PART VI

The Twister

The release of atom power has changed everything except our way of thinking . . . the solution to this problem lies in the heart of mankind. If only I had known, I should have become a watchmaker.

—ALBERT EINSTEIN

CHAPTER 17

Cyber Gauchos

Buenos Aires, Argentina

O UR CAB CHARGED through a red light, knocking the bumper off another car. I braced for a full stop and assumed we'd check on the other driver, make sure he was okay. But our driver didn't flinch. Instead, he slammed on the gas, narrowly avoiding another car and a pothole the size of a small mule as we made our way through the Buenos Aires morning rush.

I was gripped. My companion took one look at me and chuckled.

"That's why we have so many hackers in Argentina," Cesar Cerrudo said. "To get ahead, you have to work the system. Look!"

He pointed to half a dozen other cars that were all kinds of smashed and dented, their bumpers barely holding on with duct tape or wire. These were Roman drivers with a vengeance. Both Cesar and our driver were laughing at me now.

"*Atado con alambre!*" the driver chimed in.

It was the first of many times I would hear those three little words—*atado con alambre*—over the next week. It was Argentine slang for "held together with wire" and encompassed the MacGyver-like nature of so many here who managed to get ahead with so little. It was Argentina's hacker mantra.

For years, I'd heard some of the best exploits on the market hailed from Argentina. I'd run into Argentines in Miami, Vegas, and Vancouver. But it was still a hard notion to wrap your head around. And so, in late 2015, I went south to meet the Southern Hemisphere's exploit developers and to see how the world was changing.

Argentina's technology scene struck me as archaic, especially compared to Silicon Valley standards. The country's "embargo on cool shit"—as hackers put it—meant high-def televisions cost twice as much and arrived six months late. Amazon still did not ship door-to-door here. BlackBerry—or what was left of it—still had more market share than Apple. To get an iPhone, Argentines had to shell out $2,000 or more on underground auction sites.

But those setbacks, Cesar and others told me, were precisely why Argentina was such fertile ground for zero-day hunters and the brokers who now traveled long distances—from Saudi Arabia, the UAE, Iran—to buy their code. Argentines had access to free, high-quality technical education, and one of the highest literacy levels on the continent. But when it came to accessing the fruits of the modern digital economy, there were roadblocks. If Argentines wanted something that normal business channels didn't provide, they had to hack it. To access the video games and other apps we took for granted in the United States, they had to reverse-engineer a system before they could find all the ways around it.

"Cheating the system is part of the Argentine mentality," Cesar told me. "Unless you're rich, you grow up without a computer. To access new software, you have to teach yourself everything from the ground up."

It struck me, as he spoke, that the reverse was now true in the United States. Engineers who coded Silicon Valley's apps and services no longer needed to reverse-engineer a system down to its kernel, or venture far down the stack, to the metal. Increasingly, they were just skimming the surface, and in the process losing the depth of understanding required to find and develop the best zero-day exploits.

It was starting to show. Every year small teams of college students from over a hundred countries convene at the International Collegiate Programming Contest (ICPC), the oldest and most prestigious contest of its kind. Two decades ago, American teams from Berkeley, Harvard, and MIT dominated the top ten finalists. These days the winners were Russian, Polish, Chinese, South Korean, and Taiwanese. In 2019 a team from Iran beat Harvard, Stanford, and Princeton, which didn't even break into the top twenty.

America's pool of cyber talent was shrinking. U.S. intel agencies had taken a big hit in morale after Snowden and NSA analysts were leaving in droves. Some compared it to an "epidemic." Meanwhile, the most capable college graduates weren't taking jobs at the NSA anymore, especially when they could make far more money working at Google, Apple, or Facebook. It was even harder to recruit defenders to the Department of Homeland Security. It had always been more fun to be a pirate than to join the Coast Guard, and this was putting the United States at a growing disadvantage. The United States doesn't forcibly conscript talented hackers at Google and MIT to moonlight as nation-state attackers like the Russians, Iranians, North Koreans, and Chinese do. We might still be the world's most sophisticated cyber power in offense, but the scales were tipping. When I asked Gosler his thoughts on the zero-day market in one of our first conversations, he told me he didn't think it was necessary. But that was when the agencies could count on people like him to join its ranks. With the U.S. labor pool going elsewhere, the agencies were forced to purchase from outsiders exploits that they had once developed in-house.

"This is the new labor market," Cesar told me. "The new generation of young Argentine hackers have far more options than we ever did."

Cesar bore an uncanny resemblance to the actor Jason Segel from the film *Forgetting Sarah Marshall*. He was Segel's Argentine doppelgänger, and I was convinced they were twins separated at birth—one an actor famous for a nude scene in a romantic comedy, the other a small-town Argentine hacker who held the keys to the world's critical infrastructure. I pulled up a photo of Segel on my phone and held it next to Cesar's face. Our cabbie nodded in agreement. Cesar refused to acknowledge his twin.

Segel's long-lost Argentine twin came to my attention through iDefense. Fifteen years earlier, Cesar had been vying with Greg McManus, the Kiwi, for the top spot in iDefense's bounty program. Back then he was a ponytailed teenager who made $50,000 one year sending iDefense zero-days from Paraná, a small river town in northeast Argentina. Argentina's economy was collapsing, and Cesar's zero-days were making him rich. But these days he had a wife and kids, and like other Argentine hackers of his generation, he'd given up the roller-coaster ride of bug hunting and exploit development for a salaried position at an American security firm.

But he never stopped the chase. A year earlier he'd caught the world's attention with a hack straight out of *Die Hard*. He'd flown to D.C., strolled over to Capitol Hill, and pulled out his laptop. With a few points and clicks, he began turning red traffic lights green and green traffic lights red. He could have gridlocked the Capitol if he'd wanted to—but he did it just to prove he could. The company that designs traffic light sensors didn't think Cesar's little zero-day was a problem. So he'd also taken it to Manhattan and San Francisco to prove that all were vulnerable to a cyber traffic apocalypse.

I'd written about Cesar's stunts for the *Times*. You would think lawmakers would do something, but they hardly blinked. Now here we were, Cesar and I, navigating potholes in Buenos Aires, where traffic lights were apparently optional. *But at least they were offline!* So-called smart cities were dumb. Dumb cities were smart. The whole system was backward and inside out.

We made our way through Palermo, Buenos Aires' hippest boutique-and-restaurant zone. American dollars went a long way here. The country's official exchange rate was a complete fiction. The "blue dollar"—the unofficial exchange rate—was nearly twice the official rate of 9.5 pesos per dollar. Cristina Fernández de Kirchner, Argentina's soon-to-be-deposed president, refused to correct the situation. The *porteños*—as locals referred to themselves—likened Kirchner to a "female Gaddafi," and she much preferred a veneer of lies to reality. It showed in her face. She'd had more plastic surgery work than anyone in recent memory, save for Michael Jackson. To adjust exchange rates to suit Argentina's current reality would be, for her, to admit to Argentina's chronic inflation.

Argentina's hackers had been largely insulated from the financial crisis, selling exploits under the table for American dollars and putting their money toward sleek modern apartments for $1,000 a month in Palermo. For another $1,500 a month, they could rent a second home with an infinity pool thirty minutes from downtown Buenos Aires.

I asked Cesar why Argentine hackers turned to the underground exploit market when their counterparts in Brazil were making so much money in cybercrime. Brazil was gradually assuming Eastern Europe's mantle as the world leader in internet fraud. Brazil's banks were bleeding $8 billion a year to cybercriminals, much of it from its own citizens.

That was easy, Cesar told me. In Argentina, nobody, not even hackers, bothered with banks. From the moment I'd arrived in Buenos Aires, the *porteños* told me to avoid them altogether. They pointed me to various *cuevas*, illegal exchange houses, instead. After years of economic collapse and government freezes on withdrawals, Argentinians had lost their trust in banks. Online and mobile banking were still virtually unheard of, which meant there was less to be gained by hacking them. Instead, Argentina was now the "India of exploit development," the *porteño* hackers told me.

WE ARRIVED IN one piece at an old open-air oil factory on the outskirts of town. More than a thousand young Argentine hackers were lined up around the block. Some looked as young as thirteen, like teenagers at the skate park. Interspersed among them were foreigners—some Asian, some European or American, several Middle Easterners. They were here to recruit, perhaps, or broker the latest and greatest in Argentine spy code.

I had timed my visit for Latin America's largest hacking conference. Ekoparty was a mecca for hackers all over South America, and more recently zero-day brokers who came from all over the world in search of digital blood diamonds. This was my best chance of glimpsing the world's new exploit labor market. The agenda listed hacks of encrypted medical devices to e-voting systems, cars, app stores, Androids, PCs, and the Cisco and SAP business apps that could enable attackers to take remote control of computers at the world's biggest multinationals and government agencies.

Ekoparty was still dwarfed by Def Con, Black Hat, and RSA, but what it lacked in numbers and glitz, it made up for in raw creative talent. Absent were the booth babes and snake-oil salesmen that had overrun the big hacking conferences in the States. The focus here was strictly breaking-and-entering. It was an opportunity for Argentines to demo their skills on a world stage.

I saw foreign representatives from Deloitte and Ernst & Young. Avast, the Czech antivirus giant, was here. So was Synack. They were all here to recruit. I couldn't help but notice that Ekoparty's "Platinum Sponsor" was Zerodium. *Go figure.* Chaoukri Bekrar was a big topic of conversation. He'd tweeted that Zerodium had just acquired an iPhone jailbreak for $1 million.

I spotted Federico "Fede" Kirschbaum, who had cofounded Ekoparty more than a decade before. In those early days, nobody was selling exploits to governments yet; Ekoparty was mostly for fun and hijinks. It still was, but there were vast sums of money to be made on the side.

"Throw a stone," Fede said, motioning to the hundreds of Argentine hackers standing within five feet of us. "You'll hit someone selling exploits."

Once well contained within the Beltway, American mercenaries were now in Abu Dhabi, and exploits could be acquired in the barrios of Argentina. Things were getting out of control fast.

I WAS REMINDED of a Black Hat keynote address one year earlier, delivered by a CIA insider named Dan Geer. Geer was chief information security officer at In-Q-Tel, the CIA's investment arm, and a legend in the industry. He was no stranger to the zero-day market, and he'd used his turn at the lectern to encourage the U.S. government to corner the market by outbidding every other foreign buyer. Uncle Sam, Geer said, should say, "Show us a competing bid, and we'll give you 10 times" as much. That would provide the United States an opportunity to use those zero-days, then turn them over to vendors for patching, and in the process burn our enemies' stockpiles. It would keep the market lucrative, without inevitably leading to destruction. His suggestion was a provocative one, but as I stood here in this old oil factory in Buenos Aires, Geer's logic fell apart. It was too late: the United States had lost its grip on the market years ago.

Over the next few days I watched a famous Argentine hacker, Juliano Rizzo, demo a zero-day onstage that would have easily earned him six figures on the government market.

Argentine hackers demoed several hair-raising exploits that could control a car, or the grid. Foreigners—who I would later learn to be brokers—swarmed the men after their talks. I didn't understand why the brokers were approaching them *after* they'd dropped their best exploits onstage. Weren't they useless now?

"They're after whatever those researchers do next," Fede told me. "To build relationships and acquire their zero-days and weapons for a rainy day."

Foreign governments were hungrier for exploits than ever before. Stuxnet had shown what was possible. Then Snowden had given every nation a blueprint for what a truly sophisticated offensive cyber program looks like. And once Apple and Google started encrypting every last bit of their iPhones and Androids, governments had more incentive than ever to acquire the tools that could still get them inside.

The United States still had the biggest offensive cyber budgets, but compared to conventional weapons, exploits were cheap. Foreign governments were now willing to match American prices for the best zero-days and cyberweaponry. The Middle East's oil-rich monarchies would pay just about anything to monitor their critics. And in Iran and North Korea, which could never match the United States in conventional warfare, leaders saw cyber as their last hope of leveling the playing field. If the NSOs, Zerodiums, and Hacking Teams of the world wouldn't sell them their wares, well, they could just hop on a plane to Buenos Aires.

IF I REALLY wanted to know what Argentine hackers were capable of, Fede told me, I would need to visit the Cyber Gaucho, a hacker now in his forties by the name of Alfredo Ortega who had grown up in the far reaches of Patagonia. So on my third day, I set out for the Gaucho's hacking workshop of wonders, where telescopes, hacked microprocessors, and X-ray devices that could cross into nuclear power plants littered the floor.

"Almost anything you give him," Fede told me, "he will break."

The Gaucho offered me tea and cookies. "Patagonia was cold," he began. "So I never went outside."

Like so many other hackers of his generation, he'd started hacking on a Commodore 64. He'd hacked away until he could access games he couldn't otherwise get, and he joined primitive hacking forums to learn how to hack for more. It was there he met one of the godfathers of Argentina's hacking scene, a graybeard named Gerardo Richarte, or Gera. I had heard Gera's name before: he was a legend not just in Argentina but around the globe. Eren had credited Gera with helping him plan his digital Kurdish resistance movement in Turkey as a co-ed.

Two decades ago, Gera and four others started Core Security, a pen-testing business. It was impossible to divorce the history of hacking in Argentina from Core. Among their first customers were Brazilian and American banks and consultancies like Ernst & Young. The company did so well in its first few years that it set up an office in New York. The date was September 6, 2001. Five days later came 9/11, and $1 million worth of Core contracts vanished in smoke. In Argentina, the economy was imploding. Thousands of angry and impoverished *porteños* took to the streets to protest the government's handling of the crisis. They smashed windows at banks and stormed the Casa Rosada, the presidential palace. Dozens died in the protest, and Argentina's president was forced to quit.

Core's founders knew that to stay afloat, they were going to have to come up with something more compelling than just scanning banks for unpatched software. They came up with an automated attack tool, Implant, that penetrated customers' networks with exploits—some known, but many more that they discovered themselves. Analysts initially slammed the tool as unethical and dangerous. But one of Impact's first takers, NASA, helped change the industry's minds.

Core began recruiting exploit developers to write new exploits for their Implant tool, and trained up Argentine hackers like Juliano Rizzo. Gera had personally recruited the Gaucho to come to Buenos Aires and join them at Core—a big step up from where the Gaucho thought he'd be, running his dad's gas stations in Patagonia. At Core, the Gaucho specialized in hacking hardware and, increasingly "firmware," the first layer of software that touches the metal. "I became something of a firmware specialist," he told me.

Twenty-odd years later, there was not a machine the Gaucho could not defeat. Two weeks ahead of Argentina's upcoming presidential elections, the Gaucho and the Ekoparty guys had gotten their hands on Argentina's new voting machines. He defeated the whole system in less than twenty minutes. A police raid followed. Now he and others were working with lawmakers to secure the machines ahead of a critical vote.

The Gaucho gave me a little tour of his studio. As we passed the telescope, he told me he'd once hacked a satellite. In another corner was his work in progress: a Rube Goldberg–esque X-ray–emitting device that could cross air gaps and break into offline systems, like Natanz. I asked how he'd figured out how to hack the world's most protected networks.

"It's easy," he told me. "They never anticipated they would be attacked."

Chip manufacturers hired the Gaucho to make sure that their chips were secure. He'd discovered all sorts of ways one could hack chips to get into the global supply chain. He showed me how to hack a chip with a "side channel attack," sending malware via radio emissions to the copper in the chip itself. There were at least ten of these chips now in every device.

"It's hard to find one that's been compromised," he told me—albeit not impossible.

The Gaucho had come across other hackers' handiwork before. A major appliance maker—he wouldn't tell me who—had hired him to investigate its appliances. Sure enough, he confirmed that someone had compromised its firmware in the most sophisticated supply-chain attack he had ever seen, the kind Gosler told me only Tier I nation-states were capable of. "This attack wasn't the work of cybercriminals—you don't want to mess with those guys," the Gaucho said. "This was a nation-state."

That was about all he could share. But it seemed like a logical opening for my next by-now-standard question: "Have you ever sold exploits to brokers or governments?" The Gaucho was so sweet-natured that it was hard to imagine him selling exploits to Iranians in dark alleys.

"No," he said. But it was not out of some moral calculus. "I have no problem with spying," he told me. "I don't think spying is bad."

"Then why not?"

"It's no way to live," he told me. "It's not worth losing your freedom. It's like being a physicist in the 1930s with the atomic bomb. You'd get killed."

I THOUGHT ABOUT what the Gaucho said as I walked back to my hotel that day. Hackers weren't hobbyists anymore. They weren't playing a game. In short order, they had become the world's new nuclear scientists—only nuclear deterrence theory did not so neatly apply. Cyberweapons didn't require fissile material. The barrier to entry was so much lower; the potential for escalation so much swifter. Our own stockpile of cyber exploits and cyberweapons hardly deterred our adversaries from trying to acquire their own. What Iran, North Korea, and others could not develop on their own, they could now just buy off the market. The Gaucho might not sell them a way in, but there were many others here that would.

IF I HAD been ten minutes later, I might have missed them. Turning a corner, I'd stumbled on a protest march. I would learn later that this was an every-Thursday event. Argentina's grieving mothers, dressed in white head scarves, filled the city's oldest square, the Plaza de Mayo, holding up signs bearing the names of their missing children. I'd struck on their march at 3:30, just as it began.

The mothers were so old and frail now, they only managed a few loops around the obelisk before taking a seat in their chairs. One of the mothers stopped to address the crowd of onlookers. You could hear that her grief was still sharp. It was so easy to forget, walking around old-town Buenos Aires, with its hip bars and cafés, that not so long ago Argentina's military turned on its own people. Over the course of Argentina's "Dirty War" in the late seventies and early eighties, the military junta "disappeared" some thirty thousand people. Left-wing activists were accused of terrorism, tortured, raped, kidnapped, and machine-gunned to death at the edge of enormous pits, or drugged and tossed naked and half-conscious from military planes into the Río de la Plata river. So long as the *desaparecidos* were never discovered, the government could pretend they didn't exist. Forty years later, Argentina had yet to make any serious effort to identify or document its victims. This was the era in which older Argentine hackers like Gaucho and Gera had come of age.

No wonder they weren't keen to turn their digital exploits over to governments. Younger generations of Argentines were a different breed. They hadn't lived through the era of the *desaparecidos*. And with so much money to be made, there was little to restrain them from diving headfirst into the market.

THAT EVENING, FEDE and his crew invited me to dinner, but I demurred. I wanted to check out another part of town. I tucked my laptop in the hotel safe, slipped on a dress, squeezed my sore feet into heels, and hailed a cab across town to the old Puerto Madero harbor.

I'd arrived just in time to catch sunset, and I strolled along the banks of the Rio de la Plata. The streets were each named for famous women in Argentine history. Olga Cossettini. Carola Lorenzini. Juana Manso. Alicia Moreau de Justo. I crossed over the Puente de la Mujer, "the Woman's Bridge," designed by the Valencian architect Santiago Calatrava to look like a couple in the throes of the tango. It was a welcome respite from the testosterone of Ekoparty.

As I walked past elegantly dressed Argentine women, I realized it had been days since I'd spoken to a female. I wondered if maybe women were the key to keeping this market from spinning out of control. They had started wars before; perhaps they could preempt them. I was reminded of something a female hacker friend had told me late one night when I was still getting my legs under me on this beat. She told me the only way to get a hacker to stop hacking was to marry him off. *If only it were that simple.*

As I walked back across the Woman's Bridge, it was getting dark. Buenos Aires twinkled in the distance; I could see why people called this the Paris of the South. I walked up a promenade and into an airy steakhouse on the water, ordered a sturdy malbec and *asado*, and savored every bite.

By the time I got back to my hotel that evening, I was looking forward to clean sheets and a good night's sleep. I caught a glimpse of myself in the elevator mirror. My eyes were sunken; I was still adjusting to the jet lag. When I got to my room, the door was ajar. I wondered if maybe I'd left it that way in haste. Maybe a maid was still doing turn-down service? I walked inside, and no one was there. Everything was just how I had left it, except the safe that had held my laptop. It was wide open. My computer was still inside, but in a different position. I checked for any trace of an intruder in the bathroom, the closets, the balcony. Nothing. Everything else was untouched. My passport, even the cash I'd exchanged at the *cueva*. I wondered if this was some kind of warning shot. Or if I'd tripped some kind of wire.

I took a sober look at the laptop. It was a loaner. I'd left my real computer at home and stuck to pen and paper at the conference. There'd been nothing on the laptop when I'd left; I wondered what was on it now. I wrapped it in an empty garbage bag, took the elevator back down to the lobby, and threw it in the trash.

I SAVED MY last day in Buenos Aires to meet with Ivan Arce. Arce, like Gera Richarte, was one of the godfathers of Argentina's hacking scene. He was one of the five who'd started Core Security twenty years before.

Arce lacked Gera's stately gray beard and geniality—he could get carried away on Twitter—but he was as passionate about Argentina's hacking history as anyone I had met. "Our generation is responsible for where we are," he said. "We were sharing exploits as a game. Now the next generation is hoarding them for profit."

Arce, Gera, and the others had trained up the next generation of Argentine hackers. But Arce told me these men had more of a millennial mentality. They weren't company men. And there was more money to be made selling exploits on the underground market than baking them into Core's exploitation tool.

"They might stay at Core for two years, and then they're out," Arce told me. "The newer generation is driven by instant gratification. There's no more loyalty. These are the same guys who are selling exploits to foreign governments."

What the younger generation was now doing clearly weighed on Arce. "Their calculus is they can make more money and not pay taxes," he told me. "There's this James Bond element to selling exploits to spies. Before they know it, they've discovered this luxurious lifestyle, and there's no going back."

The young men at the conference may not have wanted to talk to me, but they were willing to talk to their godfather. Over the years, Arce told me he'd had many conversations with the younger generation, encounters that tended to engender a mixture of disdain but also understanding. This was the way of the world post-Stuxnet, Arce told me. Selling exploits to governments were their tickets out of poverty and corporate slogs.

I asked Arce the question I had asked so many before him. But I had been shot down so many times that I lost my step. The words fell out clumsily.

"So will they only sell their exploits to good Western governments?"

He repeated my words back to me. "*Good Western governments?*"

I slumped so low in my seat, I was sure my head wasn't even showing. The words were even more humiliating coming out of an Argentine's mouth. Relations between Argentina and the United States had been strained ever since Argentina refused to declare war on Nazi Germany in World War II and became the only Latin American nation barred from American aid. Relations improved somewhat when Argentina assisted the United States in the Gulf War, but they'd nosedived under Kirchner, who spent most of her time in office fending off American hedge funders who'd been stuck holding worthless Argentine bonds. A New York hedge fund had gone as far as to seize Kirchner's plane, an Argentine naval vessel with 220 crew members onboard, even an Argentine booth at the annual Frankfurt Book Fair. In a rambling televised speech, Kirchner had recently accused the United States of plotting to assassinate her.

"If something should happen to me, don't look to the Middle East, look to the North," she'd said.

Argentines' disdain for the United States had eased somewhat under Obama, but they were still evenly split between those who had favorable opinions of the country and those who thought we were monsters. And you couldn't really blame them. Declassified U.S. diplomatic cables showed that in 1976, Secretary of State Henry Kissinger gave Argentina's military junta the green light to engage in widespread repression, murder, kidnappings, and torture of its citizens. "We want you to succeed," Kissinger told an Argentine Admiral that year. "If there are things that have to be done, you should do them quickly." The episode was still raw for Argentines. To them, America was no democratic savior; it enabled the kidnapping of their children.

"You need to dispose of your view, Nicole," Arce told me. "In Argentina, who is good? Who is bad? The last time I checked, the country that bombed another country into oblivion wasn't China or Iran."

In the Southern Hemisphere, the whole moral calculus was flipped. Down here, the Iranians were allies. We were sponsors of state terrorism.

"Many of these guys are just teenagers," he told me. "A guy from the NSA and a guy from Iran show up with big bags of cash. Do you perform an ethical analysis? Or do you weigh the bags of cash and see which is heavier?"

My hope that there was an ethical calculus at play in this market had always been naïve. It might still be true of Americans like Desautels and former TAO operators, even if some of their allegiance to the United States was dwindling. But clearly that was not the case elsewhere.

"Not all arms dealers are ethical," he said. "In the end it all comes down to the biggest budget. Right now, that may be the NSA.

"The question is, how long can the NSA keep their weapons a secret?"

THAT NIGHT, I indulged in one final *asado* feast before making my way to an underground club in a massive industrial building in Palermo. It was the closing night of the conference, the last chance for hackers to let loose and close deals. I walked through velvet ropes, past pretty young things in miniskirts and fishnet stockings and expat banker types who'd taken their year-end bonuses to Buenos Aires. I made my way through faint green strobe lights and smoke to the VIP section to find the hackers upstairs. The deejay was remixing Talking Heads. *Home is where I want to be. Pick me up and turn me round. I feel numb, born with a weak heart. I guess I must be having fun.*

The VIP hackers were entranced by willowy young hostesses skilled at bottle service and small talk. As the booze kicked in, some danced. Some broke into limbo. Someone offered me a Red Bull and vodka. I turned off my taste buds and took a big glug. *The less we say about it the better. Make it up as we go along.* The images and conversations of the week had started to blend together, forming a single voice and image. *The new nuclear physicists.* I wondered who was going home with their code. *You may ask yourself, "Am I right, am I wrong?" You may say to yourself, "My God! What have I done?"*

I ditched the Red Bull and vodka and went to the bar to get myself a real cocktail. Through the fog, I could just make out a private conversation between two familiar faces in the corner. I recognized them from Ekoparty, two men who had studiously avoided me for much of the conference, one a buttoned-up older foreigner—a Middle Easterner, but from where? Saudi Arabia? Qatar? Certainly not American—the other an Argentine hacker in his early thirties.

With David Byrne blaring, I could not make out what they were saying. But it was clear from their body language that they were engaged in serious business. It was the intimate conversation of exploits being brokered and sold. The market America had spawned was now far beyond its control. I wondered if it would ever be contained. I wished I knew which code the hacker's keys would unlock this time, and against whom. As I took one final glug of my drink, the older gentleman, the foreigner, caught my stare. He motioned the hacker deeper into the shadows, where the fog was so dense I lost sight of them.

I walked myself out. Outside, the night was just getting started. I walked past the tipsy *porteños* and expats waiting in line. I hailed a cab. I rolled down the window. As we pulled away from the curb, I could just make out the verse. *Here comes the twister.*

CHAPTER 18

Perfect Storm

Dhahran, Saudi Arabia

I N RETROSPECT IT was the burning American flag that touched it off—the decade when every single one of America's cyber adversaries came for us.

Three years after the United States and the Israelis reached across Iran's borders and destroyed its centrifuges, Iran launched a retaliatory attack, the most destructive cyberattack the world had seen to date. On August 15, 2012, Iranian hackers hit Saudi Aramco, the world's richest oil company—a company worth more than five Apples on paper—with malware that demolished thirty thousand of its computers, wiped its data, and replaced it all with the image of the burning American flag. All the money in the world had not kept Iranian hackers from getting into Aramco's systems. Iran's hackers had waited until the eve of Islam's holiest night of the year—"The Night of Power," when Saudis were home celebrating the revelation of the Koran to the Prophet Muhammad, to flip a kill switch and detonate malware that not only destroyed Aramco's computers, data, and access to email and internet but upended the global market for hard drives. It could have been worse. As investigators from CrowdStrike, McAfee, Aramco, and others pored through the Iranians' crumbs, they discovered that the hackers had tried to cross the Rubicon between Aramco's business systems and its production systems. In that sense, they failed.

"The main target in this attack was to stop the flow of oil and gas to local and international markets and—thank God—they were not able to achieve their goals," Abdullah al-Saadan, Aramco's vice president, told Saudis' Al Ekhbariya television.

For years the generals, the suits, the spies, and the hackers had warned me that we would see a cyberattack with kinetic consequences. We just never expected Iran to get there so fast.

The United States had grossly underestimated its enemy and how quickly the Iranians would learn from our own code. The malware the Iranians used to hit Aramco was not even that sophisticated; it was essentially plagiarized from the code Americans and Israelis had used to infect and delete data on Iran's oil networks four months earlier. But the malware—called Shamoon after a word left in the code—did exactly what it needed to do: it sent Iran's chief regional rival, the Saudis, into a tailspin, and signaled to Washington that Iran now posed a formidable cyber threat of its own, that one day soon it would come for us.

"We were astounded that Iran could develop that kind of sophisticated virus," Leon Panetta, then secretary of defense, told me later. "What it told us was that they were much further along in their capability than we gave them credit for. We were dealing with a virus that could have just as easily been used against our own infrastructure. It was a weapon that could create as much havoc and destruction as 9/11 or Pearl Harbor."

While Tehran could never hope to match America in conventional weapons or military spending, Olympic Games had shown Tehran that cyberweapons had just as much potential to exact destruction. While the United States was still the top player in offense, it was woefully behind in locking up its own systems, and only becoming more vulnerable by the day. American data breaches had surged 60 percent year over year, and were now so commonplace that most barely registered as more than a blip on the eleven o'clock news. Half of Americans had to have their credit cards replaced at least once because of internet fraud, including President Obama. Breaches had hit the White House, the State Department, the top federal intelligence agencies, the largest American bank, its top hospital operator, energy companies, retailers, even the Postal Service. By the time anyone noticed, their most sensitive government secrets, trade secrets, employee and customer data, had already left the building. And Americans were now plugging in more power plants, trains, planes, air traffic control, banks, trading floors, oil pipelines, dams, buildings, hospitals, homes, and cars to the internet, oblivious to the fact that all those sensors and access points made for a soft underbelly. And lobbyists made sure U.S. regulators didn't do a damn thing about it.

Just two weeks before Iran hit Aramco, the first major congressional effort to lock down U.S. critical infrastructure failed. The bill—which started with real teeth—would have set strict cybersecurity standards for the companies that oversee America's critical infrastructure. It all seemed very promising. In closed-door briefings in a SCIF inside the Capitol, several senior administration officials—Janet Napolitano, the secretary of homeland security; Robert Mueller, then FBI director; General Martin Dempsey, chairman of the Joint Chiefs of Staff; and Mike McConnell, the director of national intelligence—tried to persuade senators that the cyber threat to the nation's critical infrastructure was dire. "For the record, if we were attacked, we would lose," McConnell told the senators. The government needed the private sector's help.

"Most of the infrastructure is in private hands," Michael Chertoff, the former secretary of homeland security, told me at the time. "The government is not going to be able to manage this like the air traffic control system. We're going to have to enlist a large number of independent actors." If that meant using regulation to force private utilities, pipeline operators, and water treatment plants to beef up their security, so be it, Chertoff said. He'd overseen the government's response to Hurricane Katrina and could see that the United States now faced a cyber threat on U.S. infrastructure that would be just as dire—if not worse.

"We are in a race against time," he told me.

But then came the lobbyists at the U.S. Chamber of Commerce—itself the target of the vicious Chinese hack that had made its way into the Chamber's thermostats and printers the previous year. The Chamber's lobbyists yelled *overregulation, big government*, etc., and soon the standards were watered down, until they were voluntary. And somehow even the voluntary standards were deemed too burdensome by the Republican senators who had filibustered and ultimately voted nay.If we couldn't get it together to agree on voluntary standards, you had to wonder if the United States even stood a chance in this new battleground. Six thousand miles away, the Ayatollahs smelled blood in the water.

Stuxnet had hit what Iran valued most—its nuclear program—and Iran soon learned it could use cyber means to hit the United States where it would hurt most: American access to cheap oil, the economy, and our own sense of safety and military superiority. Once Stuxnet had been discovered and unfurled, it became Tehran's rallying cry and the single greatest recruiting tool the Ayatollahs could have ever hoped for. The barrier to entry was so low that for the cost of three F-35 stealth bombers, Iran's Islamic Revolutionary Guard Corps built a world-class cyber army.

Before Stuxnet, the IRGC reportedly budgeted $76 million a year for its fledgling cyber force. After Stuxnet, Iran poured $1 billion into new cyber technologies, infrastructure, and expertise, and began enlisting, and conscripting, Iran's best hackers into its new digital army. Stuxnet had set Tehran's nuclear programs back years and helped stave off the Israeli bombs. But four short years later, Iran had not only recovered its uranium but also installed eighteen thousand centrifuges—more than three times the number spinning at the time of the first attack. And now Tehran claimed to have the "fourth-biggest cyber army in the world."

IRAN CAUGHT US completely off guard. The United States was already struggling to track, let alone fend off, thousands of Chinese cyberattacks. Aurora was just the tip of the iceberg. "Legion Yankee" was but one of more than two dozen Chinese hacking groups and contractors frenetically hacking into U.S. government agencies, businesses, universities, and labs. They were making off with trillions of dollars' worth of U.S. intellectual property, nuclear propulsion blueprints, and weaponry, costing the United States as many as a million jobs a year. The Obama administration dispatched delegation after delegation to Beijing to confront its Chinese counterparts. At these meetings, the Chinese would listen, deny everything, point out that they, too, were a victim of cyberattacks, and then go right back to hacking.

After Google, the *New York Times* was the first company to call the Chinese out directly for an attack on its systems. Just before I went to print with my story, my editors had done a gut check. Did we really want to publicize our attack? What would our competitors say? "Nothing," I told them. "They've all been hacked too." Sure enough, within hours of going to print, the *Post* and the *Journal* eagerly admitted that China had hacked them too. You weren't a credible news organization if you *hadn't* been hacked by China. The story opened the floodgates. Never was I prouder to say I worked at the *Times*. For years, victims had treated cyberattacks as dirty secrets better kept from customers, shareholders, and competitors, lest the disclosure sink their stock price or threaten lucrative business opportunities in China. Now victims were finally edging into the light.

Meanwhile, Chinese officials continued to deny any role. They called my story baseless and demanded "solid proof" to back it up. And so we supplied that proof. Two weeks later, my colleagues David Barboza, David Sanger, and I took the attacks right up to the PLA's doorstep. Together with Mandiant, we pinpointed the exact twelve-story white military tower in Shanghai where members of the PLA's Unit 61398 were staging thousands of attacks on American businesses, including Coca-Cola, the security company RSA, and Lockheed Martin. We could track individual hackers at specific IP locations and, in some cases, even watch what was happening on their screens. The only limitation was that we could not get inside the PLA's building.

"Either they are coming from inside Unit 61398," Kevin Mandia told us, "or the people who run the most controlled, most monitored internet networks in the world are clueless about thousands of people generating attacks from this one neighborhood."

The stories had only emboldened the White House. Five days after my story outed China's attack on the *Times*, President Obama delivered his State of the Union, in which he castigated "foreign countries and companies that swipe our corporate secrets." It was the first sign that the White House was dropping its quiet diplomacy on cyber theft and putting on the boxing gloves.

At the Justice Department, officials began to put together a legal case against the Chinese. And one year later, federal prosecutors unsealed their indictment against five members of PLA 61398, adding the men to the FBI's Most Wanted list. But it was all symbolic, given that there was zero chance the Chinese would ever turn over its soldiers. For a few weeks the Chinese hacking group abandoned their attack tools and went dark. But any optimism the White House may have had quickly fell away weeks later, when the group's tools were picked up in a slate of new attacks on U.S. targets and a new PLA hacking unit emerged, picking up where its colleagues had left off.

If the White House could not even stop a rational actor like the Chinese, from breaking into its systems, how could it possibly expect to contain an irrational actor like Iran? Nobody had a good answer when the Iranians came for our banks.

WHAT OIL IS to the Saudis, so finance is to the American economy. A little more than a month after the Aramco attacks, Iranian hackers put U.S. banks in their crosshairs. Executives at Bank of America, J.P. Morgan, Citigroup, Fifth Third Bank, Capital One, and the New York Stock Exchange could only watch helplessly as, one by one, their banking sites crumbled or were forced offline by a deluge of Iranian internet traffic.

The attacks were "denial-of-service attacks"—attacks that overwhelm a website with coordinated computer requests from thousands of computers until they collapse under the load. But there was something disturbingly different about these attacks. This was a new breed of weapon. Instead of exploiting individual computers to do their bidding, the attackers had hijacked computers at data centers around the world, transforming the online equivalent of a few yapping Chihuahuas into a pack of fire-breathing Godzillas. The flood of traffic was so powerful that no security defense was adequate to stop it. Banks that had forty gigabits of internet capacity—an incredible amount, when you stop to consider that most businesses may have only one gigabit—were brought down by a sustained fire hose of seventy gigabits of traffic, multiple times the amount of traffic that Russian hackers had fired at Estonia in 2007 in a months-long assault that nearly crippled the Baltic nation. There had never been this number of financial institutions under this much duress. For months Iranian hackers hit American banks, one after the one, in a sustained and increasingly powerful series of attacks that crippled some four dozen banks in all and marked the longest-running cyberattack in the history of the internet.

When Obama invited Wall Street executives to Washington for emergency briefings, the executives left shaking their heads. The administration had made it clear that Iran was to blame. But when it came time for solutions, officials were as hapless as the executives. The attacks had exposed the limits of America's cyber defense. The Department of Homeland Security, the very agency charged with protecting critical infrastructure—which includes the financial system—was toothless. DHS could only advise private companies of the risks to their systems and assist in the event of a breach. But in terms of covering the millions of dollars in losses and remediation costs, that would be left to the victims themselves. This is what the new era of asymmetrical cyberwarfare looked like. The United States could strike a country's critical infrastructure with cyberattacks, but when foreigners retaliated, U.S. businesses would be left holding the bag. The United States had no coherent response to the escalating nation-state cyberattacks on its systems.

"If we are going to be aggressive about using our cyberweapons against these adversaries," Panetta told me, "we have to be damn well prepared when these attacks come our way."

WHEN THE IRANIANS hit next, American officials were much quicker to respond—this time to the point of near calamity.

"Should we wake up the president?" CIA director John Brennan asked White House cybersecurity coordinator J. Michael Daniel in the middle of the night in August 2013. It was the proverbial 3:00 A.M. phone call. Brennan told Daniel that Iranian hackers were inside the Bowman Dam—in the PLC controllers—and it looked as if they might open the sluice gates.

A breach of the towering Arthur R. Bowman Dam on the Crooked River in Oregon would be catastrophic. The dam, 245 feet tall and 800 feet long, holds back 150,000 acre-feet of water from inundating 10,000 residences in downstream Pineville. If Iran's hackers were to open the floodgates all at once, they would trigger a tsunami that surely merited an equally destructive U.S. response. Daniel had never heard the usually unflappable Brennan sound so freaked.

Yet as it turned out, the Iranian hackers had misfired. They were not actually in the Arthur R. Bowman Dam in Oregon; they were inside the tiny twenty-foot Bowman Avenue Dam in New York's Westchester County, which keeps a babbling brook from flooding neighborhood basements. Not exactly Hoover Dam. And on that particular evening, its sluice gates had been disconnected for maintenance.

"It's ridiculous how little that dam is, how insignificant in the grand scheme of things," the mayor of Rye Brook, Paul Rosenberg, told my colleague Joseph

Berger after the attack was disclosed. "We're not talking about something vital to the infrastructure of the country."

Years later, Daniel still winced at just how close American officials had come that night to retaliating. "It was a critical lesson that in cyber, the first assessment is almost always wrong."

Nonetheless, the Iranian threat was becoming more serious, and more dynamic. Just a few months later, Iranian hackers managed to breach the U.S. Navy. At the Pentagon and the national energy labs in New Mexico and Idaho Falls that year, analysts and engineers started war-gaming how a real Iranian infrastructure attack would play out. These simulations involved attacks on cellular networks, the financial system, water facilities, and the power grid. The calamitous cyberattack U.S. officials had long dreaded was near. As one senior American military official put it, "There's nothing but upside for them to go after American infrastructure."

The year of the Bowman incident, it was hard to miss the young Iranian hackers popping up at industrial security hacking conferences. At the same Miami conference where I'd broken bread with the Italian mercenaries and seen them stare at the salmon, I'd watched in shock the following day as a young Iranian programmer named Ali Abbasi took the stage and hacked into the computers that controlled the power grid—*in five seconds*. The only thing more startling than Abbasi's presentation was his résumé. Early on, Iran had identified Abbasi as one of its most promising young hackers. He had been running vulnerability analysis and incident response at Iran's Shafir Technical University before being handpicked to study industrial cyberattacks in China. These days, he was looking at all the ways the world's industrial systems could be hacked, with funding from China's high-tech 863 program, a program that funds Chinese universities that, in recent years, became one of many sources for cyberattacks on American targets. I'd heard Iran was sending its best and brightest programmers to China to learn the art of the hack, but Abbasi was proof. And his specialty struck me as particularly problematic. With his access to the grid, Abbasi told us, he could do just about anything he wanted: sabotage data, turn off the lights, blow up a pipeline or chemical plant by manipulating its pressure and temperature gauges. He casually described each step as if he were telling us how to install a spare tire, instead of the world-ending cyberkinetic attack that officials feared imminent.

Just a few months earlier, Panetta had delivered the first major warning of a cyberattack by a U.S. defense secretary, an attack he said would be "as destructive as the terrorist attack of 9/11." America was once again in a "pre 9/11 moment: An aggressor nation or extremist group could use these kinds of cyber tools to gain control of critical switches," Panetta told an audience on the USS *Intrepid* in New York. "They could derail passenger trains, or even more dangerous, derail passenger trains loaded with lethal chemicals. They could contaminate the water supply in major cities or shut down the power grid across large parts of the country."

Top of mind for Panetta and everyone else paying attention that year was Iran.

"Like nuclear weapons, eventually they'll get there," Jim Lewis, a former government official and cybersecurity expert, told me in early 2014.

Back then, nobody could ever have ever anticipated that the first destructive cyberattacks to hit American soil would take out a Las Vegas casino and a Hollywood movie studio.

TWO MONTHS AFTER the Bowman attack, Iran struck Sheldon Adelson's Sands casino empire. Early on the morning of February 10, 2014, Sands casino's computers went dark. Just like Aramco before it, its computers became useless bricks. Email and phones were inaccessible. Hard drives were wiped clean. This time, the message wasn't a burning American flag; instead, Iran's hackers replaced the Sands website with a world map showing flames at the location of every Sands casino in the world, a photo of Sheldon Adelson with Prime Minister Netanyahu, and a personal message for Adelson. "Don't let your tongue cut your throat," it read. It was signed the "Anti WMD Team."

Iran's hackers had retaliated against the billionaire casino magnate for recent remarks in which he suggested that the U.S. nuke Iran. Adelson, one of the world's largest donors to Israeli causes, had told an audience at Yeshiva University that the United States should drop an atomic bomb in the middle of Iran's desert and then say: "See! The next one is in the middle of Tehran." Iran's Ayatollahs didn't like that very much. The Supreme Leader, Ayatollah Ali Khamenei, said Adelson "should receive a slap in the mouth."

Iran's cyber army took that as a direct command, striking Sands just as they had Aramco before it, but with a twist. Tehran's hackers had gone one step further this time, dumping employees' names and Social Security numbers online too. (Sands disclosed in a security filing that the cyberattack had cost the casino roughly $40 million).

In Washington the attack paralyzed U.S. officials, who had yet to formulate a clear strategy for containing the growing cyber threat from Iran, let alone how to respond when private American companies were caught in the crossfire. They were finding it hard enough to protect their own systems from attacks.

American officials didn't know it at the time, but the very same month that Sands was hacked, Chinese hackers were in the preparation phase of their attack on the Office of Personnel Management, the attack that resulted in the theft of the most sensitive data imaginable for 21.5 million Americans who had ever sought a security clearance.

We were getting hit from all sides. This twister of escalation was spiraling out of control.

SIX THOUSAND MILES away, another U.S. foe was watching the Iranian attacks—and the lack of any serious American retaliation—closely.

That December 2014, as American officials were consumed with China, Iran, and the Russians' increasing incursions into Ukraine's election systems and its grid, North Korean hackers popped out of nowhere and struck Sony Pictures in an Aramco/Sands–style attack that destroyed 70 percent of Sony's computers and reduced employees to pen and paper for months.

North Korean hackers targeted the Sony studio as revenge for a ridiculously bad James Franco–Seth Rogen film, *The Interview*, in which Franco and Rogen's characters assassinate North Korea's Dear Leader, Kim Jong-un. The North Koreans, like the Iranians before them, had wiped the data, leaked employees' Social Security numbers, and, in an escalation from previous attacks, dumped executives' embarrassing emails online.

The attackers called themselves "the Guardians of Peace," but within days it was clear that the real culprit was North Korea. Sony's hackers had used the same attack infrastructure that North Korean hackers used to attack South Korean banks and broadcasters one year earlier.

At the White House, U.S. officials saw the Sony intrusion as an assault on free speech—especially after several theater chains, under continued threat from North Korea, chose not to show *The Interview*. But what alarmed officials more was how remarkably similar the Sony attack was to the destructive hits on Aramco and Sands. Our enemies were not just learning from us; they were learning from each other.

"Destructive alarm bells should have gone off," Panetta told me.

Instead, the media coverage homed in on the leaked emails, in which Sony executives panned Adam Sandler films and called Angelina Jolie "a minimally talented spoiled brat." The emails also revealed stunning racial and gender pay gaps. The studio was the victim of the attack, and yet it was the one coming under harsh public scrutiny. Amy Pascal, Sony's cochairman, resigned in the wake of the attack, after emails showed her mocking the movie-viewing habits of President Obama. The leaked emails were derogatory enough that media outlets ate them up, while repeatedly glossing over their provenance.

"The attack went off in a big way, and yet we got no support from fellow movie studios, no support from the mayor of Los Angeles, no support from then attorney general Kamala Harris," Michael Lynton, Sony's former CEO and chairman, told me later. Five years after the event, he was still embittered, and it wasn't hard to see why. "I realized Hollywood is a 'community' in name only. Nobody lent a helping hand. But in a funny way, I really don't blame them, because from a distance, no one really understood how damaging or difficult the situation was."

The Sony leaks—and the frenzy with which American media outlets went after them—would later offer a playbook for another hack, this time on the U.S. election. "Ours brought to public light, very quickly, what a major cyberattack on U.S. soil could look like," Lynton said.

It was unnerving to see how each of these attacks evolved and built off the last, how much more destructive each was. The Sony attack, like the attack on Sands before it, was also a strike on free speech. If Americans were no longer at liberty to put out bad movies, make bad jokes, or share their darkest thoughts without the threat of a cyberattack that cost them millions of dollars' or leaked their email for all to see, this would inevitably lead to an erosion in free speech, perhaps not all at once but little by little, bit by bit.

At the White House, President Obama concluded that nonresponse was no longer an option. That December, the president announced that the United States would "respond proportionally" to North Korea's attack. He declined to give specifics, saying only that the United States would hit "in a place and time and manner of our choosing." Later, officials would tell me Obama was speaking to two audiences that day; he was addressing Iran too. Three days later, a funny thing happened. North Korea's already dim connection to the outside world went dark for an entire day.

IN 2015 THE Obama administration struck two deals: one with Tehran over its nuclear plans, the other with Beijing over cyberattacks. Both worked to curb cyberattacks on American systems for a time, but neither would stick for long.

After decades of coups, hostage-taking, terrorism, sanctions, and cyberattacks, Iranians had finally signaled their willingness to discuss limits to their nuclear weapons program. And while it was only one part of their overall calculus, the White House came to believe that bringing Tehran to the table might be the only way to stave off the next cyberattack. Sure enough, as State Department officials engaged their counterparts in Tehran, Iran's destructive attacks ceased. The attacks never stopped altogether; they just went underground. Iran's hackers pivoted from loud, destructive attacks to quiet sleuthing of American diplomats. Together, David Sanger and I uncovered a subtle Iranian hacking campaign against State Department employees' personal emails and Facebook accounts. Iran was clearly keeping close tabs on the diplomats it was negotiating with, in an apparent effort to gauge whether the United States was serious. For all the criticism of the Iran nuclear deal that was signed in July 2015—that the deal didn't go far enough, that the release of sanctions would lead to regional instability, that the United States had been duped—the cybersecurity community breathed a sigh of relief. After the deal was signed, the wrecking crew ceased.

"The nuclear deal imposes a constraint on them," Jim Lewis told me that month. But, he warned, "When the deal goes away, so goes their restraint."

*   *   *

AS SECRETARY OF state John Kerry defended the Iran deal that summer, a separate delegation of American officials was drawing a red line for their counterparts in Beijing. Chinese commercial trade theft was costing American companies billions of dollars every year. Repeatedly, American officials had pressed the Chinese to cease hacking American businesses, but despite their vigorous demands and the Justice Department indictments of Chinese military personnel, the United States had little to show for it.

"The OPM breach significantly raised the stakes," Rob Silvers, then the Department of Homeland Security's assistant secretary for cyber policy, told me. It was time to act.

Xi Jinping was due to make his first formal visit to the White House as China's president that September. Xi had—tellingly—reserved his first state visit for Moscow. He was proving himself to be China's most authoritarian leader since Chairman Mao, confiding to Putin, "We are similar in character." Xi wasn't riding bare-chested on horseback, but like Putin he sought power and prestige at all costs. Early in his first term, Xi had investigated tens of thousands of his countrymen, arresting more Chinese citizens than at any time since the mid-1990s and the aftermath of the Tiananmen Square massacre. Xi had assumed or created ten titles for himself: not only was he head of state and military, but also the leader of the Communist Party's most powerful committees—on the economy, foreign policy, and Taiwan—as well as new bodies overseeing the internet, national security, the courts, the police, and the secret police. He mocked bureaucratic apparatchiks as "eggheads" and praised the "team spirit of a group of dogs eating a lion." Abroad, he was determined not to be seen as weak.

As Xi's first official state visit neared in September 2015, American officials figured they could use China's sensitivity to any embarrassment to their strategic advantage. That August, national security advisor Susan Rice went to Beijing with a strong and clear message: "If China doesn't stop stealing our stuff, we will sanction you on the eve of President Xi's first state visit to the United States." She repeated the U.S. position to her Chinese interlocutors at every opportunity, including more diplomatically to Xi himself. She reminded the Chinese that just four months earlier, Obama had signed an executive order that gave senior officials the tools to swiftly sanction any foreign actors for cyberattacks. If the Chinese didn't cut it out, the United States would greet

Xi's visit with sanctions. As Rice went wheels-up from Beijing that August, anonymous senior officials leaked to the *Washington Post* that the administration was already preparing to sanction China. The leaks had their intended affect.

"The Chinese panicked," a senior official told me. Worried that sanctions would upend Xi's visit, China asked to send a high-level delegation to Washington immediately to head off any embarrassing surprises. They proposed a visit led by Meng Jianzhu, Xi's security czar, on September 9, the precise day White House officials planned to announce sanctions.

Inside the White House, officials debated how to proceed. On one side was Rice, who argued that they should wait and see what the Chinese envoy had to say before announcing sanctions. If they were not satisfied, they could still announce sanctions ahead of Xi's visit. Other officials argued that the United States had tried diplomacy ad nauseam. The United States should sanction China preemptively, they argued, then double down if Beijing's envoy refused to meet their demands. But there was still a question of what, exactly, U.S. officials would demand. The attack on OPM had been a bridge too far, but American intelligence officials did not actually want to demand a ban on foreign hacks of government agencies. That was a deal the United States would never keep itself.

"It was the old people-in-glass-houses problem," a senior Obama official told me. The NSA's bread and butter was hacking foreign agencies and officials. China's breach of OPM was essentially a countermeasure. "There was an inherent tension between protecting the private sector, the personal data of our citizens, and the interests of our intelligence community, which was directing the same kind of campaigns. But the real killer was commercial espionage."

Obama ultimately decided to hold off on sanctions until he had heard the Chinese out. Over the course of three days in early September 2015, Meng met with Homeland Security secretary Jeh Johnson, attorney general Loretta Lynch, and FBI director Comey, before coming to see Rice in the Roosevelt Room at the White House. Meng reiterated the old Chinese line, denying Beijing's role in Chinese cyberattacks and complaining about U.S. attacks on its own networks. Rice told Meng that unless the Chinese cut out its cyber trade theft, the United States was prepared to humiliate Xi with sanctions ahead of his visit in two weeks. Rice recapped the U.S. position in a private discussion with another member of Meng's delegation, deputy foreign minister Zhang Yesui, telling him: "We are really at a critical point. We are not bluffing, and I have no room to maneuver. If you don't agree to our proposal we are in for a rough ride."

Over thirty-six hours of nearly continuous negotiations at the White House and the Chinese embassy, Obama's deputies sketched out a proposed deal. Meng said he would take the proposal back to Xi.

THE MORNING OF September 25, 2015, began with an elaborate welcome. With the celebratory backdrop of cannon fire, the military band played "March of Volunteers," the Chinese anthem, and "The Star-Spangled Banner." Xi and Obama strolled across the South Lawn at the White House, passing rows of troops and stopping to greet children waving American and Chinese flags.

Over two hours of closed-door meetings that day, Obama took a forceful tone. The Chinese cyberattacks on American businesses had to stop, Obama told Xi. If they didn't, the United States had the next round of indictments ready to go and would move to sanctions. Xi agreed, but everyone in the room knew that by then, China had already collected enough U.S. intellectual property to last it well into the next decade. Chinese hackers had taken everything from the designs for the next F-35 fighter jet to the Google code, the U.S. smart grid, and the formulas for Coca-Cola and Benjamin Moore paint.

With Xi standing beside him in the Rose Garden that afternoon, Obama announced that the two leaders had come to a "common understanding" that neither the United States nor China would engage in state-sponsored theft of each other's intellectual property. Together, Obama told the reporters gathered that day, the two would seek new "international rules of the road for appropriate conduct in cyberspace." Together, they had pledged to set up an informal hotline through which the two could alert each other to malicious software on their respective networks, with the assumption that Chinese and American investigators would work to root out the source. The two also embraced a United Nations accord, adopted the previous July, to refrain from targeting one another's critical infrastructure—power plants, cell-phone networks, banks, and pipelines—in peacetime. There were still many unanswered questions to be worked out: What constituted critical infrastructure? Would a cyberattack on an airline or hotel constitute industrial trade theft? What if the attack was aimed at tracking foreign officials' travel?

But American officials saw the agreement as a win. That evening executives from Apple, Microsoft, Facebook, Disney, and Hollywood gathered for a lavish state dinner to celebrate. The First Lady donned an off-the-shoulder gown made by Chinese American designer Vera Wang. Asian influences were everywhere, from the Meyer lemons in the lychee sorbet to the sixteen-foot silk scroll in the East Room that depicted two roses, symbolizing "a complete meeting of the minds." In a toast, Obama said that while disagreements between the two countries were inevitable, he hoped the American and Chinese could "work together like fingers on the same hand in friendship and peace." Xi called the visit an "unforgettable journey" and praised his warm reception.

Almost immediately, the Chinese cyber theft that had ravaged American businesses over the previous decade plummeted. Security firms reported a 90 percent dropoff in Chinese industrial cyberattacks. For eighteen months, the world's first cyberarms-control agreement appeared to stick.

As ballet dancer Misty Copeland performed for the Chinese delegation and singer Ne-Yo belted out "Because of You" that evening in September, the Chinese leader smiled and clapped. Xi appeared genuine. But then came Trump, who turned the table over with tariffs and the trade war. If it weren't for that, some officials told me, Chinese industrial cyberattacks might have slowed to a trickle. But the cynics saw it differently. The agreement had always been a con job, they said. Xi was just biding his time.

Two years later, the cyberattacks resumed. Only these weren't the sloppy spearphishing attacks of the previous decade. They were vastly more stealthy, strategic, and sophisticated. And they drove up the price of a zero-day even further.

CHAPTER 19

The Grid

Washington, D.C.

S OMEBODY WAS MAPPING out the grid, they said. But nobody knew who they were, why they were doing it, or what to do about it.

This was the gist of the phone calls I started getting in late 2012, from numbers I did not recognize, people I had never met. They were analysts inside the Department of Homeland Security, the agency charged with protecting America's critical infrastructure, and there was an unmistakable urgency in their voices. The attacks marked the beginnings of a new era in cyberwar.

It started with phishing campaigns against employees at American oil and gas firms. But over the course of just a few months, the attacks expanded to include employees at electrical utilities who had direct access to the computers that controlled the power switches.

"Where is cyber at DHS right now?" a panicked analyst asked me in early 2013. "Who is minding the shop?"

There were no good answers. Over a period of four months in early 2013, at the precise moment the cyberattacks were escalating, Homeland Security's top cybersecurity officials—Jane Holl Lute, the agency's deputy secretary; Mark Weatherford, the top cybersecurity official; Michael Locatis, the assistant secretary for cybersecurity; and Richard Spires, the chief information officer—all resigned. It wasn't just at the top. DHS was failing to recruit talented engineers to its ranks. That year, Homeland Security secretary Janet Napolitano estimated the department would need to add six hundred hackers to its ranks to keep pace with the influx of new threats, but the agency was far behind. A National Sciences Foundation scholarship program was created to recruit promising high school students to federal agencies, by offering them college scholarships in exchange for a federal service commitment. But the numbers showed that the vast majority of scholarship recipients were eschewing DHS for NSA, where they could work on offense. DHS was becoming a bureaucratic afterthought. Those who did go to DHS did so as a charity case. And now, these analysts told me, the worst-case scenario was playing out. In late 2012, several decided that—for better or worse—tipping off a reporter at the *New York Times* might just be their last, best hope of pushing their bosses, or their bosses' bosses, to put their heads in the game

It wasn't that the Obama administration was negligent; it was Congress. The U.S. grid is powered by local electricity distributors who are regulated state-to-state and not held to any federal security standards. The computer systems that powered the grid were designed long before cyberattacks became the norm; they were built for access, not security. Many ran older, expired software that software companies like Microsoft no longer patched. And these local grid operators had few of the resources that big power distributors like PG&E had at their disposal.

For years the military and intelligence officials warned Congress that a foreign nation or rogue hacker could exploit software holes and access points to take down the substations that power Silicon Valley, the NASDAQ, or a swing county's voting systems on Election Day. In 2010 a bipartisan group of ten former national security, intelligence, and energy officials—including former secretaries of defense James Schlesinger and William Perry, former CIA directors R. James Woolsey and John Deutsch, and former White House national security advisors Stephen Hadley and Robert McFarlane—sent a confidential letter to the House Committee on Energy and Commerce in support of a bill to improve the cybersecurity of America's critical infrastructure. Their letter was blunt: "Virtually all of our civilian critical infrastructure—including telecommunications, water, sanitation, transportation, and health care—depend on the electric grid. The grid is extremely vulnerable to disruption caused by a cyber or other attack. Our adversaries already have the capability to carry out such an attack. The consequences of a large-scale attack on the U.S. grid would be catastrophic for our national security and economy." It went on: "Under current conditions, timely reconstitution of the grid following a carefully targeted attack if particular equipment is destroyed would be impossible; and according to government experts, would result in wide-spread outages for at least months to two years or more, depending on the nature of the attack."

The House heeded their warnings, but the bill languished in the Senate. It was national security "maverick" Senator John McCain who led Republican opposition to the bill that summer of 2012. The lobbyists had even managed to convince McCain, a senator who prioritized national security over most every-thing else, that any security regulations would be too onerous for the private companies that oversee the nation's dams, water sources, pipelines, and grid. Part of the problem was the invisibility of the threat. If intelligence officials had told senators that a well-financed nation-state was planting bombs and land mines inside American utility companies and transmission lines, this story may have ended differently. In effect, the attacks that began in late 2012 were the cyber equivalent.

"Beyond a certain community focused on grid security, nobody was treating this as a Code Red situation," a senior Obama official told me. "It was a concern, sure. But you also have to remember where we were. There was a lot going on."

Indeed. China was pillaging U.S. intellectual property. Iran was just coming online. But the escalating cyberattacks on the U.S. energy sector that began to spike in 2012 presented a far graver threat. Russia emerged as the chief suspect, but only because the hackers had gone to great lengths to hide their tools and cover their tracks. Perhaps the NSA hackers at TAO knew who they were, but if they did, they weren't sharing this intel with their defense coun-terparts at DHS. DHS analysts told me that nobody at the agency knew who was behind the attacks, and nobody had been able to crack the attackers' payload. This was telling in its own right. Iran's code was destructive but crude. If Tehran could disguise their payloads, they would have. Likewise, China's cyberattacks on American companies were brazen but relatively trivial to attri-bute. Whoever was now embedding itself in the American electrical grid was demonstrating levels of obfuscation and sophistication that analysts had only seen in America's own cyber campaigns. And they were succeeding at an alarming rate. By the end of 2012, DHS analysts had responded to 198 attacks on U.S. critical infrastructure systems—a 52 percent increase from the previous year.

Investigators at CrowdStrike, the security firm, started getting called into U.S. oil and energy firms to investigate. As CrowdStrike teased the code apart in late 2013, they began to pick up Russian-language artifacts and time stamps indicating that the attackers were working on Moscow hours. Either this was a Russian campaign, or someone taking great pains to look like a Russian campaign. CrowdStrike gave the grid hackers a deceptively affable name, Energetic Bear—Bear being the firm's code word for Russia's state-backed groups. As they unspooled the attacks, they discovered that the code dated back to 2010—the same year Stuxnet was uncovered in Iran.

Perhaps Russia's timing was coincidence. But anyone who had been closely following Moscow's reaction to Stuxnet saw a clean through-line from the American cyberweapon that had just debuted on the world stage and the timing of the Russian attacks. Not long after Stuxnet escaped, Russian officials— dismayed by what the Americans and Israelis had pulled off in the cyber realm—began agitating for an international cyberweapons ban. At a conference in Moscow the following year, Russian academics, government officials, and cybersecurity experts ranked cyber escalation the most critical threat of our time. Time and time again, Russia had proven vulnerable to cyberattacks on its critical infrastructure. For years, Kaspersky, the Russian cybersecurity firm, had run hacking competitions on the grid, and every year, one Russian hacking group or another discovered just how trivially they could take over Russian substations and short-circuit the power running through a transmission line. Russia's attack surface was vast, and only getting bigger. Russian Grids PJSC, the state-controlled utility, ran 2.35 million kilometers of transmission lines and 507,000 substations around the country, with plans to fully automate its substations and transmission lines by 2030. Each new digitized node presented another inroad for attack. Following on the discovery of Stuxnet, Russian officials feared they made for an obvious American target. In a speech in 2012, Russia's minister of telecommunications pushed for an international treaty banning computer warfare, while Russian officials back-channeled with their American counterparts to come up with a bilateral ban. But Washington dismissed Moscow's bids, believing them to be a Russian diplomatic ploy to neuter the U.S. lead in cyberwarfare.

With no treaty in sight, it appeared that Russia was now implanting itself in the American grid—and at an alarming pace. Over the next year and a half, Russian hackers made their way inside more than a thousand companies, in more than eighty-four countries, the vast majority of them American. In most cases, the Russians were hacking people—the industrial control engineers who maintain direct access to pipelines, transmission lines, and power switches. In others the Russians infected legitimate websites frequented by utility, pipeline, and grid operators with malware in what security specialists call a "watering-hole attack," because hackers poison the well and wait for prey to come. And in still others, Russian hackers were conducting man-in-the-middle attacks, redirecting victims' web traffic through Russian hackers' machines, taking American grid operators' usernames, passwords, blueprints, and emails on the way through.

This was hardly the first time a foreign actor had targeted the energy sector. China had hacked into one American energy firm after another with cyberat-tacks that American officials concluded were designed to steal U.S. fracking and renewable energy technology. As the frequency and intensity of the Russian attacks escalated in early 2013, American officials wondered if the Russians were looking to glean their own competitive edge. For decades the Russian economy had been overreliant on oil and gas, two exports for which Putin could not control the price. By gross domestic product, Russia was now under-performing Italy, even though it had twice Italy's population. By other metrics like purchasing power parity, Russia now ranked seventy-second in the world—trailing even Europe's economic problem child, Greece. With its population in rapid decline—at last count, Russia was losing a million working-age adults annually—Russia's economic growth prospects had dropped to near zero. And now, with Putin and crony capitalism at the helm, a surge in foreign investment seemed unlikely. American officials came to believe—perhaps hope is the better word—that Russia's attacks on U.S. energy firms were simply Moscow's shady way of diversifying its economy. There was simply no good reason they could think of that Moscow would want to turn off our lights.

All optimism evaporated in 2014, when the Russians took their attacks one step further. That January, CrowdStrike discovered that Russian hackers had successfully compromised industrial control software companies and Trojanized the software updates that made their way into hundreds of industrial control systems across the country. It was the same technique the Americans and Israelis had used five years earlier with Flame, when they infected computers in Iran using Trojanized Microsoft software updates. But the Russians had been far less judicious. It wasn't just U.S. oil and gas companies anymore; Russian hackers infected the software updates that reached the industrial controllers inside hydroelectric dams, nuclear power plants, pipelines, and the grid, and were now inside the very computers that could unleash the locks at the dams, trigger an explosion, or shut down power to the grid.

This was not Chinese-style industrial espionage. Moscow was preparing the battlefield. "This was the first stage in long-term preparation for an attack," John Hultquist, a top threat researcher, told me. "There's no other plausible explanation. Let's just say they're not there to collect intelligence on the price of gas."

At the same time Russia was embedding in our grid, "little green men"— armed Russian Special Forces wearing green uniforms without insignia—had started cycling into Crimea. The Kremlin was signaling to Washington that if it retaliated on behalf of its Ukraine ally, or ever dared turn off the lights in Moscow, Russia had the ability to turn around and do the same. Call it mutually assured destruction for the internet era.

AND IF RUSSIA did attack the grid, we were screwed. The Department of Homeland Security had emergency preparedness plans for natural disasters, earthquakes, hurricanes, tornadoes, heat waves, and power outages that spanned days. But there was no grand master plan for a cyberattack that denied power to millions of people for any sustained period. Intelligence officials had warned Congress, time and time again, that a carefully orchestrated cyberattack on the American grid could unleash outages for at least months, if not years.

Cybersecurity experts and hackers had always despised those in their community who so much as broached the subject of a grid attack, accusing them—rightfully, in many cases—of scaring people into buying more useless mousetraps. Fear, uncertainty, and doubt were so common a scourge in the cybersecurity industry that hackers had shortened it to a code—FUD. For years now I'd spent most of my working hours ducking a fire hose of FUD. My inbox was full of PR pitches from security types citing imminent world-ending doom. My daily commute along the forty-mile stretch of highway between Palo Alto and San Francisco was festooned with FUD billboards that screamed some variation on the following messages: *You are being watched! Do you know where your intellectual property is? Because China does! Do you know Russian cybercriminals? Because they know your Social Security number! Hide your kids. Hide your wife. The Internet is going to blow up your life. Unless, of course, you buy this one thing we are selling.* Cybersecurity marketing had become blood sport.

For twenty years the industry had screamed of these world-ending scenarios, but from late 2012 into 2014, an attack on the grid—the one they had long warned me about—was in its beginning stages. I didn't know whether to regret that we had not listened more carefully or to be furious that the cybersecurity industry's marketing tactics had made it all too easy for Americans to tune the real threats out.

Inside NSA, analysts watched Russia's hackers probe the American grid. They had tracked the group to a Russian intelligence unit. But in July of 2014, after private security researchers at CrowdStrike, FireEye, and Symantec published their findings, the group packed up its tools and vanished, leaving the analysts scratching their heads.

THE FIRST INKLING the Russians were up to something new was the zero-day.

It had been nine years since Watters sold his first baby, iDefense, to Verisign, and now he was close to selling his second firm, iSight, another Chantilly-based threat-intelligence firm. The threat landscape had drastically evolved in the previous decade. Companies weren't just up against cybercriminals and script kiddies anymore; they were now expected to fend off advanced nation-states with infinite resources. This time, Watters had set out to assemble the largest private counterintelligence firms in the world. iSight had 243 dedicated threat researchers on staff, many of them former intelligence analysts fluent in Russian, Mandarin, Portuguese, and twenty other languages. If iSight were a government counterintelligence agency, Watters claimed it would rank among the ten largest in the world. Of course, given the classified nature of such things, his claim was impossible to verify.

The goal, Watters told me, was to get "left of boom"—military jargon for the moment before a bomb detonates. iSight's analysts spent their days behind enemy lines, posing as Black Hats on the dark web, mining hacking channels for bits of information on hackers' intentions, targets, and techniques, tracking malware, exploits, and the like, to provide iSight's clients at banks, oil and gas firms, and some three hundred government agencies, with another early-warning system.

When I visited iSight in late summer 2015, Watters was still sporting Tommy Bahama shirts and crocodile cowboy boots and spitting out military analogies. "When we went into Iraq, the biggest loss of life wasn't from snipers," he told me. "It was from concealed explosive devices. We didn't get ahead of the threat until we started asking ourselves, 'Who's making the bombs? How are they getting their materials? How are they detonating them? And how do we get into that cycle before the bombs are ever placed there?' Our business is tracking the arms merchants and bomb makers so we can be left of boom and avoid the impact altogether."

At iSight's offices that summer, I found several familiar faces. Endler and James were long gone, but Greg McManus, the Kiwi, he was still disassembling code in the black room. There were new faces too. John Hultquist, an army reservist from Tennessee and a bear of a man who'd served in Afghanistan after 9/11, was now running the iSight cyberespionage division. Hultquist had closely tracked Energetic Bear until they abruptly vanished the previous year. He was still trying to understand their motives when he received an intriguing email from a colleague in iSight's Kyiv office.

The email included an innocuous-looking Microsoft PowerPoint attachment that purportedly contained a list of Kremlin sympathizers inside Ukraine. The email played on Ukrainians' worst fears. For months, Russian soldiers, supposedly on "vacation," had started popping up in the Donbass area of Eastern Ukraine. Even as Putin played dumb, Russian soldiers were filmed moving artillery, air-defense systems, and armor to their sympathizers in eastern Ukraine. Now came an email that claimed to contain a list of those complicit. It was cleverly timed phish bait.

Hultquist opened the PowerPoint attachment on computers inside iSight's virtual black laboratory, where a team of researchers, like McManus, spent their days immersed in the latest digital threats. But what iSight's researchers saw that day was perhaps the most advanced threat they had seen to date. The PowerPoint attachment unloaded malware onto their lab computers that took full control of the latest, fully patched version of Microsoft software; they were witnessing the very beginnings of a zero-day attack that would change the nature of digital warfare as we knew it. Attackers were using the zero-day to inject a highly advanced version of malware that had been bouncing around Russia for years. The malware was called BlackEnergy. It had first popped up on Russian hacking forums seven years earlier, when Russian hacker Dmytro Oleksiuk, alias "Cr4sh," started advertising his new tool on Russian forums for $40 a pop. Oleksiuk had specifically designed his BlackEnergy tool for denial-of-service attacks, but in the seven years since he'd unleashed his malware on the world, it had evolved. Hackers had crafted new variants and baked in new features. It was still used in denial-of-service attacks, but variants were being used for financial fraud.

But this BlackEnergy variant was something new entirely. The malware was trying to phone home to a command-and-control server somewhere in Europe. When Hultquist's team took a closer look at that server, they discovered the attackers had left it unprotected. It was one of those digital strokes of luck that can crunch the time it takes investigators to unspool an attack, from years to weeks. Miraculously, inside the attackers' command-and-control server was a list of BlackEnergy commands. This variant of BlackEnergy was not designed to bring websites to a halt or steal bank credentials; it was an advanced nation-state espionage tool that could extract screenshots, record keystrokes, and pilfer files and encryption keys off victims' computers. And it was no mystery who was behind it: BlackEnergy's file commands were all written in Russian.

iSight uploaded the BlackEnergy malware sample to VirusTotal, a kind of Google search engine for malware that researchers use to see where a piece of malware may have popped up before. VirusTotal revealed that four months earlier, in May 2014, attackers had used the same BlackEnergy attack to pop a Polish energy company, this time with a Microsoft Word file that purported to include the latest update on European oil and gas prices. Over the next few weeks iSight uncovered other lures: Some were clearly aimed at trying to infect computers belonging to attendees of a Ukraine-focused summit in Wales; other emails were written to entice the attendees of a NATO-focused event on

Russian espionage later that year in Slovakia. One lure was specifically aimed at an American expert in Russian foreign policy. And another targeted the engineers at Ukraine's railway agency, Ukrzaliznytsia. Some of the files dated as far back as 2010, the same year Energetic Bear started hacking the American energy sector, but this group appeared distinct. Littered throughout attackers' code were references to the 1965 science fiction epic *Dune*, a Frank Herbert science fiction novel set in a not-too-distant future in which the planet has been destroyed by nuclear war. The protagonists take refuge in the desert, where thousand-foot-long sandworms roam just beneath the surface. Hultquist called this new Russian attack group Sandworm.

Inside NSA, intelligence analysts tracked Sandworm by a different name: It was one of several departments working under Unit 74455, a division of Russian General Staff Main Intelligence Directorate, the GRU. And the NSA's analysts were increasingly alarmed by what they saw. But, of course, their findings were all highly classified. Hultquist suspected that Sandworm was a GRU unit, but without definitive proof, he couldn't just come out and say that. By the time his team published their Sandworm report six weeks later, all he knew for sure was that the Russians were five years into a zero-day-equipped espionage campaign whose true purpose would not be revealed for another year.

Hultquist's team celebrated the publication of their Sandworm findings that October inside iSight's SCIF—actually a windowless bar with Miller Lite beer on tap. If the world was going to end one day, Watters wanted to make sure his lieutenants would be guaranteed beer. But on this occasion iSight's team switched it up and toasted Hultquist with vodka, a nod to Sandworm's Russian origins. As they clinked their shot glasses together that October 2014, some 2,500 miles away two security researchers at Trend Micro, a Japanese security firm, had started digging through iSight's report from a conference in Cupertino, California. Trend Micro's researchers searched their databases and VirusTotal for the list of IP addresses Sandworm used in its attacks. They were able to trace them back to a server in Stockholm, which revealed more digital crumbs.

Included in Sandworm's files was a telling clue. Sandworm wasn't after emails and Word docs. It was targeting files used by industrial engineers. One of Trend Micro's researchers had previously worked at Peabody Energy, the world's largest coal producer. This gave him a unique window into what they were seeing. Sandworm's attackers were targeting ".cim" and ".bcl" files, two file types used by General Electric's industrial control Cimplicity software—the same software Peabody's engineers used to remotely check on their mining equipment. That very same GE software was used by industrial engineers the world over. It was a human-machine interface used to check on the PLCs that control the world's water treatments facilities, electric utilities, transportation companies, and oil and gas pipelines. Trend Micro's researchers peeled the code back further and discovered that it was designed to install itself, execute code, then immediately delete itself once the job was done. Among its various commands were the words "die" and "turnoff"—the first step in sabotaging the machinery on the other end. Russia's hackers weren't in those systems to play; they were looking to exact harm.

Two weeks after Trend Micro published its alarming addendum to Hultquist's findings, DHS sounded the sirens further. Sandworm, the agency warned in an October 29, 2014, security advisory, wasn't just after General Electric's clients. It was also targeting clients of two other industrial control software makers: Siemens, the same company the United States and Israelis had hijacked in the Stuxnet attack, and Advantech, one of the leading "Internet of Things" enablers in the world. Advantech's software was embedded in hospitals, power facilities, oil and gas pipelines, and transportation networks around the globe. DHS made clear that as early as 2011, Sandworm had started embedding itself in the computers that control the world's critical infrastructure, not just in Ukraine and Poland but in the United States as well. Sandworm had not yet used this vast access for destruction, but reading the report that October, it became clear that is what Moscow had planned.

Almost as soon as DHS published its report, Sandworm went dark. The attackers unplugged their tools and fell off the radar. When the Russians resurfaced one year later, they did so with a boom.

"I GUESS YOU could call it a crash in the night," Oleksii Yasinsky told me of the day Sandworm reared its ugly head in 2015. We were sitting in Yasinsky's offices in the heart of Kyiv's industrial zone, one of the few buildings in Kyiv that straddled two districts, each with their own power substation. So long as only one district's power went down, the lights would stay on. This was no accident.

On the eve of Ukraine's election on October 24, 2015, Yasinsky had been working as the chief information security officer at Starlight Media, a Ukrainian company, when his IT director called to wake him in the middle of the night. Two of Starlight's main servers had gone down. One server crashing was not unheard of. But two simultaneously? That was cause for concern, if not panic. Russian hackers had been shelling Ukraine's computer networks with cyber-attacks, and the timing was ominous. Just ahead of Ukraine's election the previous year, Russian hackers had taken Ukraine's Central Election Commission offline. Perhaps the crashed servers were a coincidence, Yasinsky told himself, but with voting set to begin in a few hours, he figured he should check for himself. So in the dark morning hours that October, Yasinsky got dressed, tiptoed out of his apartment, and made his way to the office. By the time he arrived, his engineers had uncovered another anomaly: The YouTube channel of a Starlight competitor, STB, was promoting a far-right candidate. Ukraine had strict rules against media outlets pushing election-related news on Election Day. Either STB had gone rogue or someone had just hijacked their competitor's YouTube channel.

Yasinsky started sifting through Starlight's server logs, when he bumped right into his attackers. Someone could see Yasinsky was starting a forensics investigation and torched a server that contained their attack commands. Yasinsky watched, in real time, as the server went dark. "It was the first sign we were under attack, and that the attackers were still inside," he told me. "We had just caught each other in a dark hallway."

Yasinsky picked up the pace, digging for any points of entry or egress. Scanning the logs, he could see one of his servers had been beaconing out to a computer in the Netherlands. As he unwound the traffic back further, he discovered the first communiqué from the Dutch server, dated back six months. Someone had sent a Starlight employee an email purporting to contain infor-mation about a Ukrainian court decision. The employee forwarded the phishing email to Starlight's legal department, where someone had opened it.

"That," Yasinsky told me, "made Starlight Patient Zero."

As soon as Starlight's lawyer clicked the attached Excel file, BlackEnergy slithered out. It had been so long that the lawyer was no longer at the company.

Later, Yasinsky and others briefly speculated whether he was some kind of Russian mole. From the attackers' beachhead that April, Yasinsky could see that they had made eighty-nine requests to Starlight's network. This was no smash-and-grab job. It was an insanely complicated break-in, carefully thought-out, and meticulously executed, until the night the servers crashed. "This wasn't capture-the-flag," said Yasinsky. "They weren't trying to get in and get out as quickly as possible."

The attackers had carefully unloaded BlackEnergy not all at once but in pieces, transferring one malware module at a time to different computers over a period of several months. It was a brilliant approach. Each bit of BlackEnergy looked perfectly harmless; only when the pieces were all in place did attackers start assembling their digital weapon. By the time Starlight's servers crashed six months later, two hundred Starlight computers were infected. What Yasinsky's team found inside the infected machines was a relatively basic KillDisk not so dissimilar from the tool Iranian hackers had used to wipe out the computers at Saudi Aramco and Sands. And like those attacks, the KillDisk had a ticking time bomb. Attackers planned to detonate their malware at 9:51 P.M. that night, just as Starlight was due to start reporting the results of Ukraine's election. Had the servers not crashed, they might have just gotten away with it.

Across town, another Starlight competitor, a Ukrainian television station named TRK, had not been so lucky. That evening nearly one hundred TRK computers were wiped clean with BlackEnergy and the KillDisk. When Yasinsky conferred with the other victims, slight differences in the attackers' tactics emerged. All discovered BlackEnergy and KillDisk on their systems, but the attackers had broke into each network using different techniques and methods, almost like they were tinkering. In one case, the attackers downloaded their tools over time—1:20 P.M. every day. In another, they downloaded them rapid-style.

"They tried one technique here, one technique there," Yasinsky told me. "This was the scientific method in action."

What Yasinsky couldn't figure out was why the attackers would go to such lengths to attack media companies? Media companies had no particularly valuable intellectual property to speak of and little in the way of customer or financial data. The assembly-level mutations their attackers had used to install and hide their tools were as advanced as Yasinsky had ever seen. Why go through all that trouble just to kill some data? It didn't add up.

"Think *Ocean's Eleven*," Yasinsky told me. "Why would they spend half a year doing all of this"—he pointed to his detailed timeline of the attacks— "only to kill two servers and wipe some data at the end? It made no sense."

The Russians were only running simulations. The KillDisk was their way of cleaning up after themselves. "It wasn't until the attacks hit three months later," Yasinsky said, "that we realized we were just their testing ground."

A few weeks after Yasinsky thwarted Sandworm's attacks on Starlight, John Hultquist was invited to the Pentagon. That November Hultquist walked senior officials through his Sandworm findings—the elaborate obfuscation techniques, the evolution of BlackEnergy from its crude origins as a tool for script kiddies to a sophisticated instrument for surveillance and, possibly, destruction. He noted Sandworm's fixation with critical infrastructure, its victims in the United States and Poland, the Ukrainian railway agency, and then the nation's two media companies. As he went through his presentation, officials looked on, expressionless. It was difficult to tell if the gravity of Hultquist's findings had set in. Then one Pentagon official asked Hultquist where he thought this was all leading.

"I think there's a good chance," Hultquist told the officials, "that they're going to turn off the lights."

ONE MONTH LATER, just before Christmas Eve of 2015, Russia's GRU hackers did just that.

In the weeks following Hultquist's Pentagon briefing, Sandworm slipped into one piece of Ukrainian infrastructure after another: the country's treasury, its pension fund, its ministries of finance and infrastructure; its railways and power operators, Ukrenegro, Ukrzaliznytsia, Kyivoblenergo, and Prykarpattyaoblenergo, which supplies power to vast sections of western Ukraine.

At 3:30 P.M., December 23, in the Ivano-Frankivsk region of western Ukraine, residents were just starting to pack up their desks and head home for the holidays when an engineer inside Prykarpattyaoblenergo's control center noticed his cursor gliding across his computer screen, as if pushed by an invisible hand.

The cursor moved to the dashboard that controlled Prykarpattyaoblenergo's circuit breakers at substations around the region. One by one, the cursor double-clicked on the box to open the breakers and take the substations offline. The engineer watched in horror as a pop-up window suddenly appeared, confirming one last time that he wanted to cut heat and power to thousands of his countrymen. He desperately tried to regain command of his mouse, but it was no use. Whoever was inside his machine had left him powerless, and now was logging him out. He tried to log back in, but the hidden hand had already changed his password, kicking him out permanently. Now he could only watch as a digital ghost moved from one breaker to another, meticulously shutting down thirty substations altogether. Meanwhile, two other Ukraine power suppliers were struck in the same way, leaving a total of 230,000 Ukrainians in the dark. Once the power was cut, the hidden hand shut down Ukraine's emergency phone lines too, just to add to the chaos. And finally, the coup de grace: The attackers shut off the backup power to the power distribution centers, forcing Ukrainians to try to put the pieces back together in the dark.

It was an act of unprecedented digital cruelty, but the Russians stopped just short of taking lives. Six hours later, they flipped the power back on in Ukraine, just long enough to send their neighbor, and Kyiv's backers in Washington, a clear message: "We can torch you."

Back in Washington, officials went into high alert. Representatives from the FBI, CIA, NSA, and Department of Energy convened inside the Department of Homeland Security's National Cybersecurity and Communications Integration Center to assess the wreckage and calculate the risk of an imminent U.S. attack. The Ukraine blackouts were the nightmare scenario officials and cybersecurity specialists had forecast for years. The Russians had stopped just short of the dreaded Cyber Pearl Harbor, but officials shuddered just to think how deadly the attacks could have been, and how much worse the damages would be if the Russians pulled the same stunt here.

By now, Russian hackers were so deeply embedded in the American grid and critical infrastructure, they were only one step from taking everything down. This was Putin's way of signaling the United States. If Washington intervened further in Ukraine, if it pulled off a Stuxnet-like attack in Russia, they would take us down. Our grid was no less vulnerable than Ukraine's; the only difference is we were far more connected, far more dependent, and in far greater denial.

"We were still stuck in legacy spy-versus-spy Cold War thinking," a senior official told me. "When we first saw those attacks, we said, 'That's what Russia does. That's what we do. It's a gentleman's thing. Nobody goes too far with it.' Then Ukraine and the election blew the lid right off that theory."

PART VII

Boomerang

The old law about an eye for an eye leaves everybody blind.

—Martin Luther King Jr.

CHAPTER 20

The Russians Are Coming

Washington, D.C.

I N LATE 2015, as the Russians were making their way through networks at
the State Department, the White House, and the Joint Chiefs of Staff and
preparing their assault on Ukraine and the 2016 election, I flew to Washington
to meet with Obama's cybersecurity czar, J. Michael Daniel.

I walked through the iron gates of the Eisenhower Executive Office
Building, the giant gray monstrosity adjacent to the White House that houses
offices for the president's staff. Once I'd been vetted, a White House staffer
escorted me to a cramped, windowless office to wait while Daniel finished up
his business in the West Wing. Printed in giant font on the office door were
the following words: I GET TIRED OF COMING UP WITH LAST-MINUTE
DESPERATE SOLUTIONS TO IMPOSSIBLE PROBLEMS CREATED BY OTHER #@%
PEOPLE. I recognized the quote from the 1992 film *Under Siege*. They belonged
to an embittered ex-CIA officer, played by Tommy Lee Jones, who had sold
Tomahawk cruise missiles and nuclear warheads to terrorists, which were now
headed toward the United States.

Below the quote was an emergency plan for cyberattacks. "Hour 0," the
notice read. "Notify White House Security Response. Hour 1: FBI and Secret
Service reach out to victim. NSA searches intelligence for information. DHS
coordinates national security response. End of Day: Send status message. If

appropriate, this message will state that: 'No further messages will be sent unless or until significant new information is obtained, which may take days or weeks.'"

I'd always imagined the White House would have some advanced, real-time map of cyberattacks, denoted in red blips, sailing toward the White House from decoy servers around the globe, and a team of responders waiting to zap them in real time. *Nope.* When it came to defense, the nation with the most advanced hacking capabilities in the world was reduced to a printout, like the rest of us.

A staffer led me to a stately wood-paneled room across the hall, where I waited for Daniel. This would be the last time we met face-to-face before the election. In a year, Daniel would be out of office, and a few years after that, Trump would eliminate the White House cybersecurity coordinator completely. Daniel and I had spoken many times before, about the Iranian attacks at Aramco and the banks, China's attacks on OPM, and the woeful state of America's cyber defenses. But this would be the first, and potentially last, time I would have the opportunity to ask him the questions I needed to ask about the government's zero-day stockpile.

ONE YEAR EARLIER, Daniel had been dragged into the zero-day debate, dragged being the operative word. A zero-day had forced the government's hand. On April Fool's Day, 2014, almost simultaneously, security researchers in Finland and at Google discovered a zero-day in a widely used encryption protocol. So critical was the zero-day that they uncovered that they spun up an entire branding campaign for the bug, complete with a memorable name, Heartbleed, a logo, and T-shirts.

"On a scale of 1 to 10, this is an 11," Bruce Schneier, a well-respected cyber-security expert, wrote at the time.

Heartbleed was a classic flaw in OpenSSL, a popular open-source software tool used to encrypt internet traffic. Everyone from Amazon to Facebook to the FBI used the free tool to encrypt their systems. It was baked into Android phones, home Wi-Fi routers, even the Pentagon's weapons systems. The Heartbleed bug was the result of a classic coding error, a case of buffer over-read, which allowed anyone to extract data, including passwords and encryption keys, from systems meant to be protected. Unlike proprietary software, which is built and maintained by only a few employees, open-source code like OpenSSL can theoretically be vetted by programmers the world over.

"Given enough eyeballs, all bugs are shallow," is how Eric S. Raymond, one of the elders of the open-source movement put it in his 1997 book, *The Cathedral & the Bazaar*, a manifesto for the open-source philosophy. But in the case of Heartbleed, Raymond told me, "there weren't any eyeballs."

The world soon learned just how neglected OpenSSL had become. The code played a critical role in securing millions of systems, and yet it was maintained by a single engineer working on a shoestring annual budget of $2,000—most of that donations from individuals—just enough to cover the electric bills. The Heartbleed bug had been introduced in a software update two years earlier and yet, nobody had bothered to notice it.

Within days of Heartbleed's discovery, Bloomberg published a thinly-sourced report claiming that the NSA had known about the bug and had been quietly exploiting it all along. The allegation was picked up by CNN, the Drudge Report, the *Wall Street Journal*, NPR, and *Politico*, forcing an official response from the secret agency. The NSA tweeted that it never knew of the bug until it was made public.

But NSA was now nine months into an endless stream of Snowden disclosures, and nobody was taking the agency at its word. The controversy forced the White House to go to lengths greater than any other nation before it to publicly address the process by which it deals with zero-days. Officials, speaking on background, told reporters that President Obama had decided, in recent months, that when NSA discovered zero-days it should—in most cases—see to it that they got fixed. But the President had apparently carved out a glaring exception for "a clear national security or law enforcement need," a loophole so big that critics deemed it worthless. It was left to Daniel to clarify.

Until Heartbleed, the government had never so much as uttered the word "zero-day" aloud. But that April, Daniel addressed the United States' zero-day policy head-on. In a statement posted to the White House website, Daniel detailed what he called a "disciplined, rigorous and high-level decision-making process for vulnerability disclosure" in which multiple government agencies weigh in on the pros and cons of withholding a zero-day. He included a list of questions agencies ask to determine whether to withhold or disclose. Among them: "Does the vulnerability, if left unpatched, impose significant risk?"; "How much harm could an adversary nation or criminal group do with knowledge of this vulnerability"; "How likely is it that someone else will discover the vulnerability?"

It was the first time any government publicly acknowledged withholding information about a security hole from the public. But Daniel had left many questions unanswered. I figured this would be my last chance to ask Daniel to answer them.

JUST PAST FIVE P.M., Daniel walked in and slumped in his chair at the long, mahogany table. With his brown hair and fatigued eyes, he actually resembled Tommy Lee Jones in *Under Siege*, with thinner hair. I wondered if Daniel was responsible for the quote on the door. After all, it was his job to come up with last-minute desperate solutions to impossible problems created by other fucking people.

It wasn't just Heartbleed. Daniel was still picking up after Snowden. North Korea had struck Sony Pictures on his watch. "I have Kim Jong-un to thank for the fact I didn't spend Christmas with my family," he told me. It was his phone that rang at 3:00 A.M. when Iranians breached the (wrong) Bowman Dam. It was Daniel who ran point on China's recent hack of OPM. And now, right below our feet, a Russian hacking unit—a division of the old KGB now known as the SVR—was winding its way through computers at the State Department, the White House, the Joint Chiefs of Staff—and, though neither of us knew it at the time, the Democratic National Committee.

"It never stops," Daniel told me. "We're doing this stuff on the fly. I'm a history buff. But on this there is no precedent."

It was Daniel's job to formulate a coherent cyber policy at the White House—a fool's errand, if there ever was one—and it was Daniel who inherited the unenviable position of leading the administrative process by which the government decides which zero-days in its arsenals to keep, and which to turn over for patching. The process Daniel oversaw had a vague bureaucratic name—the Vulnerabilities Equities Process—and of course an acronym, the VEP, and Daniel hated every last bit of it. He'd inherited ownership of the VEP from his predecessor, Howard Schmidt, a thoughtful, grandfatherly man who helped advise George W. Bush on cybersecurity and who set up the first formal cyber strategy under Obama. Schmidt, who passed away in 2017, understood the intelligence value in zero-days, but also just how vulnerable they left the rest of us.

"Governments are starting to say, 'In order to best protect my country, I need to find vulnerabilities in other countries,'" Schmidt told me before his passing. "The problem is that we all fundamentally become less secure."

Stuxnet had inspired dozens of other countries to join the zero-day hunt, and the United States was losing control over the market it had once dominated. "If someone comes to you with a bug that could affect millions of devices and says, 'You would be the only one to have this if you pay my fee,' there will always be someone inclined to pay it," Schmidt told me. What he said next never left me: "Unfortunately, dancing with the devil in cyberspace is pretty common."

The process Schmidt instituted at the White House was adapted from the NSA's own. For years the NSA had its own VEP for deciding which zero-days to keep and which to turn over, but given the agency's investment in finding its zero-days, and the role they played in the agency's most critical operations, it was only in rare circumstances that the NSA chose to turn them over, and only after the agency had put them to use. By the time Schmidt handed the reins to Daniel in 2014, the zero-day deliberations were still fairly informal, a mismatch for the gravity of the attacks slamming American networks nonstop.

In the preceding twelve months, Russian hackers had launched an aggressive cyberattack on the State Department. They'd convinced naïve staffers to click on their phishing emails and sifted through the computers of American diplomats who worked on Russian policy. By the time Kevin Mandia's team was called in, they found Russian hackers in the deepest crevices of the State Department's network. Every time they blocked one backdoor, the Russian hackers just came in through another. It was among the most brazen efforts Mandia's team had ever seen. And just as soon as Mandiant started getting a handle on the hack, Russians popped up a mile away, this time inside the White House. The Russians had been characteristically stealthy, but this time, American officials had a crystal-clear picture of who was doing the hacking. Dutch intelligence agencies had hacked into a university just off Moscow's Red

Square, where the SVR's hackers—known to private security researchers as Cozy Bear—sometimes operated. The Dutch had managed to get inside the university's security cameras and, using facial-recognition software, identified the SVR's hackers by name. It was a high-def view of America's most adept enemy, and if the White House had followed through more closely, they might have caught those very same hackers in the beginning stages of their attack on our elections.

But on the day I sat down with Daniel, Russia was no more than a tangent in our conversation. Daniel described Russia as a "wild card." Russia was in our systems; this much we all now knew. And it had the skills and the access to pull off a calamitous attack; but for now, at least, Moscow had shown restraint. The bigger worries, Daniel told me, were Iran and North Korea. He hoped the Iran nuclear deal would put Tehran on better behavior, but he wasn't optimistic. North Korea was at the gates too, but still lacked the capabilities to pull off the big one. As for ISIS, Daniel told me, its terrorists were using social media to enlist recruits and plan attacks. But in terms of cyberattacks, the most the "Islamic State Hacking Division" had managed was a dump of names and addresses for a thousand-plus U.S. military and government personnel. ISIS hackers called it a "kill list" and claimed to be inside their machines. In reality, it was just a list of people with .gov or .mil email addresses culled from a breach of an online retailer based in Illinois. The hacker responsible for the breach was now in prison; the hacker who tweeted out the "kill list" had been taken out in a drone strike the previous August. In terms of cyber capabilities, the terrorists were still years behind. And yet Daniel knew all too well that if any one of those actors got their hands on America's cyber arsenal, well, God help us.

The VEP Daniel now oversaw was designed to weigh the competing interests involved in keeping Americans safe. On the one hand, retaining a zero-day vulnerability undercuts our collective cybersecurity. On the other, disclosing a zero-day so vendors can patch it undercuts intelligence agencies' ability to conduct digital espionage, the military's ability to carry out offensive cyberattacks, and law enforcement to investigate crimes. This calculation was far simpler back in the day when we were all using different typewriters.

"It's no secret that every nation conducts espionage," Daniel told me. "In the 1970s and 1980s, Russia was using technology we did not. We were using technology that they didn't. If we found a hole in their systems, we exploited it. Period. End of story. But now it's not so cut-and-dried. We've all migrated to the same technology. You can no longer cut a hole in something without poking a hole in security for everyone."

When Daniel took over the VEP, he'd fast-tracked Schmidt's efforts, pulling together the usual suspects—representatives from the NSA, CIA, FBI, Homeland Security—with representatives from a growing number of government agencies like the Treasury, Commerce, Energy, Transportation, Health and Human Services, and others whose systems could be attacked if an American zero-day got into the wrong hands.

It was better than nothing. In fairness, the United States was one of only two countries—the other being the United Kingdom—to claim anything like it. Even in privacy-conscious Germany, officials told me that they were still a long way off from instituting a VEP of their own. And chances were slim that officials in Iran and North Korea were sitting around long mahogany tables debating whether to turn over a Windows zero-day to Microsoft.

THE PROCESS WAS, Daniel conceded, more art than science. He would never say it, but given the vast resources U.S. intelligence agencies were now pouring into offense, and the intelligence a zero-day could render on imminent terrorist attacks or North Korean missile launches, the process would always weigh more heavily on hoarding a zero-day than turning it over for patching. But as more hospitals, nuclear plants, stock exchanges, airplanes, cars, and parts of the grid came online, the VEP discussions could get ruthless.

"There are a lot of emotions involved," Daniel told me.

As transparent as he could be, the discussions were shrouded in secrecy. Daniel wouldn't even confirm which agencies, exactly, had a seat at the VEP table. "But you can imagine who they might be," he said, "when you stop and think of all the systems that would be impacted if these tools got into the wrong hands."

The calculus itself was reasonably straightforward in theory, but messier in practice. "When we make these assessments," Daniel told me, "we look at how widespread the technology is. If it's very widespread, we err on the side of disclosure. Conversely, if it's only used by our adversaries, that weighs on the other side of the table, and we're more likely to withhold. And when we with-hold, [the intelligence agencies] have to make a case why, and for how long. And we periodically review it and decide whether or not it's time to patch. If we uncover evidence our adversaries are using it, we'll patch."

Attaching a process gave the White House some semblance of account-ability, but in practice it was a high-stakes game of chicken that was hurtling out of control. I asked Daniel about the fact that the United States had spawned the zero-day market in the first place, and then shown the world what seven zero-days, strung together, could wreak. I told him the supply side of this underground economy had sputtered up elsewhere, in uncontrollable markets. I spoke of the Argentines—the men who'd told me they felt no more obligation to sell to the United States than to Iran or the deep-pocketed Gulf monarchies, arguably less.

"Listen, I'm not going to pretend we have it all figured out," Daniel said. "Sometimes," he added ruefully, "There's blood left on the table."

DANIEL NEVER WOULD address specific exploits, but among those that would have crossed his desk was a family of NSA exploits that the agency had code-named Eternal.

It was an NSA computer algorithm that had come up with the name, but Eternal ended up being a fitting moniker for a set of zero-day exploits that would haunt Daniel, the NSA, and American businesses, towns, and cities for years to come. One of those exploits, EternalBlue, targeted critical bugs in a Microsoft software protocol called the server message block (SMB). The protocol enabled computers to pass information, like files or printer services, from server to server at internet speed. At the NSA, finding the underlying flaws that EternalBlue exploited was only half the battle. The real feat, former TAO hackers told me, was finding a way to use the bugs without crashing a target's computer. Shortly after TAO first uncovered, or purchased, the flaws that made up the tool EternalBlue, they took to calling it EternalBluescreen—a reference to the eerie blue screen of death that pops up anytime a computer crashes. For a time TAO operators were under strict orders to use EternalBlue only for precision attacks. They had to get special permission from on high to fire the exploit, for fear of jeopardizing a mission. It took a team of some of the NSA's best analysts to develop the algorithm that ensured that EternalBlue would land on the target's computer without crashing screens on the other end. And once they figured it out, TAO marveled at the magic of their polished espionage tool. "It netted some of the very best counterterrorism intelligence we got," one former TAO hacker told me.

One of EternalBlue's best attributes was that it wasn't "dirty"—it left minimal logging behind. It allowed the agency's hackers to move from server to server undetected. The chance that the NSA's targets—terrorists, Russia, China, North Korea—would ever discover they'd been had with the agency's exploit was virtually null. The agency used EternalBlue for espionage. But they knew if the exploit ever got out, it could just as easily function as an intercontinental missile. If hackers in Iran, North Korea, China, Russia, or God knows where else swapped out the payload for one that could sabotage data or shut down systems on the other side, it could wreak total havoc.

"We knew it could be a weapon of mass destruction," one former TAO hacker told me.

Some officials argued that the exploit was so dangerous that they should turn over the underlying zero-days to Microsoft. But the intelligence it produced was so critical, one former intelligence analyst told me, that turning it over was never seriously considered. Instead the NSA held on to EternalBlue for seven years—over a period that saw some of the most aggressive cyberattacks on American networks in history—and prayed it would never be found.

Daniel never spoke of EternalBlue, or any other exploit, directly. But in a reflective moment years later, he conceded there were some VEP decisions he had come to regret. And once EternalBlue had been picked up by not one adversary but two and used to wreak billions of dollars of destruction around the globe, it was safe to assume the decision to withhold the Microsoft zero-days for seven years, was one of them.

DANIEL DIDN'T KNOW it then, but the Russians were already making moves to tip the scales on the 2016 election.

By June 2014 the Kremlin had already dispatched two Russian agents, Aleksandra Y. Krylova and Anna V. Bogacheva, to the United States for a three-week recon tour. The two women bought cameras, SIM cards, and burner phones and devised "evacuation scenarios" in case American officials grew wise to the real impetus for their trip. Altogether the women visited nine states—California, Colorado, Illinois, Louisiana, Michigan, Nevada, New Mexico, New York, and Texas—to "gather intelligence" on American politics. Krylova sent their findings about American partisanship and "purple states" back to their bosses in St. Petersburg that summer. Their report formed a field guide for Russia's 2016 interference.

Back in St. Petersburg, Putin's propaganda machine, known as the Internet Research Agency, was just sputtering up. The Russians code-named their creation the Translator Project, and its stated goal was to "spread distrust toward the candidates and the political system in general." Putin nominated his former chef—a burly bald man named Yevgeny Prigozhin, who spent nine years in prison for fraud before working his way from a hot-dog salesman to Putin's confidante—to oversee Russia's information warfare campaign from an unassuming four-story building just off Red Square. With a multimillion-dollar budget at its disposal—source still unknown—the Internet Research Agency (IRA) set to work recruiting twentysomething news writers, graphics designers, and "search engine-optimization specialists" with $1,400 weekly salaries, more than four times what they could make anywhere else. On one floor, Russian trolls operating in rotating twelve-hour shifts created and deployed hundreds of fake accounts on Facebook and Twitter to pummel anyone who criticized their master, Vladimir Putin. On another floor, the IRA trolls waited for their daily assignment: a list of America's political crises du jour, anything the Russians could exploit for division, distrust, and mayhem.

With Krylova's field guide in hand, Russia's trolls started in on Texas and spread out from there. In September 2014 the IRA launched a Heart of Texas Facebook group and started pumping out pro-Texan secessionist memes, #texit hashtags, and the usual scare tactics: Hillary Clinton was coming to take their guns away, and the like. Within a year the group had generated 5.5 million Facebook likes. Then, in a countermove, the IRA created a separate Facebook group, the United Muslims of America, and promoted rallies and counterrallies outside the Islamic Da'wah Center in Houston. Demonstrators from the Heart of Texas group confronted pro-Muslim protesters across the street in a terrifying real-world standoff that Russia's digital puppeteers were coordinating from five thousand miles away. Even the Russian trolls back in St. Petersburg couldn't believe the Americans were so gullible.

To add legitimacy, the Russians used the stolen identities of real Americans, whose Social Security numbers and bank and email logins were readily available on Russian dark web platforms. At the height of the Russian campaign, the IRA employed more than eighty people, who logged into Facebook and Twitter from secure virtual private networks to further mask their identities. And they started reproducing their luck in Texas all over the United States, with a special focus on purple states, like Colorado, Virginia, and Florida (FBI agents would later discover that "purple states" had become something of a Russian mantra in the 2016 interference). "Use any opportunity to criticize Hillary and the rest (except Sanders and Trump—we support them)," the bosses at the IRA told their minions in leaked memos.

The IRA used their fake personas to communicate with Trump campaign volunteers and grassroots groups who supported his cause. They bought pro-Trump and anti-Clinton Facebook ads and churned out race-baiting and xenophobic memes, aimed at suppressing minority voters' turnout and steering them toward third-party candidates like Jill Stein. The Russians put up Black Lives Matter pages and Instagram accounts with names like Woke Blacks that tried to convince African Americans, a crucial Clinton demographic, to stay home on Election Day. "Hatred for Trump is misleading the people and forcing Blacks to vote Killary," their message read. "We cannot resort to the lesser of two devils. Then we'd surely be better off without voting AT ALL." In Florida, the IRA paid an unwitting Trump supporter to build a cage on the back of a flatbed truck, and paid an actress to dress up as Clinton and sit in the cage at a rally while crowds chanted, "Lock her up." When that took off, they promoted rallies in Pennsylvania, New York, and California. By the time the IRA campaign was fully revealed, years later, Putin's trolls had reached 126 million Facebook users and received 288 million Twitter impressions—a staggering number, given that there are only 200 million registered voters in the United States, and only 139 million voted in 2016.

But the IRA campaign was only the most visible of Russian interference efforts. Starting in 2014, Russian hackers started nosing around the voter rolls in all fifty states, breaching Arizona's voter registration systems and vacuuming up voter data from a database in Illinois. They started probing American defenses and identifying weaknesses in the vast back-end election apparatus—voter registration operations, electronic poll books, and other equipment—through which American elections are run. They hacked into VR Systems, the company that provided e-pollbook check-in software to critical swing states like Florida, North Carolina, and six other states. Americans would only catch a shimmer of the Russians operation in June 2016, when they hacked the Democratic National Committee.

THAT JUNE, I was on vacation in the Sierras when the alert crossed my phone. The *Washington Post* was reporting that CrowdStrike had uncovered not one but two separate Russian hacking groups inside the computer networks of the DNC. The first, Cozy Bear—the same SVR group that had successfully hacked the State Department and the White House—had been inside the DNC's networks for over a year. The second, Fancy Bear—a group I knew all too well for its hacks on everyone from American journalists to diplomats and their wives—had breached the DNC's networks three months earlier using a simple phishing email. That March, Fancy Bear's Russian hackers had sent John Podesta, Hillary Clinton's campaign chairman, a fake Google alert, declaring that he had to change his Gmail password. Podesta had forwarded the email to the DNC's IT staff for vetting, and in what would become the most tragic typo in American election history, a campaign aide wrote back, "This is a legitimate email." He had intended to type "illegitimate," but the damage was done.

When Podesta entered his new password into the Russian's fake Gmail log-in page, it gave Russian hackers access to sixty thousand emails, stretching back a decade, and a foothold to dig even deeper into the DNC and Hillary Clinton's emails. The *Post*'s story was comprehensive, but it had also missed the mark. Full of reassurances that "no financial, donor or personal information appears to have been accessed," the *Post* reported that this was a traditional Russian espionage campaign, driven by a "desire to understand the policies, strengths and weaknesses of a potential future president—much as American spies gather similar information on foreign candidates and leaders." But who could blame them? Nobody was quite prepared for what happened next.

The minute I saw the *Post* story, I called David Sanger, who was also on vacation in Vermont. Together we had watched the escalation in Russian tactics and had a full appreciation for what the United States was up against. "This is Watergate," we both agreed. We called our editors at the *Times*, but it was difficult to get much momentum that June, in the midst of the most mind-boggling president campaign of our era. Cyberattacks had become the overplayed soundtrack to our lives, and the editors buried the story at the back of the political pages. At the White House, officials were similarly jaded, after responding to the onslaught of escalating Russian attacks on the grid, the White House, and the State Department. Reports suggested that the DNC wasn't the only victim; the Republican National Committee (RNC) had been targeted as well. Until a mysterious lone hacker appeared out of nowhere, officials wrote the hacks off as traditional Russian espionage.

A DAY AFTER news broke of the breach, an enigmatic figure calling himself Guccifer 2.0 appeared on Twitter with a link to an online screed titled "DNC Servers Hacked by a Lone Hacker."

"Worldwide known cybersecurity company CrowdStrike announced that the Democratic National Committee (DNC) servers had been hacked by 'sophisticated' hacker groups," Guccifer 2.0 wrote. "I'm very pleased the company appreciated my skills so highly))) But in fact, it was easy, very easy."

Almost immediately, U.S. officials realized that they had grossly underestimated Russia's motivations. Guccifer 2.0's post included a sample of stolen DNC emails, policy documents, the names of Democratic donors and where they lived, and the DNC's opposition research on Trump, which featured chapter headings like "Trump Has Repeatedly Proven to Be Clueless on Key Foreign Policy" and "Trump Is Loyal Only to Himself." This, Guccifer 2.0 claimed, was "just a tiny part of all docs I downloaded from the Democrats' networks." The remaining "thousands of files and mails" were now in WikiLeaks' hands. "They will publish soon," he wrote, adding: "Fuck the Illuminati and their conspiracies!!!!!!!!!"

Guccifer 2.0's hacking alias and Illuminati reference were all part of an elaborate Russian cover story. The original Guccifer (pronounced GUCCI-fer) was a real person: Marcel Lazar Lehel, a Romanian cybercriminal who used the pseudonym to hack members of the Bush family, Hillary Clinton's Benghazi memos, and Colin Powell's website. He made headlines with his leak of paintings George W. Bush had painted of himself in the shower.

government officials, including members of the Bush family, former secretary of state Colin Powell, and Sidney Blumenthal, an informal adviser to Hillary Clinton. Lehel was known for fixating on the Illuminati, a shadowy "deep state" that conspiracy theorists believe controls the world. Lehel had been arrested in Romania two years earlier and extradited to Virginia to face hacking charges. While awaiting sentencing, Lehel claimed to have hacked Clinton's private server. Now Guccifer 2.0 claimed that he was just picking up where Lehel's efforts had left off.

But immediately computer security experts honed in on the metadata from the DNC's leaked documents, which showed that they had passed through computers with Russian-language settings. Some of the files had last been marked up by a person with a telling username, written in the Cyrillic alphabet: Felix E. Dzerzhinsky, aka Iron Felix, the Soviet's first chief of the Secret Police. As sleuths laid bare their findings on Twitter, Guccifer 2.0 claimed that he was just a lone Romanian with no Russian ties whatsoever. An enterprising reporter at Motherboard, the online tech news site, interviewed Guccifer 2.0 over Twitter. Motherboard's reporter, Lorenzo Franceschi-Bicchierai, had cleverly phrased his questions in English, Romanian, and Russian. Guccifer 2.0 answered the questions in broken English and Romanian, but claimed not to understand the questions in Russian. When linguists began to tease Guccifer 2.0's responses apart, it was clear that he was no Romanian at all: he'd used Google Translate. This was a Russian influence operation through and through. The Russians had a name for this kind of operation—*kompromat*, the Russian art of spreading damaging information to discredit their enemies. Russians had perfected *kompromat* for years, and the hackers behind the DNC hack were the very same Russian group that had breached Ukraine's election reporting systems ahead of the country's critical vote just two years earlier.

But the origins of the DNC leak quickly got lost in the media maelstrom that followed. Guccifer 2.0 had passed chunks of the stolen DNC emails to reporters at *Gawker* and The Smoking Gun. Journalists and pundits from both sides of the political aisle descended on the DNC emails like flies. The *Gawker* post alone generated half a million clicks. And soon, as Guccifer 2.0 promised, WikiLeaks began to dribble out tens of thousands of emails and other stolen goods, which were immediately picked up by the *Guardian*, the *Intercept*, Buzzfeed, *Politico*, the *Washington Post*, and my colleagues at the *Times*. The Russians saved their most damaging revelations for the days ahead of the Democratic National Convention, when party members were due to come together, leaking emails that showed the DNC had secretly favored Hillary Clinton over her primary opponent Bernie Sanders. Party officials had deliberated how best to discredit Sanders. Some questioned Sanders's Jewish faith and argued that painting the candidate as an atheist "could make several points difference" this late in the primaries. Others proposed publicizing an incident in which Sanders's staff allegedly stole Clinton's campaign data. But the most damaging emails belonged to the DNC chairwoman, Debbie Wasserman Schultz, who wrote that Sanders "isn't going to be president." The leaks had their intended effect. When the convention began just days later, Wasserman Schultz was greeted with boos and jeers. Protesters—or were they planted IRA puppets?—held up signs that read "E-MAILS" and "Thanks for the 'Help' Debbie!:)." In the back and forth, Americans had lost sight of where these leaks were coming from. That July, as the convention wound down, I wrote a piece with David Sanger challenging readers to remember: "An unusual question is capturing the attention of cyber specialists, Russia experts and Democratic Party leaders in Philadelphia: 'Is Vladimir V. Putin trying to meddle in the American presidential election?'" Clinton's campaign manager, Robby Mook, insisted that the Russians were leaking data "for the purpose of helping Donald Trump," but without any evidence to back up his claims, the Clinton campaign was left twisting in the wind.

Back in Russia, Putin's hackers and trolls went into overdrive. Unsatisfied with the traction the DNC's purloined emails were getting on WikiLeaks, Russia's hackers started pushing out Democrats' stolen emails on their own channels. The DNC's emails started popping up on a new site called DCLeaks—which had conspicuously been registered back in June—a sign that the Russians had prepared to weaponize the Democrats' emails months earlier. Facebook users with American names like "Katherine Fulton" and "Alice Donovan" emerged out of thin air and pushed DCLeaks to their followers. And one month ahead of the election, WikiLeaks published the motherlode: John Podesta's personal emails, which included eighty pages worth of controversial speeches Clinton had been paid to deliver to Wall Street. In one leaked speech, Clinton told her audience that it was important for politicians to have one "public" position and one "private," playing into criticisms that she was duplicitous and not acting in the public's interests. Trump's "build the wall" hard-liners seized on one speech in which Clinton advocated for "open borders." Each leak was disseminated, maligned, and hashtagged by Russia's IRA troll army, who aimed the leaks at an already cynical American populace. Months after Sanders ended his campaign and endorsed Clinton, several activists who ran Facebook pages for Bernie Sanders began to note a suspicious flood of hostile comments aimed at Clinton. "Those who voted for Bernie will not vote for corrupt Hillary!" they read. "The Revolution must continue! #NeverHillary." "The magnitude and viciousness of it," one Facebook administrator told my colleague Scott Shane, suggested that this was the work of a cold-blooded adversary with an agenda, but the sheer idea that any of this was a Russian campaign still struck many Americans as crazy Cold War–speak.

INSIDE THE WEST Wing, officials already had a much clearer picture of who was behind the effort to discredit the Clinton campaign, though the full extent of Russian interference wouldn't be made plain until years later. The question was, what to do about it?

DNC officials pressed the White House to lay bare what they knew of the Russian hacking and information campaign. Already, the CIA had concluded with "high confidence" that the Russian government was behind the DNC hack. But the White House stayed conspicuously silent. Inside the administration a fight was brewing. The NSA said that it would not attribute the DNC hack to the Russians with anything more than "moderate confidence." Its analysts were still poring through the signals intelligence and wanted to be 100 percent sure before the NSA put out a rare public statement. Inside the CIA, officials knew with 100 percent certainty that this was the Kremlin's dirty work, but their intelligence relied on highly sensitive American spies inside Putin's network. The agency worried that making anything public would put its CIA sources in danger. And Obama worried that by definitively declaring the hacking campaign a Russian operation, he would be seen as interfering in the election too.

The usually mild-mannered Daniel argued strenuously for an in-kind response. Could we somehow seize the GRU's command-and-control servers? Or knock DCLeaks and Guccifer 2.0 offline? What about WikiLeaks? Could we knock them offline too? They considered bringing the information warfare to Putin's doorstep. What if they leaked the corrupt financial dealings of Putin and his cronies? Or could they them where it hurt most, by cutting off their access to the global banking system? But officials told me that none of these scenarios made their way to the president's desk, or were ever seriously considered.

To Obama and his most senior intelligence officials, the constant barrage of DNC leaks and headlines was a sideshow compared to the Russian hacks playing out inside each state's voter registration databases. In Arizona, officials discovered that passwords belonging to an election official had been stolen and could be used to manipulate voter registration data. In Illinois, state officials were only beginning to assess the damage from a hack of their networks, in which the Russians made off with Illinois voter data. DHS analysts started picking up Russian scans of voter registration systems across the country. If they could get inside the voter rolls, they could switch a voter's status from registered to unregistered, manipulate the data to show that someone had already voted when they had not, or delete voters from the rolls completely. Russian hackers would not necessarily even need to compromise the voting machines themselves; it would be far easier, and less visible, to simply digitally disenfranchise thousands of voters in traditionally blue urban counties in purple states. Even if they tweaked the data just a little, the Russians could cause fears of a rigged election and throw the election, and the country, into chaos. And chaos, Russian foreign policy experts said, was always the point.

With the threat snowballing that fall, the White House knew it had to do something. Obama decided it would be better to send a bipartisan message of solidarity. He dispatched his top lieutenants at Homeland and FBI to debrief top lawmakers and ask that, together, they finger Russia. But when Lisa Monaco, Jeh Johnson, and James Comey arrived on Capitol Hill in a caravan of black SUVs that September, the meeting devolved into a partisan melee. Mitch McConnell, the Senate majority leader, made it clear that he would not sign onto any bipartisan statement blaming the Russians; he dismissed the intelligence, admonished officials for playing into what he wrote off as Democrats' spin, and refused to warn Americans about efforts to undermine the 2016 election.

Into the void stepped candidate Trump. "I love WikiLeaks!" Trump declared at one of his rallies. He promoted Russia's hacks at every opportunity.

"Leaked e-mails of DNC show plans to destroy Bernie Sanders . . . Really vicious. RIGGED," he tweeted. In another tweet, he joked that he hoped the DNC hackers had breached Clinton's personal email server as well. All the while, Trump refused to call out the Russians by name. At rally after rally, he expressed doubts that Russia was involved. In a September interview with the Russian television network RT, Trump said it was "probably unlikely" that Putin had ordered the DNC hacks. "I think maybe the Democrats are putting that out. Who knows, but I think it's pretty unlikely." And at the first presidential debate, Trump said the hacks could have been the work of "somebody sitting on their bed that weighs 400 pounds."

As the election neared, the White House delivered two warnings to Putin. One came from Obama himself, who warned Putin face-to-face at a summit meeting in Hangzhou that September that if Russia persisted, America had the ability to destroy the Russian economy. The other warning came from CIA director John Brennan, who warned his Russian counterpart at the FSB, the successor agency to the Soviet-era KGB, that unless Russia backed off, it would "backfire."

And Russia did back off, or perhaps they were done. The campaign had bloodied Clinton plenty. She might emerge victorious but ineffective, they thought. By some intelligence estimates, Russia never expected that Trump would actually win. Their main goal was to bruise Clinton and throw her victory into question. When Trump won that November, it was impossible to say, statistically, what impact Russian meddling had. Disinformation experts reported that the Russian *kompromat* had little effect. But I am not so sure. The numbers show that, in fact, Trump not only lost the popular vote by three million votes but received a smaller share of the vote than Al Gore, John Kerry, and Mitt Romney in their losing campaigns. It wasn't so much that Trump won in 2016 as that Clinton lost. A number of longstanding voting trends either reversed or stalled in 2016. Black voter turnout—the very constituency Russian trolls aggressively targeted—declined sharply in 2016 for the first time in twenty years. And Trump's 2016 victory margin in swing states was smaller than the total votes for Jill Stein, the Green Party candidate buoyed by Russian trolls. In Wisconsin, where Clinton lost by 23,000 votes, Stein won 31,000; in Michigan, where Clinton lost by 10,704 votes, Stein garnered 50,000. Of course, conservative political strategists argue Democrats vastly underestimated how deeply disliked Clinton was to begin with. But we'll likely never know how much Russia's daily barrage of anti-Clinton memes, simulated rallies, and bots kept would-be Clinton voters at home or created such a dark cloud over her candidacy that it pushed them to vote third-party.

Obama officials had planned on handling the Russia question after the election—after Clinton won. But Trump's victory that November 2016 threw everything into doubt. The Obama administration exacted punishing sanctions on Russia that December. They kicked out thirty-five Russian "diplomats," many of them spies, and closed two secret Russian diplomatic properties, one a forty-nine-room mansion on a fourteen-acre compound in Long Island, the other a waterfront spy nest in Maryland, where neighbors noted, with some alarm, that the Russians next door didn't do crab boils like the locals. "They stab them with a screwdriver, break the back shell off, clean them and then boil the body," one neighbor told the Associated Press.

Altogether, it was barely a spanking for burning down the house. And speaking of burning, when the Trump administration finally ordered Russia to close its San Francisco consulate nine months later, conspicuous plumes of black smoke started pouring out of the building's chimney on moving day. Inside, the Russians were burning who-knows-what. Locals gathered on the sidewalk to gawk; the fire department was called to investigate; local environmental officials sent an inspector. A local news reporter approached a Russian man and woman exiting the building to inquire about the burning. With acrid black smoke billowing all around them, the woman replied: "There is no burning."

CHAPTER 21

The Shadow Brokers

Location unknown

T HE FIRST SIGN that the NSA's cyberweapon stockpile had gotten out was a dribble of barely coherent tweets from the Twitter account @shadowbrokerss.

In August 2016, just two weeks into the first DNC leaks, as Russian trolls were battering Hillary Clinton on social media and probing America's election systems state-by-state, the new Twitter account appeared out of nowhere. The Shadow Brokers, whoever they were, claimed to have hacked the NSA, and were now gleefully auctioning the agency's cyberweapons online.

"!!! Attention government sponsors of cyber warfare and those who profit from it!!!" the message—written in a kind of mock Russian, ersatz English— began. "How much you pay for enemies cyber weapons?"

The Twitter account claimed to have intercepted cyberweapons belonging to the "The Equation Group." The Equation Group—like CrowdStrike's silly naming convention for Russian hacking units Cozy Bear and Fancy Bear—was the Russian firm Kaspersky's name for the NSA's elite hacking squad, TAO.

> We follow Equation Group traffic. We find Equation Group source range. We hack Equation Group. We find many many Equation Group cyberweapons. You see pictures. We give you some Equation Group files

free, you see. This is good proof no? You enjoy!!! You break many things. You find many intrusions. You write many words. But not all, we are auction the best files.

At first the screed appeared to be an elaborate hoax, another Guccifer wannabe trying to steal the limelight or distract public attention from the minute-by-minute election meddling playing out on our screens. But the cache of hacking tools the Shadow Brokers posted online appeared legitimate. Attached to the message was a link to 300 megabytes of data—the equivalent of text in three hundred novels—only in this case the files contained hacking tools with code names like Epicbanana, Buzzdirection, Egregiousblunder, and Eligiblebombshell. A few figured that some idiot with way too much time on his hands had simply gone through the Snowden documents and the TAO ANT catalog *Der Spiegel* posted years earlier, come up with his own silly names, and slapped them onto hacking tools plucked from the dark web.

But as NSA operators, security researchers, and hackers all over the world started teasing the file apart, it became clear this was the real deal. The trove contained zero-day exploits that could invisibly break through the firewalls sold by Cisco, Fortinet, and some of the most widely used firewalls in China. I immediately called up every former TAO employee who would pick up their phone.

What is this?

"These are the keys to the kingdom," one put it bluntly. He had already combed through the sample cache and recognized the tools as TAO's. They were all a cyberterrorist would need to break into government agencies, labs, and corporate networks all over the world. If Snowden had dribbled out descriptions of NSA programs and capabilities, the Shadow Brokers had just unleashed the capabilities themselves. The code and algorithms to exact mass destruction were now freely available to anyone with an ax to grind, or data to steal—the NSA's worst nightmare essentially, the very scenario the VEP was designed to impede.

But the cache was just a teaser, an advertisement for a much larger trove of NSA tools they planned to release to the highest bidder. The Shadow Brokers followed up with another encrypted file—"better than Stuxnet!" they wrote—and this time they offered to decrypt it for anyone who bid the most

Bitcoin. This time, the Shadow Brokers added a twist. If the bidding reached one million Bitcoin—worth well over half a billion dollars at the time—they would dump the entire contents of their stolen trove online. Chaos came with a hefty price tag.

The Shadow Brokers concluded with a strange screed about "elites."

"Let us spell out for Elites. Your wealth and control depends on electronic data," they wrote. "If electronic data go bye bye where leave Wealthy Elites? Maybe with dumb cattle? 'Do you feel in charge?' Wealthy elites, you send bitcoin, you bid in auction, maybe big advantage for you?"

To my ear, and to others in other newsrooms, and to Russian experts all over the world, the Shadow Brokers' mock-Russian broken English sounded like a native English speaker trying to sound Russian, a false flag of sorts. This did not sound like the sophisticated Russian hacking units we all were becoming intimately acquainted with. But that August—on the heels of Russian hacks of the DNC—nobody put it past them.

JAKE WILLIAMS, THIRTY-NINE, was sitting in a nondescript command center in Ohio, helping yet another company clean up from a vicious cyberattack. His team had been working insane hours trying to get cybercriminals off his client's network when he saw the Shadow Broker's tweets.

He downloaded the Shadow Brokers sample cache and immediately recognized the handiwork. Williams did not advertise it, but four years earlier he had left TAO. A former paramedic who served in military intelligence before joining the NSA, Williams had served as a TAO exploitation specialist from 2008 to 2013, a longer tenure than most these days. And while he could not tell me whether any of the tools the Shadow Brokers dumped were his, he could vouch for their authenticity.

Williams traded looks with his partner, another former TAO analyst, who also recognized the tools. *This is not fucking happening. Holy fucking shit. This cannot be happening.*

Back in the newsroom, I was thinking the same. By now I had tracked the government's zero-day stockpile from its infancy—Gunman, Gosler, the agency, the hackers, the brokers, the spies, the market, the factories. The dilemma had always been: What if our adversaries or cybercriminals discover these same bugs in the wild? Few, if anyone, had ever paused to consider what might happen if the government's stockpile was stolen. I could not believe what I was seeing. The Snowden leaks were disastrous from a diplomatic perspective, and the release of the NSA's ANT catalog had shut down NSA operations around the world. *But the code? The actual exploits?* The cache contained exploits that had taken months, in some cases years, for America's preeminent hackers and cryptographers to get just right. No doubt they were deployed in active TAO operations on American adversaries, and even some allies, around the world. The Snowden leaks were bad; this was far worse.

Williams pored through the files. His team was finally getting close to knocking attackers off their client's network for good, and he was ready to get back home to Georgia. Then he saw that his client's firewalls were exposed to the same zero-day exploits the Shadow Brokers had just dumped online. Nobody would be going home anytime soon.

His team spent the next several hours making changes to their client's network, to protect them from the blow to follow. By the time they were anywhere close to setting up a buffer, it was late. The building had been emptied of its workers. Williams headed back to his hotel. There, over several extra-strength Long Island iced teas, he read through the Shadow Brokers' release, dissecting the mock Russian-English, the taunts, trying to figure out who in their right minds would do this. America's worst enemies and best allies would soon be scanning their own networks for any trace of the NSA's stolen code, he knew. If, or when they found them, the fallout wasn't going to be pretty. This promised to be a whole new level of torture.

Diplomatically speaking, the most damaging Snowden leak had been the revelation that the NSA was hacking German chancellor Angela Merkel's cell phone. Three years later, American diplomats were still trying to mend relations with Berlin. Now which NSA operations would our allies uncover? Over more Long Islands than Williams cared to admit, his mind ran through a list of U.S. adversaries that would do this, and which most stood to benefit. For years Iran and North Korea had demonstrated the will to do us harm. But as damaging as their attacks had been, and as quickly as they had proven themselves formidable cyber enemies, the United States was still light-years ahead of them when it came to cyber capabilities. Now someone—*who the hell was it?*—had just handed them our cyberweapons and sealed the gap.

Williams shuddered to think of the colossal havoc these tools could wreak. His clients were in for a wild ride. Cybercriminals the world over would surely use them for profit. But nation-states could just as easily bolt digital bombs and data wipers onto the tools, detonate data, and take America's government agencies, corporations, and critical infrastructure offline.

The next morning, Williams woke up and pounded his hangover antidote: Monster Rehab energy drinks. As he made his way back to his client's offices, the question kept coming back: Who the hell *could* do this?

The timing couldn't be coincidence. The Russians had just hacked the DNC and were oozing *kompromat* all over the place. Surely the Pentagon was now reckoning with options for possible retaliation. Williams wondered if the Shadow Brokers' release was a preemptive strike. Perhaps it was the Kremlin reminding the world that it wasn't the only nation that played in this game. Or perhaps Russia was warning the United States that, should they choose to retaliate with cyberattacks, the Kremlin was already onto their methods.

Among those who pushed that view was Edward J. Snowden himself. Tweeting from Moscow, Snowden wrote, "Circumstantial evidence and conventional wisdom indicates Russian responsibility"; the Shadow Brokers leaks were "likely a warning that someone can prove U.S. responsibility for any attacks that originated from this malware server. That could have significant foreign policy consequences," he added in another tweet. "Particularly if any of those operations targeted U.S. allies" or their elections. "Accordingly," he went on, "this may be an effort to influence the calculus of decision-makers wondering how sharply to respond to the DNC hacks."

In other words, Snowden said, "somebody is sending a message" that retaliating against Russian election interference "could get messy fast."

AT CISCO HEADQUARTERS in Silicon Valley and the company's satellite offices in Maryland—just ten miles from Fort Meade—threat analysts and security engineers ripped the NSA code apart. This was Day Zero. The zero-days the Shadow Brokers exposed in Cisco's firewall were the nightmare scenario—not just for Cisco but for millions of their customers around the world. Now Cisco's engineers were racing to come up with a patch, or a workaround. Until they did, anyone with the digital wherewithal could secretly break into their customers' networks untethered.

The Cisco zero-days could have each fetched five figures on the underground market, but the damages from the Shadow Brokers' public release easily exceeded hundreds of millions of dollars, my sources told me. The Shadow Brokers' sample files dated as far back as 2013, but some of the code dated all the way back to 2010. *The NSA held on to these for this long?* The zero-days didn't just neutralize one firewall but eleven different security products. At Fortinet, another security corporation just up Highway 101, the same nightmare scenario was playing out. Fortinet had been steadily capturing more share abroad. Now, executives worried foreign customers would read the leaks as complicity. Engineers cursed their own government.

The Shadow Brokers' leaks made liars of American officials who, in recent years, had talked up the VEP and "NOBUS"—the "nobody but us" calculus NSA officials used to decide whether or not to turn over a zero-day. But the exploits Shadow Brokers leaked were hardly NOBUS-level. They exploited bugs that anyone—our adversaries, cybercriminals, amateurs—could have found and developed on their own. These firewalls were used to secure American networks, and yet the NSA had held onto these zero-days for years. If the VEP worked the way Daniel and others said it did, the bugs would have been turned over and fixed long ago.

For years U.S. officials agonized that the nation's own cyber operations—the Stuxnet strikes—would inspire its enemies to develop their own. That one day, with enough money and training, they might catch up. Now America's own hacking tools were hanging out there on the open web, free for anyone to pick up and fire back at us. Back at Fort Meade, the spies began to sweat.

AS LUCRATIVE AS the Shadow Brokers' zero-days would have been on the underground market, their public auction never amounted to much. Perhaps that's because would-be bidders were afraid—for good reason—that a bid would make them targets of the world's preeminent spies. Twenty-four hours after their auction began, the NSA's torturers only had one measly $900 bid.

But none of the investigators at the FBI or the NSA's counterintelligence arm, known as the Q Group, believed that the Shadow Brokers had done this for profit. Whoever had done this had done it at great risk to the agency, to their country, to every affected computer network in the world, and to themselves, if they ever were caught. Investigators came to believe this was a terror plot, unfolding in slow motion.

The media did not flock to the Shadow Brokers release the way they had to the Snowden dumps or the DNC leaks. At the *Times*, my colleagues David Sanger, Scott Shane, and I covered the leaks in story after story, several on the front page, but, with their technical aspect, they did not land the same way as previous leaks. Still, the damages to the NSA's operations were far greater.

Inside the Fort, and its satellite campuses across the country, the agency was feverishly shutting down any operation impacted by the leaked code, swapping out tools, and trying to anticipate what the Shadow Brokers might leak next. Meanwhile, anyone who had ever so much as glanced at the stolen code was dragged in for questioning. In the hunt for turncoats allied with the Shadow Brokers, some were given lie-detector tests, others suspended indefinitely. NSA morale, already badly bruised by the Snowden leaks, fell to all-time lows. Some of the agency's specialists, even graybeards who had spent their entire careers at the agency, started looking for jobs in the private sector, where they could expect higher salaries, less red tape, and fewer polygraphs.

That summer, as the agency tried to track down the source of the leaks, the NSA alerted the FBI to a Twitter post made by an NSA contractor named Harold "Hal" Martin III. Martin had used Twitter to contact Kaspersky Lab, which in turn alerted the NSA. The FBI used the tip to obtain a search warrant for Martin's residence, where they found 50 terabytes of data—six full boxes' worth of classified code and documents, some containing the names of covert intelligence officers—littered throughout Martin's car, trunk, house, garden, and shed. But as it turned out, Martin was a hoarder, not a leaker. There was no evidence that he'd ever actually accessed the files he'd stolen, or shared them. And if the NSA thought they had their guy, they were quickly proven wrong.

With Martin in custody that October, the Shadow Brokers reappeared the day before Halloween with a new blog post: "Trick or Treat?" This time it wasn't code they leaked but the web addresses of NSA decoy servers around the world, giving U.S. allies and adversaries a comprehensive map of secret NSA hacking operations around the globe, in North Korea, China, India, Mexico, Egypt, Russia, Venezuela, the UK, Taiwan, and Germany. The leak included a taunt toward Vice President Joe Biden, who just days earlier had told

NBC's *Meet the Press* that Russia was responsible for the hack of the DNC, and that U.S. intelligence agencies planned to retaliate. "We're sending a message," Biden had said. "It will be at the time of our choosing—and under circumstances that have the greatest impact."

The Shadow Brokers didn't like that very much. "Why is DirtyGrandpa threating [*sic*] CIA cyberwar with Russia?" they tweeted. "Oldest control trick in book, yes? Waving flag, blaming problems on external sources, not taking responsibility for failures. But neverminding, hacking DNC is way way most important than EquationGroup losing capabilities. Amerikanskis is not knowing USA cyber capabilities is being screwed? Where is being 'free press'?" The post ended with a more ominous threat: "On November 8th, instead of not voting, maybe be stopping the vote all together? Maybe being Grinch who stopped election from coming? Maybe hacking election is being the best idea? #hackelection2016." The password to the Shadow Brokers dump reminded bidders that its ongoing auction of NSA tools might soon come to a close. It was "payus."

Six weeks later the Shadow Brokers emerged again, this time on a different tack, more along the lines of what Netragard, Vupen, NSO, and others had done. "The ShadowBrokers is trying auction. People no like. The Shadow Brokers is trying crowdfunding. Peoples is no liking. Now the Shadow Brokers is trying direct sales." Alongside their Boratesque rant was a screenshot of various files that the group said were worth between 1 and 100 Bitcoin each—$780—$78,000. If buyers wanted to purchase the NSA's hacking tools à la carte, they could make a direct bid for each exploit. Whether would-be buyers found the risk that they would become an NSA target too great, or the entire auction was a farce, nobody bit. That January the Shadow Brokers announced that they were quitting the cyberarms market altogether.

"So long, farewell peoples. The ShadowBrokers is going dark, making exit; Continuing is being much risk and bullshit, not many bitcoins. Despite theories, it always being about bitcoins for the ShadowBrokers. Free dumps and bullshit political talk was being for marketing attention."

For three months, the Shadow Brokers disappeared. Meanwhile another leak, this time a CIA vault—the leakers called it Vault7—of Langley's hacking tools dated between 2013 and 2016 were published online. The vault detailed how the CIA could hack into cars, smart TVs, web browsers, and the operating systems of Apple and Android phones and Windows, Mac, and Linux computers. Essentially, the motherlode. But the Shadow Brokers did not take credit for the leaks. And based on the tools, it appeared Vault7 was the work of a second leaker. Two years later, the CIA would pin the Vault7 leaks on a former elite CIA programmer by the name of John Schulte, who claimed innocence. A jury determined Schulte was guilty of making false statements to investigators, but as to whether Schulte was the source of the leaks, the jury was deadlocked, forcing the judge to declare a mistrial.

As the NSA and FBI raced to find out who was behind the Shadow Brokers' leaks, the leading theory was that an NSA hacker had left the agency's arsenal on a computer or server that was breached by Russian hackers. Some NSA insiders were not convinced. The leaks included a large share of TAO's zero-day collection, including tools that the agency kept on physical disks. Investigators suspected an agency insider may have pocketed a thumb drive and walked out. But that explanation still did not explain why some of the files taken by the Shadow Brokers were not on the disks and appeared to have been stolen from different systems at different times. The dumps included PowerPoint presentations and other files, making it unlikely that the Shadow Brokers had simply grabbed tools left on the internet by a sloppy TAO operator.

One lead, from Israelis, led investigators to an NSA employee's home computer. The employee had installed antivirus software made by Kaspersky, the Russian cybersecurity company. The Israelis, I learned from sources, had hacked into Kaspersky's systems and discovered the firm was using its antivirus software's access to computers all over the world, to search for and pull back Top Secret documents. The Israelis shared screenshots taken inside Kaspersky's systems with their American counterparts, proving as much. And now it appeared that Kaspersky's software may have stolen Top Secret NSA documents from an employee's home computer. It was a dizzying story of spies hacking spies hacking spies, but by then, nothing surprised me anymore. After our story published in the *Times*, Kaspersky claimed an internal investigation determined that the firm's antivirus software had simply been doing its job: searching for a particular string of malware that contained the word "secret" in the code. It was an admission that Kaspersky's software had taken Top Secret data off an NSA employee's computer, but Kaspersky said it destroyed the NSA's data as soon as it realized what, exactly, had been caught in its dragnet. Some found Kaspersky's explanation plausible, others laughable. For years, American officials suspected Kaspersky was a front for Russian intelligence. Now the episode cast a dark cloud of innuendo over the firm, adding fuel to the theory that Russia was somehow involved in theft of NSA tools.

If the Shadow Brokers were Russian operatives, several in the security industry surmised that with Trump's election that November, they had accomplished their mission. For three months the group dropped off the map. But any relief proved fleeting. In April 2017 the Shadow Brokers reemerged, posting the password to the first encrypted file they had posted some eight months earlier, the trove they had advertised as "better than Stuxnet." That proved to be false advertising. The deciphered file included exploits that affected older versions of Linux, Unix, and Solaris, hardly the cyberweapons of mass destruction they promised.

If their goal had all been to help elect Trump, then the Shadow Brokers had grown disillusioned with their candidate. Attached to their leak was a long list of political grievances: With the ease of a seasoned American pundit, the Shadow Brokers addressed Trump directly. They wanted the president to know they were upset about Steve Bannon's recent removal from the National Security Council; the Pentagon's strike on Syria one day earlier; the "deep state"; the Freedom Caucus in Congress; and white privilege.

"TheShadowBrokers is wanting to see you succeed," the Shadow Brokers told Trump. "TheShadowBrokers is wanting America to be great again."

WILLIAMS HAD FOLLOWED the leaks with equal measures awe and angst. He had his own theories. He was convinced that the leaks were the work of the Russians, timed to embarrass the United States and shift the news cycle from Russian election meddling, and later to protest the U.S. incursions in Syria. Sitting in an Orlando hotel room the night before he was to lead an all-day security training session, Williams decided to make his case. In a blog post, he called the Shadow Brokers a classic Kremlin influence operation, noting that the timing of their latest post—just a day after the United States struck fifty-nine Tomahawk cruise missiles at a Syrian airport base—was a Russian effort to embarrass the United States.

"This is a seminal event. Russia is using a cyber operation (the theft of data from a likely hack) to influence real world policy," Williams wrote. "Russia is quickly responding to the missile attacks on Syria with the release of the dump file password that was previously withheld. This was a nuclear option for the Shadow Brokers."

Williams hit publish and went to bed. The following morning, he woke up at 7:30 A.M., rolled over, and checked his phone. It was blowing up with messages and Twitter mentions. The Shadow Brokers had responded directly to his blog post. His worst nightmare was about to be realized. The Shadow Brokers had just correctly outed Williams as a former member of TAO, a role Williams had never publicly disclosed. When clients or colleagues inquired about his career background, he told them only that he worked at the Defense Department. There was not much about his work at NSA that he could speak to, and he worried that if word got out, it would limit his travel. The United States had started indicting nation-state hackers in China, Russia, and Iran. He worried if his previous work became public, he might get picked up in the course of his travels, hit with lawsuits, or forced to share his tradecraft.

Staring at his phone that morning, "I felt like I'd been kicked in the gut," he said.

The Shadow Brokers screed against Williams was littered with odd references to "OddJob," "CCI," "Windows BITS persistence," and the investigation involving "Q Group." Those weren't the usual Shadow Broker babble; they were agency code words. Whoever the Shadow Brokers were, they were much deeper into TAO's operation than Williams had given them credit for. This was some kind of insider.

"They had operational insight that even most of my fellow operators at TAO did not have," Williams said. "Whoever wrote this either was a well-placed insider or had stolen a lot of operational data."

The jolt from the Shadow Brokers' riposte changed Williams's life. He canceled business trips to Singapore, Hong Kong, even the Czech Republic. He always thought that if someone outed him like this, the agency would have his back. But since the Shadow Brokers post, he hadn't gotten so much as a phone call.

"That feels like a betrayal," he said. "I was targeted by the Shadow Brokers because of that work. I do not feel my government has my back."

The NSA, meanwhile, had been shaken to its core. The agency regarded as the world's leader in breaking into foreign computer networks had failed to protect its own. And just when it thought things couldn't get any worse, it turned out that the Shadow Brokers had saved the best tools for last.

A FEW DAYS later, on April 14, 2017, the Shadow Brokers unleashed their most damaging leak yet. The tally of damages to the NSA, to tech companies, and their customers, would go from millions to tens of billions of dollars, and counting.

"Last week, the shadowbrokers be trying to help people," the message read. "This week theshadowbrokers be thinking fuck peoples."

And there they were: the crown jewels, the code for twenty of the NSA's most coveted zero-day exploits, exploits they had spent months building and honing, tools that netted the best counterintelligence the agency could get. But these weren't just espionage tools, they had the power to inflict incalculable destruction. Some of the exploits were "wormable," meaning that anyone could pick them up and bolt on code that would self-replicate malware around the world. The cyberweapon of mass destruction.

Former NSA director Michael Hayden, who had defended the agency for years in the wake of the Snowden release, was unusually speechless. "I cannot defend an agency having powerful tools if it cannot protect the tools and keep them in its own hands," he told my colleague Scott Shane. The loss of those tools, and the damage they could cause, Hayden said, "poses a very serious threat to the future of the agency."

As hackers and security experts began to parse through the latest leaks, one TAO exploit stood above the rest: EternalBlue, the exploit that could invisibly penetrate millions upon millions of Windows machines and leave barely a speck of digital dust behind.

"Hard to detect and easy to use. It was pretty much point-and-shoot," one former TAO operator told me. It was the tool that, some VEP representatives argued, would be far too dangerous if it ever got out, but it had been kept anyway for its sheer intelligence value.

As it turned out, the zero-days underlying EternalBlue weren't zero-days after all. A month earlier, Microsoft had quietly rolled out a patch for its underlying bugs. Typically, Microsoft credits anyone who turns over a bug in its system, but this time the attribution was left blank. The NSA had tipped off Microsoft to the flaw in its systems before the Shadow Brokers had a chance to unleash into the wild. Now, as researchers tried to understand how widely EternalBlue had been used, they discovered just how nebulous a tool it was. The only trace that it had been used was a second, complementary NSA exploit, code-named DoublePulsar, that was often used to implant EternalBlue into machines.

When researchers scanned the web, tens of thousands of infected machines the world over pinged back. Now, with the NSA's tools in everyone's hands, the number of infected systems would explode. One week later, the number of infected machines topped 100,000. Two weeks later, 400,000 victims were infected.

Back at Fort Meade, the agency braced for impact.

CHAPTER 22

The Attacks

London, England

T HE FIRST SIGN that our cyberweapons were boomeranging back was the commotion outside London's hospitals on May 12, 2017. Ambulances were getting diverted. Emergency rooms were turning people away. Patients were being rolled out of the operating room on gurneys, told their surgeries would have to be postponed to another day. Nearly fifty British hospitals had come under assault from the most vicious ransomware attack to hit the internet.

It was the middle of the night when my phone started rattling. "Are you seeing this?!" the messages read. "British health system down!!" By the time I was vertical, ransomware attacks were detonating across the globe. Russian railroads and banks, Germany's railway, French automaker Renault, Indian airlines, four thousand universities in China, Spain's largest telecom, Telefonica, Hitachi and Nissan in Japan, the Japanese police, a hospital in Taiwan, movie theater chains in South Korea, nearly every gas station run by PetroChina, China's state owned oil company, and, in the United States, FedEx and small electrical utilities scattered around the country—were all held hostage by a red screen with a ticking countdown clock demanding $300 in ransom to decrypt their data. If they didn't pay in three days, victims were told, the ransom would double. In seven days their data would be deleted for good. "Your important files are encrypted," the ransom note read. "Maybe you are busy looking for a way to recover your files, but do not waste your time. Nobody can recover your files without our decryption service."

Across the world, people started ripping their computers out of the wall. But often it was too late. The speed of the ransomware was like nothing security researchers had ever seen. Some started live-tracking the infections on a map. Within twenty-four hours, 200,000 organizations in 150 countries had been infested. Only Antarctica, Alaska, Siberia, mid-Africa, Canada, New Zealand, North Korea, and a wide swath of the American West were spared. Hardest hit were China and Russia, both known for using pirated software. In China, 40,000 institutions were infected. And though Russian officials at the state's powerful Interior Ministry initially denied it, over a thousand Ministry computers had been roasted.

As analysts started dissecting the ransomware code, they dubbed the attacks WannaCry—not because the word perfectly encapsulated the way so many victims felt—but because of a tiny snippet left in the code: ".wncry." As they teased the code further still, they discovered why the attacks had spread so quickly. The attackers had used a powerful catalyst, the stolen NSA exploit EternalBlue.

It was an inconvenient detail that Trump officials were careful to omit from their talking points over the next several days, as the tally of damages climbed. Three days into the biggest attack to hit the web, Tom Bossert, Trump's home-land security advisor, told the hosts of *Good Morning America* that the attacks signaled an "urgent call for collective action" by governments throughout the world. When Bossert was asked in a press conference the same day whether WannaCry's code had indeed originated at the NSA, he deployed a clever deflection strategy. It "was not a tool developed by the NSA to hold ransom data. This was a tool developed by culpable parties, potentially criminals or foreign nation-states." *Sure, the tools were ours, but we're not responsible for how others use them* became the official government line.

In this case, it helped that the attackers were an almost universally despised enemy: North Korea. Despite their speed and element of surprise, WannaCry's authors had made several careless mistakes. For one, they had used recycled tools. It didn't take long for researchers to trace the WannaCry attacks back to the same command-and-control servers North Korean hackers had used in their 2014 attack on Sony Pictures. Other hard links to

North Korea emerged. The attackers had barely bothered to tweak the back-door programs and data-wiping tools that previously had only been only seen in attacks by Pyongyang. Some surmised that the recycling of North Korea's tools, the blatantness of it, was itself an artful deflection, a false flag to throw investigators off.

But within a few hours I was on a call with researchers at Symantec, who concluded that the WannaCry attacks were in fact the work of the group they called Lazarus, their code name for North Korea's notorious hacking unit. It wasn't just Sony; North Korea's hackers had used the same attack tools in an impressive list of bank heists over the previous year and a half. Pyongyang was learning that cyberattacks were a far easier way to get around sanctions than North Korea's usual methods of counterfeiting and illicit wildlife trafficking. North Korea's hackers had been caught—but never punished—for major cyber heists at banks in the Philippines, Vietnam, and at the Bangladesh Central Bank, where they'd made a $1 billion transfer request from the New York Federal Bank. Only a spelling error (they'd misspelled foundation as "fanda-tion") had kept bankers from transferring the full billion, but they'd still made off with $81 million, among the largest bank heists in history. WannaCry was the next evolution in North Korea's efforts to generate badly needed income.

"Cyber is a tailor-made instrument of power for them," former NSA deputy director Chris Inglis said after North Korea's role in the WannaCry attacks became clear. "There's a low cost of entry, it's largely asymmetrical, there's some degree of anonymity and stealth in its use. It can hold large swaths of nation-state infrastructure and private-sector infrastructure at risk. It's a source of income." In fact, Inglis said, "You could argue that they have one of the most successful cyber programs on the planet, not because it's technically sophisti-cated, but because it has achieved all their aims at very low cost."

But like the spelling error before it, the attack had been sloppy. The North Koreans failed to develop a way to give victims the decryption keys if they paid the ransom. Even if victims paid, they had no real way to retrieve their data. Once that became clear, victims stopped paying. WannaCry netted less than $200,000 in payouts, a pittance compared to the millions of dollars professional ransomware cybercriminals were making a month. Secondly, and fortunately for the victims, the attackers had also unwittingly baked a kill switch into their code. Within hours of the attack, a twenty-two-year-old British college dropout named Marcus Hutchins discovered that he could neuter the attacks by redirecting victims' servers away from the attackers' command-and-control server toward a web address he bought for less than $11. By redirecting WannaCry's victims to his own benign site, Hutchins stopped the attacks cold. An attack that could have held millions more hostage was now immunized—not because of some great intelligence coup, but because one hacker dared to hack a way out of total chaos. Hutchins' last-second heroics made him a target for U.S. feds, who picked him up a few months later at the Las Vegas airport, en route home from Def Con, and charged him with writing malware early on his career. The case was a reminder to hackers everywhere that no good deed goes unpunished.

The WannaCry attacks appeared to have been planned in such haste that some questioned whether North Korea's hackers had accidentally let their attack code escape before it was ready. Or perhaps they had simply been testing their tools, clueless to the potency of their newfound NSA weapon. Whatever the explanation, they not only failed to generate income or cover their tracks but also managed to piss off their biggest backer and benefactor, China. China's addiction to bootlegged software left its systems among the hardest hit.

The White House response to the WannaCry attacks was notable not only for how casually it deflected any responsibility for losing America's own cyberweapons but because of how quick it was to call out North Korea. By contrast, more than a year after intelligence officials concluded that Russia had aggressively meddled in the 2016 elections, Trump was still reluctant to name and shame Russia for its attacks. In a one-on-one meeting with Putin that year, Trump told reporters he believed Putin when he said Russia was not responsible. "He did not do what they're saying he did," Trump said aboard Air Force One. "That whole thing was set up by Democrats."

One month after Trump's tête-à-tête with Putin, the White House leaped at the chance to blame North Korea for the WannaCry attacks. In a *Wall Street Journal* op-ed headlined: IT'S OFFICIAL: NORTH KOREA IS BEHIND WANNACRY, Bossert, the Homeland Security advisor, wrote, "North Korea has acted especially badly, largely unchecked, for more than a decade, and its malicious behavior is growing more egregious. WannaCry was indiscriminately reckless. . . . [North Korea] is increasingly using cyberattacks to fund its reckless behavior and cause disruption around the world."

Absent from his op-ed was any mention of the role the NSA's tools had played in abetting the attack.

INSIDE MICROSOFT'S HEADQUARTERS in Redmond, Brad Smith, Microsoft's president, was seething. The company had better telemetry into the attacks than anyone—given that EternalBlue exploited Windows—and now Microsoft was witnessing firsthand just how destructive the NSA zero-days in its software could be.

Microsoft's security engineers and executives convened in the company's war room. The NSA had only given the company a few weeks heads-up to patch the bugs in its software before the Shadow Brokers dumped them online. This was a marginal improvement from Flame, when engineers were called in from their vacations because the NSA had abused Microsoft's software update mechanism to infect computers all over Iran. But realistically it took Microsoft customers months, sometime years, to install patches—a reality that was now in stark display as hundreds of thousands of unpatched systems were held hostage by North Korea's adaptation of the NSA's cyberweapon. Given how many affected systems were running older, expired versions of Microsoft's XP software, executives decided they could no longer ignore older software. A staggering number of computers that controlled the world's critical infrastructure—hospitals, patient records, and utilities—still ran Microsoft XP, even though Microsoft had stopped patching the software in 2014. As easy as it was to blame operators for not keeping their systems up to date, patching and updating the software that runs large-scaled industrial machinery or touches the grid is no easy thing. Automated patches were still big no-nos inside critical infrastructure networks. Often any software updates to these systems need to be approved at high levels, and often only occur during narrow maintenance windows or when when it is safe to pull systems offline—which can easily mean once or twice a year. Even critical patches, like the one Microsoft had rolled out for EternalBlue's underlying bugs that March, are not applied if there is even the smallest chance of disruption. Now engineers at Microsoft were stuck working around the clock to figure out how to patch these older, vulnerable systems. Once again, Microsoft was working overtime to clean up the government's mess.

In many ways, the United States had dodged a bullet. Unlike Russia and China, U.S. companies were at least cognizant enough to not use pirated software. Save for FedEx and smaller electric utilities and manufacturing facilities around the country, most U.S. networks had been spared the damage. But Smith was already bracing for the next attack. Each new attack had a way of building on the last, and chances were the next attack wouldn't be so reckless. There would be no quick kill switch; no twenty-two-year-old hacker to save them.

For years Smith had stayed silent as the NSA weaponized Microsoft's software to spy on, then destroy, targets in Iran. Snowden was the breaking point. After Snowden's leaks suggested the NSA had direct access to Microsoft's systems, Smith began to get his voice. He assailed the U.S. secret surveillance courts for prohibiting the companies from telling their side of the story. And when negotiations between the companies and the government stalled, he led a team of Microsoft lawyers to court, successfully getting a judge to rule that Microsoft and others could publish data related to the number of requests the company received from governments around the world. It wasn't much, but at least it helped Microsoft show that it wasn't giving the NSA a direct pipeline to customers' data. But the WannaCry attacks were different. The NSA had withheld Microsoft's vulnerabilities for years, allowed its customers to get hacked, and once again, left it to Redmond to clean up the mess. Smith was apoplectic. It was time to hold the agency to account. And so, he took direct aim at the NSA in a manifesto.

"This attack provides yet another example of why stockpiling of vulnerabilities by governments is such a problem," Smith wrote. "This is an emerging pattern in 2017. We have seen vulnerabilities stored by the CIA show up in Wikileaks, and now this vulnerability stolen from the NSA has affected customers around the world," he continued. "The governments of the world should treat this attack as a wake-up call . . . We need governments to consider the damage to civilians that comes from hoarding these vulnerabilities and the use of exploits."

Back at Fort Meade, the NSA acted like it couldn't be bothered. The agency had yet to publicly comment on the Shadow Brokers, or even confirm that the leaked cyberweapons were theirs. When pressed off-record, senior intelligence officials told me to stop focusing on the tools, but instead on how our adversaries had used them. There was not even the slightest touch of remorse or responsibility for the role its stolen cybermunitions had played.

Meanwhile, back in Moscow, Russia's GRU hackers watched the WannaCry attacks with a mixture of disdain and bemusement. By the time they were ready to launch their own attacks, they took care not to repeat Pyongyang's mistakes.

DYMTRO SHYMKIV WAS running through the Catskills two months later when his phone started buzzing. The forty-one-year-old Ukrainian had been dropping his kids off at a French summer camp in upstate New York—his family's annual respite from the day-to-day cyber skirmishes playing out in Kyiv. Three years earlier Shymkiv abandoned his cushy job running Microsoft's offices in Ukraine, walked to Kyiv's Independence Square, and joined his people's revolution. It was the first time any major Ukrainian executive had so publicly joined the 2014 protests, and the press made much of the fact that a Microsoft executive had left his job to shovel snow at the site of protests.

Three years later, Shymkiv was now in the people's government. Petro Poroshenko, Ukraine's new president, had personally asked Shymkiv to join him as his deputy and to help the country defend itself against nonstop Russian cyberattacks.

With the Ukrainian independence holiday coming up that June, he assumed all would be quiet back home. But when he returned from his run in the Catskills, he glanced at his text messages.

"The machines are dying," one read.

All of Ukraine had seized up. It wasn't a power-grid attack this time; but something equally sinister. Computers at Kyiv's two major airports were down. Ukraine's shipping and logistics systems were frozen. Ukrainians couldn't take money out of ATMs. Or pay for gas; the payment machines were no longer functioning. The same Ukrainian energy companies that had been taken out in the blackouts were paralyzed once more. Computers at bus stations, banks, railways, the postal service, and media companies were all displaying a familiar ransom message.

In the first hours of the campaign, researchers believed that the attack was from ransomware known as Petya—a reference to the James Bond film *GoldenEye*, in which top-secret Soviet satellites armed with nuclear warheads, one nicknamed Petya, the other Mischa, prepare to trigger a nuclear electromagnetic pulse to take out power worldwide. But it did not take long before researchers could see the attack was vastly more sophisticated than that of Petya. It used not one but two stolen NSA tools—EternalBlue and another called EternalRomance—to spread. And it had baked in yet another formidable exploit, MimiKatz, a password-stealing tool developed by a French researcher five years earlier as a proof-of-concept exploit, to crawl as deep into victims' networks as it could.

In their haste, researchers dubbed the attack NotPetya. Despite the cover story, NotPetya was not ransomware at all. The encryption in the ransomware could not be reversed. This was no for-profit venture; it was an attack designed to wreak maximum destruction. The timing of the attack, on Ukraine's equivalent of the Fourth of July, was no coincidence either. Shymkiv knew this was Moscow sending Kyiv a message that Mother Russia was still in charge.

From his makeshift operations center at his kids' summer camp, the first thing Shymkiv did was post to the administration's Facebook account. "We are under attack," he wrote. "But the Presidential Office is still standing."

"It was critical to let the country know someone was alive," Shymkiv told me. "You need to maintain the narrative whenever there is an attack."

His team back home started sharing the Microsoft patch and posting a recovery plan. He started dialing his counterparts at Ukraine's Ministry of Infrastructure, former colleagues at Microsoft, and trading messages with contacts on Facebook. There wasn't one company or government agency at home that had not been badly affected. What he didn't yet know was just how far beyond Ukraine the virus would spread.

At Merck, the pharma giant, factory floors stopped. The multinational law firm DLA Piper could not access a single email. The British consumer goods company Reckitt Benckiser would be offline for weeks. So would subsidiaries of FedEx. Maersk, the world's largest shipping operator, was paralyzed and would sustain hundreds of millions of dollars in damages. India's largest container port was turning shipments away. In the United States, doctors at hospitals in rural Virginia and across Pennsylvania were locked out of patient records and prescription systems. NotPetya had even spread to the far reaches of Tasmania, where factory workers at the Cadbury's chocolate factory in Hobart watched in horror as their machines froze up with the same ransom message flashing across computers all over the world. The attack even backfired on Moscow. Computers at Rosneft, the Russian oil giant, went down too.

Over the next few days, researchers learned just how meticulously the Russians had planned their attack. Six weeks earlier they had broken into a small family-owned Ukrainian software company on the outskirts of Kyiv, Linkos Group. Linkos sold a Ukrainian version of TurboTax, tax reporting software called M.E. Doc that was required at most Ukraine government agencies and many of its largest companies. The company was the perfect foil for the Russian attacks. Russian hackers had cleverly Trojanized an M.E. Doc software update to infect the whole country. As soon as investigators traced the infections back to Linkos' software, Ukrainian soldiers descended on the company, guns drawn. Reporters gathered outside by the hundreds to ask the mom-and-pop tax software company if they were Russian agents. But they were just unwitting accomplices, no more culpable than the hundreds of thousands of victims who'd left their systems vulnerable to the NSA's weapons.

Linkos was Patient Zero. By infecting M.E. Doc, Russia's hackers apparently believed they could limit their blast radius to Ukraine. But that proved wishful thinking. The internet has no borders. No cyberattack can be confined to one nation's citizens anymore. That had been the short-lived lesson from Stuxnet's escape. These attacks were transnational. Any company that did any business in Ukraine—even those with a single employee working remotely from Ukraine—got hit. Once that employee was infected, EternalBlue and MimiKatz finished the job; breaking through the rest of their networks and encrypting anything in their path. The speed at which NotPetya traveled from Ukraine's Ministry of Health to Chernobyl's radiation monitors and on to Russia, Copenhagen, the United States, China, and Tasmania was stunning. And this time too, American officials were eager to name and shame those responsible in Russia. At the White House, Tom Bossert penned another op-ed for the *Wall Street Journal* blasting Russia for the attack and outlining a new American strategy for cyber deterrence. But Bossert's op-ed never saw the light of day. The president ultimately kiboshed it—out of fear it might anger Trump's friend, Putin.

This, even as Russia's attack became the most destructive in world history. Months later, Bossert would pin NotPetya's damages at $10 billion. But some still believe that is a gross underestimate. The world has only been able to tally damages reported by public companies and government agencies. Many smaller organizations surveyed their wreckage in quiet, and publicly denied they'd been hit. Shymkiv laughed as he recalled the phone calls he received from IT executives all over Ukraine, many of whom had publicly claimed to have been spared. "They would call and say, 'Um, do you know how to install six thousand PCs?'"

The damage to Merck and Mondelez alone topped $1 billion. Their insurers would later refuse to pay out damages relating to NotPetya, citing a widely written but rarely invoked "war exemption" clause in their policies. The Russian attack, insurers concluded, qualified as an act of war; while no lives were lost directly that June, it was a demonstration of how a stolen NSA weapon and some cleanly written code could do as much damage as a hostile military force.

When I flew to Ukraine in 2019 to visit Ground Zero for myself, the country was still reeling. Shymkiv met me at my hotel for breakfast. He had blond hair, piercing blue eyes, and the look of a sailor: blue blazer, collared shirt, and a tan that seemed incongruous in the dead of winter. He had just returned from the ends of the earth. He'd left his job in government and had signed on for a weeks' long sailing excursion from Argentina to Antarctica. Like my trek across the Maasai Mara before this journey ever began, it was the only way he knew to escape the digital hellscape all around. He's sailed with a crew of Israeli, German, and even Russian sailors across the Southern Ocean to Antarctica's research stations.

"We avoided politics," he said with a chuckle.

On their return, his crew sailed through Drake's Passage, the point where the Atlantic, Pacific, and Southern Oceans converge. Breaking waves knocked their boat in all directions. As his crew fought to keep their boat upright, Shymkiv told me, he glanced upwards. The skies in the Southern Hemisphere were as clear and calm as he'd ever seen them. "I'd never seen anything like it," he told me. For one brief moment in time, like a disembodied spirit, he could see clearly past the mayhem all around him.

For five years he had been fighting one Russian cyberattack after another, but he knew that as much as Russian meddling in Ukraine would continue in one form or another, the country was its digital test kitchen, not its end target.

"They were experimenting with us," he told me over bacon and eggs. "They couldn't have even imagined what kind of collateral impact NotPetya was going to have. Someone in Russia got a star for that operation."

Two years later, Ukraine was still picking through the rubble.

"The question we should all be asking ourselves," he continued, "is what they will do next."

FIVE MONTHS AFTER the NotPetya attack, Brad Smith took the stage at the United Nations headquarters in Geneva. A century and a half earlier, in 1949, he reminded the crowd, a dozen countries had come together to agree on basic rules of warfare. Hospitals and medical personnel were off limits, the countries agreed. It took three more diplomatic summits over the next century before 169 nation-states signed on to the Fourth Geneva Convention, agreeing to basic protections for wounded or captured military personnel, medical personnel, and nonmilitary civilians during wartime—rules that still hold today.

"It was here in Geneva in 1949 that the world's governments came together and pledged that they would protect civilians even in times of war," Smith told government officials from around the globe. "And yet let's look at what is happening. We're seeing nations attack civilians even in times of peace."

Smith pointed to the nonstop cyberattacks. Data breaches had become so commonplace that we now accepted them as our way of life. Hardly a news cycle went by when we did not hear of some new hack. We were all inured to what happened next: an offer of a year's worth of free credit monitoring, a weak public apology from a CEO. If the breach was really terrible, he or she might get fired; but more often than not, after a temporary dip in stock price, we all moved on.

The latest attacks were different. The destructive back-to-back cyberattacks that assailed the globe in 2017—WannaCry, followed by NotPetya—were defining the post-Stuxnet era. In the absence of any universally accepted cyber rules, or even definitions, the United States had set the rules itself, making it permissible to attack a country's critical infrastructure in peacetime. Now North Korea and Russia were using American cyberweapons for their own attacks and proving just how vulnerable the world's infrastructure had become. Patients had been turned away from hospitals. Merck's production of a critical vaccine had been interrupted; the company was now dipping into emergency stockpiles at the Centers for Disease Control to meet demand. Global shipments froze as Maersk tried desperately to get its inventory systems back up and running. By the time Mondelez, the snack food conglomerate, tallied the losses of Oreo cookies, crackers, fried laptops, and vanished invoices, the hit to its business was more than \$100 million. At Chernobyl, with radiation systems down, engineers in hazmat suits were monitoring radiation levels above the old nuclear blast site manually with hand-held counters. And if North Korea had properly vetted its code. If Russia had kept the power off in Ukraine a little longer. If it had taken the NotPetya attacks even one step further, the financial and human cost would have been unfathomably worse. "It's clear where the world is going," Smith told the crowd of diplomats. "We're entering a world where every thermostat, every electrical heater, every air conditioner, every power plant, every medical device, every hospital, every traffic light, every automobile will be connected to the internet. Think about what it will mean for the world when those devices are the subject of attack."

And though he didn't call out the agency by name, Smith took direct aim at the NSA, and the market the United States had created for cyberweapons. "Nation-state attacks are growing because of increasing investments that are leading to increasingly sophisticated cyberweapons," Smith said. "We simply cannot live in a safe and secure world unless there are new rules for the world." The twenty-first century required new rules of wartime and peace, Smith proposed. "The world needs a new, digital Geneva Convention . . . What we need is an approach that governments will adopt that says they will not attack civilians in times of peace. They will not attack hospitals. They will not attack the electrical grid. They will not attack the political processes of other countries; that they will not use cyberweapons to steal the intellectual property of private companies. That they instead will work together to help each other and the private sector respond when there are cyberattacks. In fact, what we really need is not only to recognize the need for rules but, frankly, to know when others are violating them."

The idea of an international cyber treaty had been pitched before, by Europeans and Russians, especially in the wake of Stuxnet. A small handful of former American officials with classified insights into the pace, scale, and destructive nature of cyberattacks had proposed similar ideas. The year Stuxnet was uncovered in 2010, Richard Clarke, counterterrorism czar under Reagan, Clinton, and Bush, proposed a policy whereby nations would agree not to attack civilian infrastructure. For years the United States had failed to engage in these discussions, in large part because it was the world's supreme cyber superpower, with offensive capabilities it assumed would take adversaries years, decades even, to develop. But the theft of its tools, and the WannaCry and NotPetya attacks, made clear that the gap was closing. Scores of new nation-states were moving into this invisible battlespace. The United States had, for two decades, been laying the groundwork for cyberwar, and it was now American businesses, infrastructure, and civilians who were bearing the brunt of its escalation and collective inaction.

And yet, instead of a multilateral, or even bilateral, treaty, the United States went the other way. At the very moment Smith was wrapping up his speech in Geneva that November 9, 2017, the Pentagon's hackers—unbeknownst to the commander-in-chief—were busy laying trapdoors and logic bombs in the Russian grid.

CHAPTER 23

The Backyard

Baltimore, Maryland

B Y THE TIME the NSA's exploits boomeranged back on American towns, cities, hospitals, and universities, there was no one to guide Americans over the threshold, to advise them, or even to tell them that this was the threshold they would be crossing.

For decades the United States had conducted cyberwarfare in stealth, without any meaningful consideration for what might happen when those same attacks, zero-day exploits, and surveillance capabilities circled back on us. And in the decade after Stuxnet, invisible armies had lined up at our gates; many had seeped inside our machines, our political process, and our grid already, waiting for their own impetus to pull the trigger. For all the internet's promise of efficiency and social connectivity, it was now a ticking time bomb.

Under Trump, things unraveled much more quickly, in a dimension few Americans could truly grasp.

The agreement Obama had reached with Xi Jinping to cease industrial espionage ended the day Trump kicked off his trade war with China.

Trump's abandonment of the Iran nuclear deal—the only thing keeping Iran's hackers on good behavior—unleashed more Iranian cyberattacks on American interests than ever before.

The Kremlin—which had yet to feel much of any pain for its 2016 election interference or its hacks on the Ukraine and U.S. grids—never stopped hacking our election systems, our discourse or our infrastructure.

Our fickle allies in the Gulf—the Saudis and Emiratis—only became more brazen in their choice of targets. Having avoided so much as a slap on the wrist for their brutal murder of Saudi journalist Jamal Khashoggi, the Saudis blinked and kept going.

And cybercriminals continued to ravage American towns and cities with attacks, steadily raising their ransom demands from a few hundred bucks to $14 million. And desperate local officials were paying.

In fact, the only adversary that appeared to back off under Trump was North Korea, but only because its hackers were too busy hacking cryptocurrency exchanges. Pyongyang had figured out that by hacking the exchanges that convert Bitcoin into cash, it could generate hundreds of millions of dollars in revenues, mitigate sanctions, and get back to its nuclear weapons.

And increasingly, the most destructive threat to our civil discourse, to truth and fact, was coming from inside the White House itself.

By 2020, the United States was in the most precarious position it had ever been in the digital realm.

THREE YEARS AFTER the NSA lost control of its tools, the long tail of EternalBlue was everywhere. The underlying Microsoft bugs were no longer zero-days—a Microsoft patch had been available for two years—and yet EternalBlue had become a permanent feature in cyberattacks on American towns, cities, and universities, where local IT administrators oversee tangled, cross-woven networks made up of older, expired software that stopped getting patched long ago. Not a day went by in 2019, Microsoft's security engineers told me, when they did not encounter the NSA's cyberweapons in a new attack.

"Eternal is the perfect name," Jen Miller-Osborn, a threat researcher, told me in early 2019, "for a weapon so useful, it will not go away."

In Allentown, Pennsylvania, city services were disrupted for weeks after malware spread like wildfire across city networks, stealing passwords, wiping police databases and case files, and freezing the city's 185-camera surveillance network along the way.

"This particular virus actually is unlike any other virus," Allentown's mayor told local reporters. "It has intelligence built in."

Nobody had even bothered to tell the mayor that the virus hitting his city had been traveling on a digital missile built by the nation's premier intelligence agency.

A few months after the attack on Allentown, federal agents stormed a jail in San Antonio, Texas, in the dead of night. Malware was spreading from a computer inside the jail at a speed officials had never seen—thanks to EternalBlue—and they feared it might be an attempt to hijack their upcoming election.

"It could have been anything," the Bexar County sheriff told the local news. "A terrorist organization, a hostile foreign government."

By May 2019 the NSA's exploits were popping up in its own backyard. Just a short drive from the Fort down the Baltimore-Washington Parkway, residents in Baltimore awoke to discover that they could no longer pay their water bills, property taxes, or parking fines. Homes drifted into foreclosure because their owners simply couldn't access the system to pay back bills. Epidemiologists had no way to warn city health officials about spreading illnesses. Even the database that tracked bad batches of street drugs had been knocked offline. Baltimore's data had been replaced with a ransom message—one becoming all too familiar to cities and towns around the country—demanding Bitcoin to unlock their data. Over the next few weeks, as Baltimore officials refused to pay their extortionists, the price of a single Bitcoin—which had nosedived the previous year—rose by half, raising Baltimore's ransom to more than $100,000. But that was nothing compared to the $18 million in damages Baltimore would end up having to pay in cleanup costs.

Baltimore called in a handful of incident response teams, including security engineers from Microsoft, to help recover their data. And there again, Microsoft uncovered EternalBlue.

My colleague Scott Shane and I broke the Baltimore story in the *Times*, but the NSA was having none of it: "NSA shares the concerns of all the law-abiding citizens around the world about the threat posed by criminal, malicious cyber activity, but the characterization that there's an indefensible nation-state tool propagating ransomware is simply untrue," Rob Joyce, who headed the NSA's hacking programs, told an audience in the days after our story broke.

Joyce was wordsmithing. As investigators would soon discover, Baltimore had actually been hit by multiple attacks. One assailant locked up its systems with ransomware; another detonated EternalBlue to steal data. Joyce and others in the exploit trade placed the onus on Baltimore for not patching their systems, and seized on the technical detail that in this particular case, the ransomware attack had not spread with EternalBlue. Absent was any mention of the fact that attackers had used NSA's tool for another purpose, or of NSA's role in dropping the world's most advanced hacking tools in our enemies' laps. Inside Microsoft, engineers and executives were livid. The agency had seized on this one technical detail to avoid responsibility. Meanwhile, Microsoft was mopping up the wreckage from EternalBlue in towns and cities across the country.

By now I was used to the NSA's semantic twisters. Weeks earlier I'd sat down with Admiral Michael Rogers, the gruff former NSA director whose tenure oversaw the Shadow Brokers leaks and the destructive attacks that followed. It was under Rogers' tenure that the NSA had made a remarkable, now dubious, admission. Seeking to rebut accusations that his agency was hoarding zero-days, Rogers signed off on a rare public statement in November 2016 which claimed the NSA turned over 91 percent of the zero-days it found. The remaining 9 percent were withheld, the agency said, because the vendor had already fixed it, or for "national security reasons." What struck me was how remarkably specific, but meaningless, the NSA's percentages were. It was unclear if the remaining 9 percent represented ten zero-days or ten thousand—not that those numbers would have clarified much more given that a single zero-day, like Heartbleed, had the ability to impact millions of systems.

Parsing the agency's statement was a waste. The Shadow Brokers leaks, the zero-days the NSA was hoarding, the years they withheld them, their severity, the ubiquity of the systems they impacted, and the wreckage from the WannaCry and NotPetya attacks confirmed just how misleading Rogers' agency had been.

None of this seemed to bother Rogers much, the day we met face to face in a San Francisco hotel in early 2019. Rogers had only been out of the agency for nine months. Gone was the uniform. In its place was a grandfatherly sweater and a beard. But the swagger was still there. I asked Rogers that day what his reaction had been when North Korea and Russia used the NSA's stolen exploits to hold the world's computers hostage. "My reaction," Rogers told me, "was, 'Our lawyers would never have let us get away with that.'"

I don't know what I expected, but Rogers' realpolitik took me aback.

"You don't lose any sleep over those attacks?" I stuttered.

"I sleep just fine," he told me. There was not one whiff of regret or self-doubt.

Then I asked him point-blank what responsibility his agency bore for WannaCry, NotPetya, and the attacks now roiling American towns and cities, to which the admiral leaned back and crossed his arms.

"If Toyota makes pickup trucks and someone takes a pickup truck, welds an explosive device onto the front, crashes it through a perimeter and into a crowd of people, is that Toyota's responsibility?"

I couldn't tell if he was being pedantic or if he expected an answer. But then he'd answered it himself. "The NSA wrote an exploit that was never designed to do what was done."

It was the first time anyone from the agency had so much as acknowledged that the NSA authored the stolen tools. It was also a toothless analogy, one that made crystal clear the agency felt no responsibly for misleading the American public or leaving them vulnerable to the cyberattacks that hit—and that keep hitting—American networks.

When I relayed Rogers' analogy to Microsoft weeks later, executives nearly lost their heads. "That analogy presupposes that exploits have any benefit to society," Tom Burt, an amateur race car driver who oversaw Microsoft's customer security teams, told me. "These exploits are developed and kept secret by governments for the express purpose of using them as weapons or espionage tools. They're inherently dangerous. When someone takes that, they're not strapping a bomb to it. It's already a bomb."

As we spoke, Burt's engineers were quietly dismantling those bombs across the country.

AS IT TURNED out, the shadow of the NSA's stolen exploits was longer and stranger than any of us knew. Months before the Shadow Brokers first leaked the NSA's tools in 2016—and more than one year before North Korea and Russia used them to wreak global havoc—China had discovered the NSA's exploits on their own systems, snatched them, and used them for their own stealth attacks. It took three years for anyone to sort this out. If the NSA knew China was hacking American allies using its tools, that intelligence never made it into the hallowed halls of the VEP, where deliberators might have seized the opportunity to get the bugs fixed long before the Shadow Brokers, North Korea, or Russia could use them for chaos.

Symantec's discovery was clear evidence that even when the NSA used its tools in stealth, there were no guarantees that our adversaries wouldn't detect them and—like a gunslinger who grabs an enemy's rifle and starts firing away—turn them back on us. It was another sign that NOBUS—the presumption that "nobody but us" had the sophistication to find and exploit zero-days—was an arrogant one. Not only that, it was obsolete. The NSA's advantage had hugely eroded over the last decade—not just because of Snowden and the Shadow Brokers and what had been learned from Stuxnet, but because we had grossly underestimated our enemies.

More unsettling was the Chinese hacking group behind the extraction and redeployment of the NSA's exploits. The group, code-named Legion Amber, was based in Guangzhou, an ancient city in southern China, but even the agency struggled to make sense of their ties to the state. Legion Amber's "operators appear to be private or contract hackers, but little is known about their affiliation," one classified NSA assessment concluded. "However, their heavy targeting of Five Eyes and global government and industrial entities suggests they are operating on behalf of elements of the Chinese government."

NSA analysts came to believe that Legion Amber's members were part of a digital reserve army comprised of China's top security engineers, who worked for private internet companies by day but were tapped by China's main spy agency, the Ministry of State Security, for sensitive operations by night. Among Legion Amber's early targets were U.S. defense contractors. But its hit list expanded over the years to include American weapons developers and scientific research labs, where they stole aerospace, satellite, and—most alarming of all—nuclear propulsion technologies. Symantec could not or would not say what, exactly, the Chinese had used the NSA's exploits to steal, but based on Legion Amber's rap sheet, we weren't talking paint formulas.

China had, for the last half century, pursued a no-first-use policy when it came to nuclear weapons. But since coming into office in 2012, Xi Jinping had backtracked. In his first presidential speech to China's Second Artillery Force—the division responsible for China's nuclear weapons—Xi told his soldiers that nuclear weapons were critical to China's status as a global super-power. Absent from the speech was any mention of no first use.

China was decades behind the United States in nuclear weapons development, but thanks to Legion Amber, it had stolen everything it needed to catch up. In 2018, U.S. officials watched in horror as Beijing successfully tested a new submarine-launched ballistic missile and began moving ahead with a new class of subs that could be equipped with nuclear-armed missiles. Meanwhile, the game of chicken between American and Chinese jets and warships in the South China Sea was coming dangerously close to triggering a broader conflict. Both countries abandoned long-standing lines of communication, the kind of measures needed to prevent minor incidents from escalating into war. By 2019 their ships and jets had faced off in eighteen near-collisions. And by 2020, American officials accused China of secretly testing nuclear weapons in violation of longstanding nonproliferation agreements.

Add to this Trump's trade war, and it was safe to assume that the 2015 agreement Xi had struck with Obama to cease commercially motivated attacks was off. By the time Trump was ensconced in the West Wing, Chinese hackers were back to popping American companies with renewed gusto. In early 2019, I discovered that Boeing, General Electric Aviation, and T-Mobile had all been targeted. Within a year, China's hit list had expanded to include as many telecom, manufacturing, health care, oil and gas, pharma, high tech, transportation, construction, petrochemical, travel, and utilities companies and universities as they could break into. Only this time, rather than brute-forcing their way into victims directly, China's hackers were coming in through side doors, breaking into companies via the software employees use to work remotely. They'd abandoned malware commonly attributed to China and started encrypting their traffic. They cleaned up after themselves, erasing server logs and moving files to Dropbox rather than sucking them directly back to China's command-and-control servers.

"The fingerprint of Chinese operations today is much different," Priscilla Moriuchi told me in early 2019. Moriuchi previously ran the NSA's cyber operation in the Pacific, where her duties included determining whether Beijing was abiding by the terms of the 2015 agreement. For years she determined that the deal was, remarkably, working. But under Trump, the gloves came off. Rather than brute-forcing or spearphishing victims one at a time, Chinese hackers were now pulling off their own variation of NSA's Huawei hack, hitting Cisco routers, Citrix applications, and telecom companies to gain access to hundreds of thousands, if not millions, more victims still. Moriuchi's team watched as Chinese hackers siphoned cargo ships worth of American intellectual property back to China for Beijing's state-owned enterprises to rip off.

Skeptics argue that Xi never planned to stick to the 2015 agreement in the first place. Former Obama officials maintain that Xi was sincere, that the deal would have stuck had Trump not flipped the tables over. What we do know is that in the three years since Xi signed the deal, he had consolidated PLA hacking divisions under a new Strategic Support Force, similar to the Pentagon's own Cyber Command, and moved much of the country's hacking operations away from the PLA's scattershot hacking units to the stealthier and more strategic Ministry of State Security.

Beijing started hoarding its own zero-days, eliminating any above- or below-ground market for them in China. Authorities abruptly shuttered China's best-known private platform for reporting zero-days and arrested its founder. Chinese police announced that they would start enforcing laws banning the "unauthorized disclosure" of vulnerabilities. Chinese hackers were forced to give authorities right of first refusal for their zero-days before making them public. The same Chinese hacking teams that had so dominated the big international hacking competitions over the previous five years stopped showing up, on state's orders. Uncle Sam had no such luxury; it couldn't force American hackers into conscription. If the U.S. government wanted exclusive access to American hackers' zero-days, its agencies—well, really the American taxpayer—were going to have to pay for them. And the market prices just kept going up, up, up.

In August 2019 I got my first glimpse at where all those Chinese zero-days were going. That month, security researchers at Google's Project Zero discovered that a number of websites that catered to China's Uighur Muslim minority were invisibly planting spyware onto the iPhones of anyone who visited the sites, using a series of iOS zero-days. It was as slick a surveillance operation as any Google's Project Zero had seen. Anyone who visited the sites, not just in China but all over the world, unwittingly invited Chinese spies into their digital lives. In the weeks that followed, a second group of researchers discovered a parallel effort to hijack Uighurs' Android phones. And not long after that, Citizen Lab discovered a separate but similar campaign aimed at Tibetans.

The targets came as no surprise. Under Xi, China was cracking down on the Five Poisons—Uighurs, Tibetans, pro-independence Taiwanese, the Falun

Gong, and prodemocracy activists—as never before. In Xinjiang, the western province that borders India and Central Asia, China's Uighur Muslims now lived in a virtual cage. As Ukraine was to Russia, Xinjiang was now to China—an incubator for every new piece of surveillance technology. Uighurs were forced to download compulsory spyware that monitored their calls and messages. Surveillance cameras now hung from every doorway, shop, mosque, and street in Xinjiang. Facial recognition algorithms had been trained to identify Uighurs by their unique facial features. And when they snagged one, China's minders inspected every pixel of the footage, sniffing for any whiff of dissent. If they found anything remotely suspicious, Uighurs were detained in "job-training schools" that were, in effect, torture chambers.

Now China was exporting its surveillance abroad. Google's researchers determined that every week for two years, thousands of people around the world—Uighurs, journalists, hell, even American high-school students interested in their plight—had visited China's infected sites and downloaded Beijing's implant.

It was a watering-hole attack that upended everything we thought we knew about mobile surveillance. For one, finding iOS and Android zero-days was supposed to be hard. There was a reason the FBI had paid $1.3 million for a single iPhone jailbreak. Those capabilities now fetched $2 million on the underground market. Considering the cost, we assumed governments would use these capabilities sparingly, for fear of blowing their access. But the Chinese had managed to hide a chain of fourteen zero-day exploits in plain sight for two years. And China wasn't using these capabilities to hunt the next Bin Laden. They'd aimed for Uighurs and their empathizers all over the world. Many were not surprised that China would test these tools on its own people first. The question was: How long before Beijing aimed these capabilities at Americans directly?

"The Chinese use their best tools against their own people first because that's who they're most afraid of," Jim Lewis, the former government official who tracked cyber threats, told me. "Then they turn those tools on us."

AT ALMOST EXACTLY the same time that the United States was reckoning with China's reemergence as IP thief and global voyeur, the Pentagon and officials at Homeland Security were reckoning with another longtime foe.

Almost as soon as Trump nullified the Iran nuclear deal, sensors all over the world lit up with Iranian cyberattacks. Initially these were phishing attacks aimed at European diplomats, in an apparent effort to gauge how likely our allies were to follow Trump out the door. But by the end of 2018 Iran's hackers were slamming into U.S. government agencies, telecoms, and critical infrastructure at a rate we had never seen. They were now the most active nation-state hackers in our digital orbit, more prolific even than China.

Even Keith Alexander, the brains behind Stuxnet, was bracing for impact: "We're probably one of the most automated technology countries in the world and we have a very good offense, but so do they," General Alexander told me the week Trump walked away from the deal. "And unfortunately, we have more to lose."

In the early months of 2019, the threat escalated. The very same Iranian hackers that had wiped out data in Saudi Arabia took aim at the U.S. Department of Energy, oil and gas firms, the national energy labs. The attacks looked to be standard intelligence collection, but as hostilities between Washington and Tehran picked up that summer, those in the know suspected Iran's hackers were "preparing the battlefield" for something more destructive.

To be fair, we were doing the same in Iran. We had been for years, actually. Under a highly classified program conceived under Bush but accelerated under Obama—code name Nitro Zeus—U.S. Cyber Command started planting time bombs in Iran's communications systems, air defenses, and critical parts of its grid. By June 2019 it was safe to assume that Iran's attacks on U.S. critical infrastructure was Tehran responding in kind. What the security community witnessed that summer was, in effect, mutually assured destruction in real time.

That summer spark plugs were everywhere. A series of escalating military skirmishes spiraled into the cyber domain. That May and June, the United States blamed Iran for snapping mines onto the hulls of several oil tankers as they passed through the Gulf of Oman—a vital shipping lane for a third of the world's oil—and then detonating them nearly all at once. Tehran called the explosions an American "false flag." To which, U.S. officials released video footage of an Iranian patrol boat pulling up to one of the stricken ships, hours after an initial explosion, to retrieve one of their unexploded mines from its hull. One week later Iran took out a U.S. surveillance drone. Trump ordered up a strike on Iran's radar and missile sites, before reversing course with just ten minutes to go. He instead ordered Cyber Command to detonate the Iranian computers they believed had been used to plan the attacks on the oil tankers. Again, Tehran responded in kind, hacking more than two hundred oil-and-gas and heavy machinery companies headquartered in the Middle East and in the United States, stealing trade secrets and wiping data at a cost of hundreds of millions of dollars.

Trump's about-face on the missile strikes suggested that the typically impulsive president, who often threatened to "totally destroy" enemies with "fire and fury," was a more cautious commander-in-chief than critics assumed. With the United States already applying maximum pressure with sanctions, and Trump unwilling to launch missile strikes, Trump pursued the third option: cyber strikes. It had its benefits but, as Keith Alexander put it, "we have more to lose."

That summer, the officials I spoke with sounded relieved that Iran's attacks had not been more destructive. With the United States more digitized and vulnerable than ever, it wasn't our defenses that stopped them. Perhaps Tehran had held some fire on the chance that Americans might vote Trump out of office in 2020, officials posited, and his successor would reverse course. To help its prospects, Iran's hackers took aim at Trump's 2020 reelection campaign. Over a thirty-day period between August and September 2019, Iran's hackers made no less than 2,700 hacking attempts against Trump's campaign and everyone in its orbit. It was the first sign that—even with Russia sucking up most of the oxygen on election interference—other nations were meddling in 2020, for very different reasons.

This, you see, was the precarious state of the Iranian cyber threat when—on January 2, 2020—Trump ordered up a drone strike on General Qassim Suleimani. The United States could have killed Suleimani a thousand times before. But previous administrations had never dared pull the trigger for fear his death would provoke the kind of large-scale retaliation that would lead to war. Iran's powerful security and intelligence commander was like a second son to Iran's Supreme Leader, Ali Khamenei. He led the Revolutionary Guards' Quds Force, and was responsible for the killing of hundreds, if not thousands, of Americans in Iraq over the years. No doubt he had more planned. But inside Iran, he was a hero.

"Buckle your seatbelt," one senior official texted me the night Suleimani was blown to smithereens.

Almost immediately, an Iranian hashtag—#OperationHardRevenge—appeared on Facebook, Twitter, and Instagram. Hackers defaced websites in Minneapolis and Tulsa with images honoring Suleimani. For a brief period one Saturday, any high-school history student looking for an annotated version of the Constitution on the U.S. federal library website would have instead been met with an image of a bloodied Trump getting punched in the face. An Iranian official tweeted out the addresses of various Trump hotel properties as possible targets.

"You cut off the hand of Qassim Suleimani from his body, and we will cut off your feet from the region," an Iranian official tweeted. It was a reference, apparently, to the fact that Suleimani's hand had been severed in the U.S. strike.

A few days later the Iranians made good on their promise, bombing a series of joint U.S.-Iraqi military bases with twenty-two missiles. By luck or design, the missiles took out infrastructure, not people. A few hours later, Iran announced the strikes had concluded its response.

Trump believed this would be the last word on the matter.

"All is well," the president tweeted.

But inside Homeland Security, officials did not share his relief. Iran's military response may have concluded, but Chris Krebs, the senior-most cybersecurity official at DHS, warned that the threat of cyberwarfare had only just begun. Iran, he told us, had the ability "to burn down the system."

The day of Iran's strike, Krebs urged 1,700 members of the U.S. private sector and state and local governments to lock up their systems, upgrade software, back up their data, and move anything precious offline. "You need to get in the head space that the next breach could be your last," he told them.

As of this writing, Iran's hackers were pushing deeper into U.S. critical infrastructure and the companies that control the American grid. And showed no signs of leaving anytime soon. It is Iran's way of saying, "We're sitting here with a gun to your head," said Suzanne Spaulding, the former undersecretary for cybersecurity and critical infrastructure at Homeland Security.

MEANWHILE A NEW zero-day broker has quietly surfaced online and started outbidding everyone else on the market. The broker calls itself Crowdfense, and I have learned it works exclusively for the Emiratis and their closest ally, the Saudis. Crowdfense is ponying up $3 million for the same iPhone exploits everyone else is offering—at most—$2 million for.

The Gulf monarchies are busy cutting middlemen out left and right. In 2019 Twitter discovered that two unassuming engineers within its ranks were actually Saudi spies. The two had stolen data on more than six thousand accounts— the bulk of them Saudi dissidents, but some Americans too—on behalf of Bader al-Asaker, chief henchman for Saudi Crown Prince Mohammed bin Salman, known as MbS. If these were the lengths the monarchies were willing to go to spy on, and silence, their critics, Silicon Valley didn't stand a chance of stopping them.

I no longer had to guess how that Twitter data was used. In the UAE, Ahmed Mansoor was still in solitary confinement for "defaming" its rulers on Twitter. And by then, the CIA concluded that MbS had personally ordered the hit on slain *Washington Post* journalist Jamal Khashoggi. None of this came as a surprise. But the White House was unrecognizable. Trump and his son-in-law, Jared Kushner, excused their oil-rich ally for the atrocity. Kushner and MbS continued to use WhatsApp even after grisly recordings emerged of MbS's thugs dismembering Khashoggi's body.

But the journalists would not let it go, none more so than Khashoggi's colleagues at the *Post*. Already Trump had taken aim at the *Post* and its owner, Amazon founder Jeff Bezos, for its coverage of his administration. Trump had taken to calling the *Post* the #AmazonWashingtonPost on Twitter, where he decried the paper as a "lobbyist weapon" and "big tax shelter" for Amazon and its CEO, "Jeff Bozo." So when the Saudis took their fight directly to Bezos, hardly anyone in the Trump White House bothered to stop them.

The Saudis saw to it that Bezos paid for the paper's commitment to the Khashoggi story. After three months of nonstop *Post* coverage of Khashoggi's murder, the *National Enquirer*—the supermarket tabloid owned by Trump's longtime friend and fixer David Pecker—published eleven pages of photos and intimate text messages showing Bezos having an extramarital affair. Somehow the tabloid had secured access to Bezos's phone.

In a blog post, Bezos hinted that Saudi Arabia had hacked his phone and said he had hired a private security team to determine as much. As it turned out, the source for the *Enquirer* exposé was Bezos's mistress's brother, who had shared his sister's private texts and photos with the tabloid for $200,000. But in the course of their investigation, Bezos's investigators determined that the Saudis had simultaneously hacked Bezos's phone using a WhatsApp zero-day exploit. It wasn't difficult to tie the hack back. MbS himself had sent Bezos a WhatsApp video, and soon Bezos's phone was sending three hundred times his usual data through a maze of servers to the Gulf.

A few weeks later I got a call from a source. That WhatsApp exploit that gave the Saudis entre to Bezos's phone? Well, it was the same one his buddy had sold to Crowdfense, the Saudi-Emirati front company.

"How can you be sure?" I asked.

"Either it's the exploit, or a WhatsApp exploit that does the same exact thing."

"Will your contact talk? Off the record? On background?"

"Never," he told me.

That *fucking salmon.*

I couldn't really blame him. I'd heard the stories. The mercenaries returning to the States told me they had started getting threatening phone calls from their former employer, warning there would be "consequences" if they ever discussed their work back in Abu Dhabi.

They knew how far the state was willing to go to silence its agitators. And Trump's transactional brand of Middle Eastern diplomacy had removed the few remaining barriers. In Trump, the Gulf monarchies had hit the jackpot. The president was willing to overlook their human rights abuses in the name of economic prosperity and the hope of a UAE-Israeli peace deal that the president's son-in-law, Jared Kushner, could one day call his own. Kushner, officials told me, bragged about his regular correspondence with UAE crown prince Mohammed bin Zayed (known as MbZ) and MbS. (Their preferred method of communication? WhatsApp.)

Having seen just how much the Gulf monarchies had gotten away with under Trump, some suspected they might try to leave their own mark on the next election.

"You don't think they're going to just let their prince walk off into the night, do you?" one former White House official asked me in late 2018.

It took me a moment to realize the "prince" in this case wasn't MbZ or MbS. It was their WhatsApp buddy, their best asset in the White House, Kushner.

Meanwhile, the price the Gulf was willing to pay hackers to surveil their enemies just kept going up.

"This cyber game is going to the highest bidder," Tom Bossert, Trump's first Homeland Security Advisor told me after he left the White House. If ever there was a need for a moral compass in this market, it was now. Maybe, as more abuses emerged, more people would refuse to sell their zero-days to autocrats. But who was I kidding anymore? At the price tags regimes were now willing to pay, there would always be hackers and brokers who wouldn't think twice.

In the years since I left Buenos Aires, Argentina's economy only contracted. Unemployment reached a thirteen-year high. The peso was swinging wildly again. The country's up-and-coming young hackers had more incentive than ever to insulate themselves in the under-the-table, inflation-free cyberarms trade. If it was dictators and despots who wanted their code, so be it.

THIS WAS ALL happening under an America First president who was temperamentally uninterested in complexity, who romanticized authoritarianism, and who dismissed any talk of Russian election interference as an elaborate "hoax." That his trade war with China, his abandonment of the Iran nuclear deal, and his refusal to confront Putin directly might have unintended and dangerous consequences seemed to matter little in the Old Western Trump had written for himself. In his retelling, he was Wyatt Earp restoring law and order, securing the border, and blazing his way to glory.

Early on, Trump grew bored with daily intelligence briefings and canceled them. His top cybersecurity officials were also shown the door. And when it came time to address Russia's election interference, he balked: What interference? Any discussion of it cast doubt on his legitimacy. By late 2018 the Trump administration was already easing some sanctions it had placed on Russian oligarchs and their companies. Over the next few months, any senior official who tried to broach election security with the White House was "given the Heisman," as one official put it. The job of cybersecurity coordinator, the very person whose job it was to coordinate American cyber policy and to oversee the VEP by which the United States decided which zero-days to keep, and which to fix, was eliminated outright.   When then homeland security secretary Kirstjen Nielsen repeatedly tried, to focus the administration on preventing a repeat of 2016, Trump's then chief of staff Mick Mulvaney told Nielsen never to mention election interference in front of the president again. A few months later, she too was shown the door.

In fact, Trump said he would welcome interference. Asked point-blank in June 2019 whether he would accept damaging information from a foreign government on an opponent in the future, Trump responded: "I think I'd take it." And a few weeks later, there were Trump and Putin making light of it all. Asked whether he would tell Putin not to interfere in 2020, Trump had mock-scolded his buddy. "Don't meddle in the election, President," he told Putin, wagging his finger with a smile. As for the journalists doing the questioning that day, "Get rid of them," Trump told Putin, whose tenure oversaw the murder of dozens of Russian journalists. "Fake news is a great term, isn't it? You don't have this problem in Russia, but we do."

In reality, Putin never stopped meddling. In fact, 2016 was just their dry run. As one expert told the Senate Intelligence Committee, hacks of state voter databases and back-end election systems in 2016 was "the reconnaissance to do the network mapping, to do the topology mapping, so that you could actually understand the network, establish a presence so you could come back later and actually execute an operation." What stopped them from going further, then? Perhaps it was Obama's one-on-one with Putin, or Brennan's phone call to the head of the FSB. But with the 2020 elections fast approaching—and evidence piling up that the Kremlin was back to its old tricks—you had to wonder who in the White House would confront Putin this time.

In the four intervening years, the Kremlin only grew more emboldened—albeit stealthier. In 2016 Russia's influence operation stood out for its brazenness. Social media posts were written in broken English. Facebook ads were paid out in rubles, and self-proclaimed Texas secessionists and Black Lives Matter protesters logged into servers from Red Square. Now Russians were setting up offshore bank accounts, paying real Facebook users to rent their accounts, and obscuring their real locations using Tor, the anonymizing software.

Inside IRA headquarters in Saint Petersburg, Russian trolls had far greater command of America's politics. They switched out obvious Russian bots for human-scripted chatbots that searched for keywords in discussions and chimed in with incendiary scripted responses. They seized every opportunity to pour fuel on America's culture wars, weighing in on guns, immigration, feminism, race, even the NFL players who took a knee during the national anthem.

Russia's trolls grew mindful of the time difference, posting anything that would make liberals' blood boil in the early Russian hours, so it would stir them up at night, and posting anything aimed at conservatives early in the morning, just as they were settling in to *Fox & Friends*. The IRA continued to create fictitious Facebook accounts with monikers like "Bertha Malone" and "Rachell Edison," seeded false stories that Obama had ties to the Muslim Brotherhood, and echoed NRA narratives that Democrats were out to take everyone's guns. The IRA's 2018 midterm election interference project had a code name: Project Lakhta. And in the six months leading up to the midterms, they poured $10 million into their efforts. Once again, it was nearly impossible to measure their impact on the American psyche.

But this time U.S. officials managed to work around Trump and respond decisively.

IN SEPTEMBER 2018 THE president ceded decision-making for U.S. offensive cyberattacks to the Pentagon, where those decisions now fell to General Paul M. Nakasone, the new director of the NSA, who also ran Cyber Command. The previous month, the president's hawkish national security advisor John Bolton had drafted a new cyber strategy giving Cyber Command far more leeway to conduct offensive cyberattacks than they had under Obama, when every attack required express presidential approval. Trump signed the secret, still-classified order known as National Security Presidential Memorandum 13 in September. And with Cyber Command's reins loosened and Trump effectively out of the way, the nation's elite warriors took the fight right back to Russia's servers.

That October, one month ahead of the midterms, Cyber Command delivered the Kremlin a message. They posted warnings directly to the IRA's computer screens, threatening the agency's trolls with indictments and sanctions if they meddled in the elections. It was the digital equivalent of the leaflets American pilots dropped over Japan in 1945, urging evacuation before the coming bombs. On Election Day, Cyber Command took the IRA's servers offline and kept them there for several days, while election officials certified the vote. We may never know what, if anything, the Russians had planned for that day, but the 2018 midterms proceeded relatively unscathed.

Cyber Command's triumph proved illusory. A few weeks later the Russian hacking unit known as Cozy Bear—one of the same units that hacked the DNC in 2016—resurfaced from a year of dormancy. For several weeks they barraged Democrats, journalists, law enforcement, defense contractors, even the Pentagon with phishing attacks before going eerily quiet again in early 2019. Either they ceased their attacks—unlikely—or they got better at hiding.

Over the next few months, the NSA and the Brits over at GCHQ caught Russian intelligence units boring into the networks of an elite Iranian hacking unit, piggybacking on Iran's systems to attack governments and private companies around the world. In a rare joint public message, the agencies laid out the Russian plot. It was a warning, as 2020 neared, that the Russian threat was rapidly evolving. Nothing could be taken at face value anymore.

Back in Washington, a nascent cybersecurity agency within the Department of Homeland Security was charged with the most thankless and incongruously controversial job in Washington: protecting the 2020 election. Just after the 2018 midterms, Trump signed into law the Cybersecurity and Infrastructure Security Agency Act, elevating a dedicated cybersecurity agency within DHS that would now be known by the acronym CISA. Chris Krebs, a forty-something, former Microsoft executive with a man-boyish charm, was tapped to lead the agency defending an election the president did not want to defend and helping states that did not want his help. It is states, after all, that run U.S. elections. To provide states with any federal election security assistance—even so much as a vulnerability scan—federal agencies must be invited. States—particularly red ones—had long interpreted federal assistance in their elections as its own kind of interference. But in 2019, all one had to do was look at the record number of ransomware attacks crippling American counties, towns, and cities to understand just how vulnerable they were.

More than six hundred American towns, cities, and counties were held hostage by ransomware attacks between 2019 and 2020. Cybercriminals were not just hitting big cities like Albany and New Orleans, but smaller counties in swing states like Michigan, Pennsylvania, and Ohio. In Texas, a new battleground, twenty-three towns were hit simultaneously. In Georgia, the tally of victims was stunning: The city of Atlanta. The state's Department of Public Safety. State and local court systems. A major hospital. A county government. A police department for a city of thirty thousand people. In each case, networks crashed; public records disappeared; email went down; laptops had to be foren-sically examined, reconfigured, and tossed out; police departments were rele-gated to pen and paper.

Officials and security experts shuddered to think about the impact a well-timed ransomware attack on voter lists, registration databases, or secretaries of state could have come November 3.

"The chance of a local government not being hit while attempting to manage the upcoming and already ridiculously messy election would seem to be very slim," Brett Callow, a threat analyst, warned me.

Florida—America's enduring election problem child—seemed to be getting the brunt of it. Palm Beach County—the same county that decided the 2000 election—hid the fact that in the weeks leading up to 2016 its election offices were taken out by ransomware. County officials did not even bother telling federal officials about the attack until 2019 when, once again, hackers held two of its towns hostage. In Riviera Beach, officials paid $600,000 to undo a vicious attack that brought down email, water utilities, and pump stations. In another attack just south, in the village of Palm Springs, officials paid an undisclosed sum to their extortionists, who never returned their data.

At first glance, the attacks hitting American towns and cities appeared to be run-of-the-mill ransomware. But starting in the fall of 2019, it was clear many were multistage attacks. Hackers were not just locking up victims' data, they were stealing it—dumping it online in some cases, peddling access to victims' systems on the dark web, and, in one especially disturbing case, got caught selling their access to North Korea. Just as Ukraine's investigators determined the NotPetya ransomware attack was a political hit job, officials at CISA, the FBI, and intelligence agencies feared the ransomware attacks hitting the heartland weren't just motivated by profit, but by politics.

There was little question where many of these attacks came from. A striking number were conducted between 9:00 A.M. and 5:00 P.M. Moscow time. Many were deployed through a botnet, a vast network of infected computers, called TrickBot, whose developers were based in Moscow and St. Petersburg. TrickBot's operators sold access to infected computers to cybercriminals and ransomware groups all over Eastern Europe. But the ransomware group that was using Trickbot to ransom American targets left telling clues behind. Scattered among attackers' code were Russian language artifacts. And perhaps most telling of all was that they specifically designed their ransomware to avoid infecting Russian computers. The code searched for Cyrillic keyboard settings and when it found them, moved right along—technical proof they were abiding by Putin's first rule: no hacking inside the Motherland.

By 2019, ransomware attacks were generating billions of dollars for Russian cybercriminals and were becoming more lucrative. Even as cybercriminals raised their ransom demands to unlock victims' data from three figures to six, to millions of dollars, local officials—and their insurers—calculated it was still cheaper to pay their digital extortionists than to rebuild their systems and data from scratch. The ransomware industry was booming and—with all that loot pouring into Russia—intelligence officials found it inconceivable that the Kremlin was not aware of, exploiting, or coercing criminals' access for their own political ends.

It was a leap, to be clear. But with the 2020 elections imminent, officials could not afford to ignore the long-standing partnership between Russia's cybercriminals and the Kremlin. Seared into memory was Russia's hack of some five hundred million email accounts at Yahoo five years earlier. It took investigators years to unravel that attack. Ultimately they traced it to two cybercriminals working hand-in-hand with two Russian agents at the FSB. The FSB agents had allowed the cybercriminals to profit from the theft of personal data, while mining their access to spy on the personal emails of American officials, dissidents, and journalists. More recently, intelligence analysts determined that a prominent Russian cybercriminal—the leader of an elite cybercrime group that called itself Evil Corp—wasn't just working hand-in-hand with the FSB. He *was* FSB.

"There's a pax mafiosa between the Russian regime and its cyber cartels," is how Tom Kellermann, a Russian cybercrime expert, put it to me as we inched closer to the 2020 election. "Russia's cybercriminals are treated as a national asset who provide the regime free access to victims of ransomware and financial crime. And in exchange, they get untouchable status. It's a protection racket and it works both ways."

Officials furnished no proof. But that fall, as ransomware attacks took down one American town after another, they came to fear the ransom elements were smokescreens for deeper probes of counties that might make for ideal targets closer to the 2020 election. That November they watched their worst nightmare almost realized. The week of Louisiana's governor election, cybercriminals held Louisiana's secretary of state's office hostage in a ransomware attack that would have upended the election had local officials not had the foresight to separate Louisiana's voter rolls from their broader network. Louisiana's election proceeded unscathed, but forensics revealed just how premeditated attackers had been. Timestamps showed that Louisiana's assailants had breached its systems three months earlier but patiently waited to time their attack for the election. It was, the FBI feared, a preview for 2020.

In the months that followed, the FBI sent confidential missives to field agents across the country, warning that ransomware would "likely" take out America's election infrastructure. As to whether those attacks were the work of opportunistic profiteers, a more calculated state adversary, or some convergence of the two, the government still had no clear answers.

BACK IN WASHINGTON, the election was shaping up to be open season for hackers. A raft of election security bills faced a one-man roadblock in Senate Majority Leader Mitch McConnell. McConnell made clear that he would not advance any election security bill, no matter how bipartisan. Even measures deemed critical by election integrity experts—paper trails for every ballot and rigorous post-election audits, bills that blocked voting machines from reaching out to the web and required campaigns to report foreign outreach—died on McConnell's desk. It was only after critics took to calling him "Moscow Mitch" that McConnell begrudgingly approved $250 million to help states guard against interference. But even then, he refused to tack on federal requirements for paper backups and audits that the experts deemed critical. Publicly, McConnell did this out of ideological purity: he had long advocated against what he called a Washington "takeover" of state elections. Privately, colleagues suspected his animus toward election bills stemmed from fears of provoking the president.

And it took very little to provoke the president. In his zeal to replace what he called the "Russia hoax" with an alternate explanation, Trump eagerly embraced a Kremlin-backed conspiracy theory that placed Ukraine at the center of the 2016 election interference, which in turn set in motion his impeachment. In the Kremlin's—and now Trump's—revisionism, the DNC was not hacked by Russia but by Ukraine; the company that the DNC hired to investigate its breach, CrowdStrike, was owned by a wealthy Ukrainian; and

CrowdStrike was hiding the DNC's compromised server in Ukraine, where the FBI could not access it, to hide Ukraine's role in hacking it. Not one iota of this was true. All seventeen U.S. intelligence agencies concluded, early on, that Russia was behind the DNC hack. The CrowdStrike theory had no basis to it whatsoever: One of its two cofounders was American; the other was a Russian exile whose family escaped to the United States when he was a child. CrowdStrike never had physical possession of the DNC servers. Like every other security company that responds to breaches in America, they conducted their investigation through a process called "imaging," which involves copying compromised computer hard drives and memory and working their way back from there. CrowdStrike had shared their findings with the FBI. The FBI also had imaged copies of the DNC hard drives, and through their own analysis agreed with CrowdStrike that Russia's hacking units Cozy Bear and Fancy Bear were to blame.

In any other political climate, anyone pushing this fringe theory would be diagnosed as certifiably insane. Not in the age of Trump. This was the Kremlin and Trump's last-ditch effort to malign Ukraine and the Democrats at the same time. And as with his earlier "birtherism," the president would not relent. He held up nearly $400 million in Congressionally approved military aid to Ukraine. And when Ukraine's new president, Volodymyr Zelensky, tried to ingratiate himself to Trump and shake the funds free, Trump told Zelensky, "I would like you to do us a favor though," in their now infamous July 2019 call, followed shortly by, "The server, they say Ukraine has it." A quid pro quo.

Another conspiracy theory emerged in that call. It involved Burisma, the Ukrainian gas company on whose board Hunter Biden, the son of Joe Biden, served. Burisma had brought on Biden's son while his father was leading Obama's Ukraine policy as vice president. At the time, Biden—backed by our European allies—was pushing Ukraine to fire its top prosecutor, who was failing to pursue corruption cases. But with Biden surging in head-to-head polls against Trump in the summer of 2019—and emerging as Trump's likeliest opponent in 2020—Trump flipped the narrative upside down: He accused Biden of messing with Ukraine's criminal justice system solely because Burisma was under investigation and his son could be implicated. The president's personal attorney, Rudy Giuliani, made pushing this theory his personal mission. He took it upon himself to travel to Ukraine where he phoned home regularly with claims to have evidence of wrongdoing. (One year later, the Treasury Department outed Giuliani's Ukrainian source as an "active Russian agent" and sanctioned the Ukrainian for waging a covert influence campaign to interfere in the 2020 election.)

In that infamous Zelensky call, Trump had asked a second favor: that the Ukrainian president announce a public investigation of Burisma and the Bidens, which Trump referred to as "the other thing." Had it not been for a whistleblower in the White House—legally required to listen in on the call—Ukraine might have complied with Trump's request and locked in its millions in aid. Instead, the whistleblower's report set off an impeachment inquiry, freed Zelensky from his vise, and stopped the sham Burisma investigation dead in its tracks.

Or so we all thought. In reality, Russian hackers picked up where Trump's pressure campaign left off. As closed-door impeachment inquiries into Trump's Ukraine intervention wrapped up on Capitol Hill in November 2019, and with the public hearings set to begin, I learned that Fancy Bear, one of the same Russian units that hacked the DNC in 2016, began boring into Burisma in an eerily similar phishing campaign. And this time too, it appeared Russia's hackers successfully stole the credentials necessary to roam free through Burisma's inboxes, in search of any emails that might play into Trump's muddy brew of conspiracy theories and perhaps sully the Bidens in the process. I broke news of Russia's Burisma hack that January. And as we entered the final months of the 2020 campaign, a hack-and-leak of Burisma materials seemed all but inevitable. (It was: The *New York Post* eventually took that baton and ran with it.) Whether it would bolster Trump and Giuliani's baseless claims of the Bidens' "corruption" no longer seemed to matter. False investigations and innuendo—not truth—had killed Hillary Clinton's campaign. That, Trump seemed to figure, might be all he needed to win again. And it appeared Russia was, once again, willing and well-positioned to help.

U.S. intelligence officials warned lawmakers and the White House, one month later, that Russian hackers and trolls were once again working overtime to elect Trump. Trump erupted at the briefings. Not because Russia was interfering in America's democracy. The president was upset intelligence officials had shared their findings with Democrats. Trump was so incensed that he replaced his acting director of national intelligence with a vocal supporter and decried the intelligence as "misinformation" on Twitter. Republicans dismissed the findings, too. Chris Stewart, the Utah Republican, told the *Times* that Moscow had no reason to support Trump in 2020.

"I'd challenge anyone to give me a real-world argument where Putin would rather have President Trump and not Bernie Sanders."

As it turned out, Russia was bolstering Sanders too. In a separate briefing, intelligence officials told Sanders that Russia was trying to boost his chances in the Democratic primary fight against Biden, in an apparent bet that Sanders would be Trump's weakest opponent.

"If Sanders wins the Democratic nomination, then Trump wins the White House," one former Kremlin advisor told a reporter. "The ideal scenario is to maintain the schism and uncertainty in the States 'til the end. Our candidate is chaos."

Over the next few months, I tracked one Russian campaign after another. But with their 2016 playbook exposed, the Kremlin was changing tack. Russian hackers were after campaign emails, again, but they weren't getting them through rudimentary phishing attacks anymore. NSA analysts discovered that Sandworm, the same Russian hackers who shut off Ukraine's power, unleashed NotPetya, and probed voter registration systems in all fifty states in 2016, was exploiting a vulnerability in an email program that, the NSA warned in a public advisory, gave Russia "dream access."

After they were caught hacking Burisma, Fancy Bear took greater care to hide their tracks. They shifted their operations to Tor, the anonymity software that concealed their true whereabouts. Microsoft revealed that in just one two-week period, Fancy Bear had taken aim at more than 6,900 personal email accounts belonging to American campaign workers, consultants, and politicians on both sides of the aisle.

And while it still was not clear what, if any, role the Kremlin was playing in the ransomware attacks, the attacks were getting worse. TrickBot's developers were now cataloging American municipalities they had access to, selling anyone who wanted it a paint-by-number approach to hacking our election.

When it came to disinformation, Russia's goal was still the same: divide and conquer. But this time, the Kremlin's trolls didn't have to spin up "fake news." Americans—perhaps nobody more so than our president—were generating plenty of false, misleading, and divisive content every single day. In 2016, Russians spun up fictitious tales of Democrats practicing witchcraft. With Americans more divided than at any time in recent history, Russia's trolls and state news outlets found it far more efficient to amplify American-made disinformation than create their own. This time they weren't looking to go viral—that would draw too much attention—they simply searched for sparks wherever they flew and offered up a little kindling.

When a mobile app Democrats used to report results from their Iowa primary caucus imploded in public view in February, I watched as Russian trolls retweeted and stoked Americans who falsely believed the app was a ploy by Hillary Clinton's inner circle to wrest the election from Bernie Sanders. When the coronavirus pandemic took hold, I watched those same Russian accounts retweet Americans who surmised Covid-19 was an American-made bioweapon or an insidious plot by Bill Gates to profit off the eventual vaccine. And as the world stood still waiting for that vaccine, Russian trolls worked overtime to legitimize the vaccination debate, just as they had during the worst of Ukraine's measles outbreak one year earlier. They retweeted Americans who challenged official Covid-19 statistics, protested the lockdowns, and doubted the benefits of wearing a mask. And when thousands of Americans took to the streets to protest the murders of African Americans at the hands of police, I watched those same Russian accounts retweet Americans, including the president, who dismissed the Black Lives Matter movement as a Trojan horse for violent left-wing radicals. With each new campaign, it got harder to pinpoint where exactly American-made disinformation ended and Russia's active measures began. We had become Putin's "useful idiots." And so long as Americans were tangled up in our own infighting, Putin could maneuver the world unchecked.

"The mantra of Russian active measures is this: 'Win through force of politics rather than the politics of force,'" is how Clint Watts, a former FBI agent who specializes in Russian disinformation, explained it to me. "What that means is go into your adversary and tie them up in politics to the point where they are in such disarray that you are free to do what you will."

Some days, American officials seemed downright determined to clear Putin's way. It wasn't just Moscow Mitch and his refusal to pass a single election-security bill. That August, the nation's newly installed spy chief, John

Ratcliffe, ended the in-person intelligence briefings to Congress about election interference. Ratcliffe blamed the decision on too many leaks, but by reverting to written assessments, it allowed the director of national intelligence to twist intelligence in a way that back-and-forth, in-person briefings to Congress could not. Over the next weeks and months, intelligence analysts and officials watched in horror as Ratcliffe twisted intelligence to Trump's tastes, stuck minders on intelligence briefers to keep them from straying from approved topics, selectively declassified intelligence to score political points, offered Trump supporters top spots in his ranks, and contradicted intelligence findings at every turn by claiming that China and Iran, not Putin, posed the gravest threats to the election.

It was a propaganda fantasy designed for one man and it had little correlation to reality. China and Iran were active too, but not in the way that Trump and his henchmen led Americans to believe. Iran's cyber corps were more aggressive than ever, but they were hardly a success story. They bombarded Trump's campaign with phishing attacks but never actually managed to break in. Closer to the election, they spoofed emails to thousands of Americans in which they claimed to be Proud Boys, the far-right white supremacist group, and warned voters to vote Republican or "we will come after you." But they did not use stolen voter data as some suspected; they used publicly available data. And because of a careless error, American officials immediately traced the emails to Iran. It was the quickest attribution in history and, perhaps, should have even been celebrated. Instead, Ratcliffe seized the moment to spin it in Trump's favor. Standing at a podium, Ratcliffe told reporters Iran's little email gimmick was not only intended to intimidate voters and sow social unrest; it was designed to "damage Trump."

Trump and his advisers continued to dilute the threat from Russia by playing up the threat from Iran and China. At a September campaign rally, the president claimed, again, that Russian interference was a hoax. "What about China? What about other countries? It's always Russia, Russia, Russia. They're at it again." Trump's deputies were happy to toe that line. Asked in television interviews which country posed the gravest threat to the upcoming election, Trump's national security adviser, Robert O'Brien, and his pugnacious attorney general, Bill Barr, each repeated that it was China—not Russia—that posed the most serious threat.

Complicating their narrative was the fact that, at that very moment, China was actively targeting Biden, not Trump. Intelligence officials would not say so publicly, but security teams at both Google and Microsoft discovered China's hacking campaign was aimed squarely at the Biden campaign. And China's efforts did not appear to be the preliminary stages of a Russian hack-and-leak, but standard espionage—not so different from China's hack of the McCain and Obama campaigns in the 2008 election, when Chinese spies gained access to the policy papers and emails of top advisers. This time too, security researchers and intelligence analysts believed China was reading the tea leaves, looking to assess Biden's policy plans for Beijing.

The mission of the intelligence community had always been nonpartisan. With each passing day, the president and his political appointees were contorting intelligence, in ways big and small, for the president's political ends. Of Trump's senior team, only the FBI director, Christopher Wray, was willing to break through the White House's alternate reality in public remarks. The same month Barr and O'Brien were on television calling out China, Wray testified to lawmakers that Russia was interfering in the election through "malign foreign influence in an effort to hurt Biden's campaign." He spoke the words matter-of-factly but, given the truth famine we found ourselves in, they landed like the words of a renegade soldier, and Trump and his minions punished him for it.

"Chris, you don't see any activity from China, even though it is a FAR greater threat than Russia, Russia, Russia. They will both, plus others be able to interfere in our 2020 Election with our totally vulnerable Unsolicited (Counterfeit?) Ballot Scam," Trump tweeted.

Steve Bannon, the president's far-right-hand man, later called for Wray—and Dr. Anthony Fauci, the infectious disease expert—to be beheaded as a warning to federal workers who dared question the president's propaganda.

We had all spent the past four years worried what our foreign adversaries were planning. But as the election neared, it was clear that the real interference was coming from within. Even before a single vote was cast, Trump was delegitimizing the 2020 election and with it, democracy itself, calling it "rigged," "a fraud and a sham." Polls began to show that a majority of Americans, some 55 percent, agreed.

"I disagree with a lot of things the president has done, but this is the worst," Senator Angus King, the independent from Maine, told me in the days leading up to the election. "Undermining the confidence of the American people in the functioning of their democratic system is profoundly dangerous. And it happens to line up with what Russia and other countries are doing."

SUCH WERE THE political landmines that American officials had to work around as they made the final push to shore up state and county election defenses going into November 2020. With McConnell unwilling to pass a single election security bill, Senator Mark Warner of Virginia, the top Democrat on the Senate Intelligence Committee, made it his personal quest to convince secretaries of states around the country, even on the far right, that the threat of foreign interference was real and to accept CISA help.

At CISA, Krebs dispatched his deputy, Matt Masterson, the former commissioner of the Election Assistance Commission, to go state to state, hat in hand, to plead with states and counties to scan and patch their systems for vulnerabilities, lock up voter registration databases and voter rolls, change passwords, block malicious IP addresses, turn on two-factor authentication, and print out paper backups. The pandemic had upended the election, shuttered polling places, and pushed millions more Americans to vote by mail. In some ways this made the election more secure: Mailed ballots had a built-in paper trail, but it also made the voter registration databases that much more precious. An attack that sabotaged voter-registration data—by switching voter addresses, marking registered voters as unregistered, or deleting voters entirely—risked the digital disenfranchisement of thousands, millions even, if pulled off at scale. The shift to mail-in voting also meant that, barring a landslide, the election would not be called in one night but would be slow-rolled over several days, weeks even, forming an ever-widening attack surface. And it made the threat of a ransomware attack on the registration systems, the post office, voter signature verification, and tabulation and reporting systems that much more chilling.

In Redmond that September, Microsoft's Tom Burt stewed over the ransomware attacks on American towns and cities. Just that month, a Texas company that sells software that some cities and states used to display election results was hit by ransomware. The company, Tyler Technologies, did not actually tally votes, but it was used to aggregate and report results in at least twenty places around the country—making it exactly the kind of soft target officials worried could be struck by anyone trying to sow chaos and uncertainty in the election. The attack was just a blip among more than a thousand ransomware attacks on American towns and cities over the previous year, but it was also precisely the sweet spot that Russian hackers took advantage of in Ukraine's election in 2014, when they dropped malware that would have claimed victory for a far-right candidate had Ukrainians not caught it in the nick of time. The attack was also disturbing for another reason: In the days that followed, Tyler's clients—the company would not say which ones—saw outsiders trying to gain access to their clients systems, raising fears that their assailants were out for something more than just a quick profit.

Burt had been watching the ransomware attacks with growing unease. The catalyst was seeing that TrickBot's operators had added surveillance capabilities that allowed them to spy on infected officials and note which belonged to election officials. From there, it would be a breeze for cybercriminals, or state actors, to freeze up election systems in the days leading up to the election and after.

"We don't know if this is Russian intelligence," Burt told me, "but what we know is TrickBot is, by volume, the key distribution pipeline for ransomware and that it would be really easy for state actors to contract with TrickBot to distribute ransomware with the goal of hacking election systems. That risk is real particularly given that so much of the ransomware is targeting municipalities. Just imagine that four or five precincts were hit with ransomware on Election Day. Talk about throwing kerosene on this unbelievable discussion of whether the results are valid or not. It would be a huge story. It would churn on forever. And it would be a huge win for Russia. They would be toasting with vodka well into the next year."

"That," he told his deputies, "is a risk I want to take out."

The obvious place to start was TrickBot. It had been used as the conduit for ransomware attacks in Florida, courts in Georgia, the *Los Angeles Times*, New Orleans, state agencies in Louisiana, and just that month, one of the largest medical cyberattacks in history after ransomware delivered via TrickBot hijacked more than four hundred hospitals in the middle of the pandemic.

Burt had put together a team of security executives and lawyers that sprung to action to figure out how to cut TrickBot off at the knees. Together they decided legal action, through the courts, was the best way to go. By bringing their case to federal court, they could force web-hosting providers to take TrickBot's operators offline, arguing that cybercriminals were violating U.S. copyright acts by using Microsoft code for malicious purposes. They strategized for months but ultimately decided to wait until October to act, fearing that an earlier move would give Russia's hackers time to regroup by November.

But when it was go time that month, Microsoft discovered someone else was already there. Starting in late September, U.S. Cyber Command started hacking into TrickBot's command and control, feeding its infected computers a set of instructions that sent them into an endless loop. It was the equivalent of a phone dialing its own number over and over and over again so that anyone else trying to get through would get a busy signal. TrickBot's operators were able to reclaim their computers in half a day. But roughly a week later, Cyber Command struck again, hitting TrickBot's systems with the same attack as before. This time too the disruption was temporary. But it was also a message, similar to the one General Nakasone's cyber troops sent the Internet Research Agency in the days leading up to the midterms: We are watching you, we are inside, and we will take you out if you come for our election. (As the election approached, we learned that Cyber Command launched a similar attack on the Iranian hackers behind the Proud Boys campaign, hindering their ability to conduct more attacks.)

In a briefing with reporters, Nakasone declined to discuss the attacks on Iran and TrickBot, but made clear Cyber Command was prepared to do more. "When authorized we have been prepared and poised to do other operations against our adversaries." As for the weeks ahead, the general said, "I am confident in the actions taken against adversaries in the last several weeks and several months to ensure they are not going to interfere in our elections."

Microsoft's TrickBot takedown in federal court in the weeks following Cyber Command's attacks formed the second part of a one-two punch that relegated TrickBot's operators to "wounded animals," as one security executive put it to me in late October. With more than 90 percent of their infrastructure

376 THIS IS HOW THEY TELL ME THE WORLD ENDS

down, TrickBot's Russian operators lashed out, shifted to new tools, and retaliated on American hospitals. They traded lists of some four hundred American hospitals they planned to target with ransomware, and slowly started hitting them one by one. This, with less than a week before the election, when hospitals were seeing a record spike in coronavirus cases.

"We expect panic," one Russian hacker told his comrades in a private message exchange captured by a threat researcher.

The FBI, CISA, and the Department of Health and Human Services set up an emergency call with hospital administrators and security researchers to debrief them on the urgent "credible threat." Already hospitals in California, Oregon, and New York were reporting attacks. They were not yet life-threatening, but they relegated hospitals to pen and paper, interrupted chemotherapy treatments, and forced hospitals that were already facing record shortfalls in staff to divert patients elsewhere. Officials feared mayhem. But with one week before the election to go, they had to keep focus.

AT CYBER COMMAND, CISA, the FBI, and NSA, officials went into high alert. With voting well under way, attacks were already surfacing: In Georgia, a database that verified voter signatures on their mailed ballots was locked up by Russian hackers in a ransomware attack that also dumped voters' registration data online; in Louisiana, the National Guard was called in to stop cyberattacks on smaller government offices that used tools that previously had only been seen in North Korea. Russia's A Team was caught probing systems in Indiana and California. And someone had briefly hacked the Trump campaign, defacing its website with threats in broken English warning that the worst was yet to come.

On their own, none of these attacks mounted to much. But taken together, they fed into what CISA's Krebs and cybersecurity executives in Silicon Valley dubbed a "perception hack": smaller attacks, perhaps concentrated in battleground states, that could be easily amplified into something bigger and seized upon as evidence that the whole election is "rigged," as the president continued to claim, ad nauseum, it would be. At CISA, Krebs' team set up a "rumor control" site to debunk conspiracy theories and exaggerated claims of election fraud. It put them squarely in the president's crosshairs. Krebs and his deputies accepted that they would likely get fired for it—just as soon as the election concluded.

For months now, in Silicon Valley, cybersecurity executives at Facebook, Twitter, and Google spent more time securing their networks and trading intelligence back and forth with one another, one Twitter executive told me, than they had with their spouses. The bulk of their efforts were aimed at defending against foreign interference, whether in the form of an attack or an influence campaign, and readying themselves to slap warnings and labels on false or misleading posts that undermined the integrity of the election. They, too, knew this would not play well with the president.

When Election Day arrived there were the expected hiccups across the country. In Georgia, a water main break in Fulton County delayed vote counts in Atlanta for several hours, then days. In two other Georgia counties, different software issues delayed poll workers when they tried to check in voters. In a third, a different software issue delayed officials' reporting of vote tallies but did not affect the final count. In Michigan, officials in one county mistakenly counted votes from one city twice but quickly corrected the error. In another, in the Republican stronghold of Antrim County, the unofficial results initially showed Biden beating Trump by roughly three thousand votes—a sharp reversal from Trump's 2016 performance there. An election worker had mistakenly configured ballot scanners and reporting systems with slightly different versions of the ballot, which meant results initially did not line up with the right candidate—a human error that was quickly caught and fixed.

But, miraculously perhaps, there was no evidence of outside interference, no fraud or even a single ransomware attack that day. Every three hours, CISA officials debriefed reporters on what they were witnessing, and while they stressed that "we are not out of the woods yet," the attacks so many feared from cybercriminals in Russia, Iran, and China never materialized. It was, Chris Krebs said, "just another Tuesday on the Internet."

We may never know what our adversaries had planned for Election Day, or what might have stopped them in the days that followed as the "red mirage" of in-person votes for Trump swung blue as record turnout in mail-in votes yielded a win for our next president, Joe Biden.

I would like to think it was the coordinated attacks by Cyber Command, the unsung heroes at CISA who zipped up state and county systems, the TrickBot takedowns, the quick attribution of the Iranian attacks, or the naming and shaming by federal prosecutors who, in the weeks leading up to

the election, unsealed charges against the Russian military intelligence officers behind NotPetya, the Ukraine grid attacks, the attack on the 2018
Olympics, the French election, and the probes of our voter registration databases in 2016. I would like to think that collectively all of it amounted to
successful deterrence that we could improve upon and deploy again and again.

There was a moment, in the weeks ahead of the election, when Putin
appeared to drop his poker face and momentarily fold. In a Kremlin statement,
Putin called for a cyber "reset" with the United States. "(I propose) . . .
exchanging guarantees of non-interference in each other's internal affairs,
including electoral processes," Putin began. "One of the main strategic challenges of our time is the risk of a large-scale confrontation in the digital sphere.
We would like to once again appeal to the United States," he said, "to reset our
relations in the use of information and communication technologies."

Perhaps it was sincere. But American officials dismissed Putin's proposal
outright. The top national security official at the Justice Department disparaged
Putin's reset call as "dishonest rhetoric and cynical and cheap propaganda."

But something tells me the reason we did not see more Russian interference
in 2020 was not because Putin was deterred, but perhaps because he concluded
his work here was already complete. These days, Russian trolls barely have to
lift a finger to sow discord and chaos when Americans, and our outgoing president, do it for them. As of this writing, we are exactly one week out from
November 3 and Trump has yet to concede. Trump's calls that the election was
"rigged," of widespread voter "fraud," of suspicious "glitches" have only intensified. Even when Twitter slaps a warning on the president's tweets, conservative sites like Breitbart, the Federalist, and newer, conservative-friendly social
media platforms like Parler carry his message loud and far. On one side of the
country, hundreds of Trump's supporters are lined up outside a poll center in
Arizona yelling, "Count the votes!" while maskless protesters in Michigan
scream, "Stop the count!" In Atlanta, the president's son is calling on supporters
to fight "to the death." In Philadelphia, our democracy's hub, poll workers are
getting death threats for counting votes. The torrent of disinformation is unlike
anything Americans witnessed in the previous four years. And it is clear that
the "perception hack" American officials feared is coming from within the
White House.

Perhaps, very soon, we will learn that Iranian and Russian trolls are
bouncing the president's messages around social media echo chambers. But

even if they are, they are getting drowned out by real Americans. If the goal of Putin's 2016 interference was to sow chaos and undermine democracy, then what is now playing out is beyond his wildest dreams.

I HAD ALWAYS been warned we were headed for the mushroom cloud. In fact, behind all the fast-breaking stories I'd been tracking—the election meddling, the disinformation, China's commercial trade theft and creeping surveillance, the coming train wreck from Iran—I'd been covering one story that seemed to foretell the worst.

Those calls I started getting years ago from DHS? The ones warning that Russia was penetrating our energy networks? Our grid? Well, as our country dithered back and forth on the "Russian "hoax," the reality was that Russian hackers were up to something far worse.

I suppose it's fitting that I would get the call on the Fourth of July. I'd been driving through the Colorado Rockies that holiday weekend in 2017, with my husband and our dog, when my phone rang.

"THEY'RE IN," the voice on the line told me. "THEY'RE FUCKING IN."

I told my husband to pull over and let me out. My source had gotten his hands on an urgent DHS-FBI alert. It was meant solely for the utilities, the water suppliers, the nuclear plants. The bureaucrats were trying to bury it on a holiday weekend. And as soon as I got eyes on it, I could see why: the Russians were inside our nuclear plants.

The report didn't spell it out. But buried in its technical indicators, analysts had included a snippet of code from one of the attacks. The code made clear that Russia's hackers had breached the most alarming target of all: Wolf Creek, the 1200-megawatt nuclear power plant near Burlington, Kansas. This was no espionage attack. The Russians were mapping out the plant's networks for a future attack; they had already compromised the industrial engineers who maintain direct access to the reactor controls and radiation monitors that could affect the kind of nuclear meltdowns the world had only witnessed in Chernobyl, Three Mile Island, and Fukushima. This was Stuxnet. Only it wasn't the United States doing the hacking. It was Russia. And the goal wasn't to stop the boom. It was to trigger one.

The Russians had been shamelessly meddling in our politics. But when it came to our infrastructure, they had probed and prodded, lurked, fired off their warning shots in Ukraine, then vanished. Now they were inside our nuclear plants, lying in wait for the day Putin yelled "FIRE." And if we had any lingering doubts about what the Russians were capable of that July, we had only look to Ukraine, or to the cyberattack they pulled off one month later at Saudi Arabia's Petro Rabigh refinery. Using zero-days, Russia's hackers had leapt from an engineer's computer into the plant controls and switched off the safety locks—the last step before triggering an explosion. The technical hurdles to exact a cyberattack of mass destruction had been cleared. Now we were all stuck in a waiting game to the point of no return.

In another joint DHS-FBI warning the following March, the agencies officially named Russia as the perpetrator behind the assaults on our grid and our nuclear plants. Included in their report was a chilling illustration of our new predicament: a screenshot showing the Russians' fingers on the switches. "We now have evidence they're sitting on the machines," Eric Chien, Symantec's director, told me. "That allows them to effectively turn the power off or effect sabotage. From what we can see, they were there. They have the ability to shut the power off. All that's missing is some political motivation."

The report also included a telling timeline. The Russians had accelerated their strikes on America's grid in March 2016, the same month Russia hacked Podesta and the DNC. Eight months later, even the Kremlin was surprised when their man was voted into the Oval Office. But instead of causing them to back off, Trump's election only emboldened them. Under his watch, Russia invisibly worked their way into an untold number of nuclear and power plants around the country.

"I would say right now they do not think very much will happen to them," General Nakasone told the Senate in the days before he was confirmed as NSA director and head of U.S. Cyber Command in May 2018. "They don't fear us."

As Nakasone assumed his new duties, his staff was still assessing the Russian attacks on our systems. It was not just Wolf Creek; the Russians had also targeted Cooper Nuclear Station in Nebraska, and an untold number of other operators whose identities we still do not know. They also discovered that the same Russian hackers that successfully dismantled the safety guards at the Saudi refinery had been doing "digital drive-bys" of our own chemical, oil, and gas operators in the United States. Russia was inching dangerously closer to attack.

It had long been Nakasone's position that the United States needed to "defend forward" in the cyber domain. The son of a Japanese American linguist who experienced Pearl Harbor firsthand, he believed that the only way to prevent the Big One was to meet the enemy on the battlefield. It was Nakasone who played a critical role in leading Nitro Zeus, the U.S. operation to plant land mines in Iran's grid. And it was Nakasone who argued that Russia's attacks on our critical infrastructure could not go unanswered. Now, under his new authorities, Cyber Command started plotting its response.

In the months that would follow, Cyber Command began planting crippling malware inside the Russian system at a depth and with an aggressiveness that had never been tried before. For years the United States had been among the stealthiest players in the digital realm, but now we were making a show of our power, letting Russia know that if they dared flip the switch here, we would reciprocate. There were some who believed that after years of getting beaten up and blacked out in the digital domain, these attacks were long overdue. Others worried that the United States was effectively enshrining the grid as a legitimate target, which, of course, it was.

For three months David Sanger and I tried to learn as much as we could about the escalating digital cold war between Washington and Moscow. The attacks were highly classified, but then national security advisor John Bolton started dropping public hints. The week we prepared to go to press, Bolton, speaking at a conference, said: "We thought the response in cyberspace against electoral meddling was the highest priority last year, and so that's what we focused on. But we're now opening the aperture, broadening the areas we're prepared to act in." Referring to Russia, he added, "We will impose costs on you until you get the point."

Over the next few days we reached out to Bolton and Nakasone through their spokesmen, both of whom declined to answer our questions about the U.S. grid attacks. But when David went to the National Security Council and presented them with the details we were prepared to publish, something curious happened. Typically, with sensitive national security stories, there is pushback. This time, there was none. They had no national security concerns whatsoever about the publication of our story, officials said. It was the clearest evidence yet that our attacks on Russia's grid were intended to be noticed.

In fact, Pentagon officials' only hesitation about our going public, we learned, was that Trump had not been briefed on the grid attacks in detail. In part, this was because Cyber Command's new authorities didn't require his knowledge or approval. But that was just the cover story. In reality, officials were hesitant to brief Trump for fear that he might countermand the attacks or discuss them with Russian officials, as he had done two years earlier, when he casually disclosed a highly classified U.S. operation—an operation considered so highly sensitive that many senior officials in our own government were still in the dark—to Russia's foreign minister.

When our story went to print that June 2019, Trump went ballistic. He took to his favorite medium, Twitter, to demand that we immediately release our sources, and to accuse us of "a virtual act of treason." It was the first time the president had ever dropped the word *treason*.

For years, we had become inured to his attacks—"fake news," "the enemy of the people," "the failing *New York Times*"—but now he was accusing us of a crime punishable by death. It was a serious escalation in his war on the press, an attack that until now had been reserved for autocrats and dictators. To his eternal credit, our publisher, A. G. Sulzberger, came to our swift defense, writing in an op-ed in the *Wall Street Journal* that the president had crossed a "dangerous line."

"Having already reached for the most incendiary language available," Sulzberger asked, "what is left but putting his threats into action?"

In that moment, it wasn't me I was worried about, but my colleagues abroad. From the moment I started on this beat, I'd found myself drifting into dangerous territory. But I had always taken heart in something my mentor, Phil Taubman, had told me. Taubman had served as the *Times'* Moscow bureau chief during the Cold War. The day I published the story about China's hack of the *New York Times*, we met for lunch. He asked me that day if I was afraid. It was a question I had deliberately refrained from asking myself, and I responded only with a nervous laugh. To which Taubman told me of the days when the KGB would follow him as he drove his young children to school every day in Moscow. The Kremlin often made a loud show of it. They wanted him to know they were watching his every move.

"You should assume the same," Taubman told me that afternoon. But he also wanted me to know this: one of the great privileges of working for the *Times* is that you carry with you an invisible armor. Should anything happen to me, Taubman told me, it would be an international incident. There was a reason the KGB had followed him so aggressively, yet never stepped over the line. For seven years, I had taken comfort in his words. *Invisible armor. An international incident.*

But these days I had begun to question whether my invisible armor existed at all. Khashoggi's death, and the lack of a serious U.S. response, had been a wake-up call. I never believed Trump would make good on his threats, but I worried about the tacit encouragement he was giving to other governments in China, Turkey, Mexico, Myanmar, Russia, the Gulf. We were already seeing this play out in distressing ways. The Mexican hacks. The jailing of journalists in Turkey. The Turkish thugs who'd been given carte blanche to viciously beat up American protesters on Recep Tayyip Erdogan's trip to D.C. China's expulsion of American journalists. The Saudi hack of Bezos. In Egypt, authorities had not thought twice before arresting my colleague David Kirkpatrick in Cairo and deporting him in 2019. And in a separate incident two years earlier that did not emerge until Trump singled us out for "treason," the *Times* received a call from a concerned American citizen. The caller made clear he was acting on his own volition. Egyptian authorities were preparing an imminent arrest of another colleague, Declan Walsh, who had recently published an investigation into Egypt's role in the torture and murder of an Italian student, whose body had been dumped along a highway in Cairo. Alarming as the call was, it was also fairly standard. The *Times* had received many such warnings from American diplomats over the years. But this one was different. The official was distressed. They told the *Times* that their bosses at the American embassy had already signaled to the Egyptians that they would not intervene. The Trump administration was going to let the arrest be carried out. When Walsh called the embassy for help, officials there feigned concern but suggested that, as an Irish citizen, he call the Irish embassy instead. In the end, it was Ireland—not the United States—who got him out safely. The episode made clear, Walsh would later write, "that journalists can't rely on the United States government to have our back as it once did." In other words, our invisible armor was gone. It vanished the day Trump was inaugurated.

THEY—THE HACKERS, the officials, the Ukrainians, the voices in the wilderness—had always warned me that a cyber-enabled cataclysmic boom would take us down. The Cyber Pearl Harbor. And when I started on this beat nearly a decade ago, I always made a point of asking, "Okay, then, when?" Their answers were, almost comically, the same: "Eighteen to twenty-four months." 18 TO 24 ✓ was scribbled throughout my spiral-bound notebooks. Just close enough to add urgency to their prediction, but just far enough away that I might not hold them to their projections if they didn't come true.

It has now been more than one hundred months, and though we have yet to see the mushroom cloud, we are closer than we have ever been. In the weeks just before the election, the same Russian hackers who penetrated our nuclear plants started boring into American local networks. The timing of the attacks, so close to the election; the actor, believed to be a unit of Russia's FSB; and the potential for disruption set off alarm bells back at Fort Meade. But Election Day came and went, and nothing came of it. As of this writing, we still do not know what they were doing there or why. Some suspect that by deputizing its stealthiest hackers to target state and local systems, Russia was hedging its bets. If Putin believed Trump would be reelected and wanted to forge a better relationship with the United States, perhaps he wanted to limit the degree to which Russia was seen as interfering. Now that Biden is elected, Russia may try to use its foothold in our systems to weaken or delegitimize him, or hold back so as not to provoke his new administration, or just sit there with the digital equivalent of a gun to his head.

"One possible explanation is that they are calling in the real pros—the A Team—who is used to operating in this really sensitive critical infrastructure where you want to keep quiet until you don't," Suzanne Spaulding, the former DHS undersecretary for cybersecurity, told me. "By doing this more quietly you give yourself more options."

The truth is, I do not know if or when we will see the kind of cyber-enabled boom I have been warned about for years. But the analogy to Pearl Harbor is a deeply flawed one. America didn't see that attack coming; we've seen the cyber equivalent coming for a decade. What we are experiencing instead is not one attack but a plague, invisible to the naked eye, that ripples across our country at an extraordinary rate, reaching ever deeper into our infrastructure, our democracy, our elections, our freedom, our privacy, and our psyche, with no end in sight. American computers are attacked every thirty-nine seconds. Only when there are highly visible mishaps do we pause for reflection. But the lessons from even the most destructive attacks tend to be forgotten too quickly. We have normalized them, even as the stakes grow higher, as the threats mutate into ever deadlier versions, as they hit us quicker than ever before. These unfolding crises play out in a dimension few of us can see, in a language few understand, and they are shutting down our cities, towns, and hospitals almost every other day. Occasionally we respond with indictments or sanctions, and, increasingly, escalating cyberattacks of our own. We too have forgotten that the internet is borderless. There are no red lines. We are not immune from our own attacks. The enemy is indeed a very good teacher. The cyberarms market is no longer ours to monopolize. We can no longer keep our cyberweapons safe. They can, and have, been turned on us. The vulnerabilities are ours, too. We just have more of them

A FEW MONTHS before the November election, I called up the godfather of American cyberwar himself. I'd caught Jim Gosler at his home in the Nevada desert. He was taking apart slot machines, searching, as ever, for new vulnerabilities. I'd called looking for reassurance, I suppose.

"This was simply inevitable," Gosler told me. "For a long time, people didn't believe the problem was big enough."

He reminded me that there were as many vulnerabilities now as there were stars in the sky. It was only a matter of time before a patient adversary exploited them against us. And now it was all happening, so frequently in fact that most attacks never even made the headlines. They were hitting our nuclear plants, our hospitals, nursing homes, our brightest research labs and companies, and somehow, no matter how much I wrote, this all seemed to escape the consciousness of the average American, of the people now plugging in their Nests, Alexas, thermostats, baby monitors, pacemakers, lightbulbs, cars, stoves, and insulin pumps to the internet.

In truth, there is no one running point. There is no cavalry. And now, the pandemic had virtualized our lives at speeds we never imagined, exposing us to cyberattacks like never before. It was no surprise when hackers seized on the coronavirus to take aim at our hospitals, our vaccine labs, and the federal agencies leading the Covid-19 response. It is not clear how successful Russia's retaliatory strikes on American hospitals will be. Ten days out from the election, more hospitals were reporting cyberattacks. And with coronavirus cases spiking to record levels, and a shortage of healthy hospital workers, I fear it is only a matter of time before cyberattacks cost us lives.

As of this writing, foreign states and cybercriminals are hitting American networks from so many sides that, from my quarantined perch, it has become nearly impossible to keep track.

"We've seen this coming for a long time," Gosler told me. "It's death by a thousand hacks. Our adversaries are basically seeing that we have systems of interest that are vulnerable. The tools to exploit them have been thrown in their lap, and they're willing to take some modest level of risk to use them because of the anonymity of the internet. You're only going to see a growing level of these attacks as time goes on."

It was so easy to forget that it was only forty years ago that humans dispatched the first message over the internet. I imagined what the internet would look like in another ten years, then twenty. I thought about how much more dependent we will have become on this web, how much more of our infrastructure will have moved online. And then, for one moment, I allowed my thoughts to drift to the potential for information mayhem and mass destruction.

"Listen, Nicole," Gosler said. "You'd have to be a cloistered monk on a mountain in Africa to not be concerned about cyber vulnerabilities."

With that, I left the godfather of American cyberwar to his slot machines. I wondered how many new attacks I'd missed in the span of our short phone call. I longed to swap places with that cloistered monk on that mountain in Africa.

I had never missed the elephants more.

EPILOGUE

Just a mile down the road from where I grew up in Portola Valley, California, lies an old wooden roadhouse. There on the banks of a shady creek lies the Alpine Inn Beer Garden, a place us locals still call "Zott's," short for the Rossotti's, the previous owners. Zott's has been around since the 1850s, first as a gambling house, then a saloon, and later a roadhouse serving burgers and beer, much to the chagrin of its prestigious neighbor to the east, Stanford University.

Stanford's campus was dry. Under the deed of Leland Stanford, no alcohol could be served on campus or even in Palo Alto, and administrators worried about the flocks of students getting drunk down the road. Stanford's first president had lobbied unsuccessfully to shut Zott's down, calling it "unusually vile even for a roadhouse."

It was, indeed, a scuzzy place for hellraisers of all types and, looking back, a fitting birthplace for the internet.

Few of the customers today know it, but the entire digital universe is in orbit around one picnic table out back where computer scientists relayed the first message over the internet one summer afternoon in 1976. That August, scientists from SRI International—the research institute in nearby Menlo Park—pulled up to the Zott's parking lot in an old bread truck to perform a demo for Pentagon officials who'd flown in for the occasion. The choice of locale was an inside joke; the SRI geeks had hoped there'd be some Hells Angels bikers in the mix. Sure enough, when they greeted the generals that day, one asked: "What are the hell are we doing in the parking lot of a biker bar?"

"We thought you'd ask that," one of the scientists had replied. "We wanted to do this demo in a hostile environment."

The scientists proceeded to haul a clunky Texas Instruments computer terminal out to the far picnic table and, under the watchful gaze of cowboys and bikers, they hooked up a cable from the terminal to their bread truck in the parking lot. The SRI team had spent months retrofitting the truck into a giant mobile radio unit with $50,000 radios inside. Once everything was wired, they ordered up a round of beers and fired off the first email over the internet.

Within milliseconds, it left Zott's via the bread truck's mobile radio unit and traveled to a second network—the Pentagon's Advanced Research Projects Agency Network, ARPANET—and on to its final destination in Boston. The dispatch was the first time two distinct computer networks were linked. In another year, three networks would be "internetworked" and the web as we know it would be well on its way.

There's still a plaque commemorating the BEGINNING OF THE INTERNET AGE on the wall, and a picture of the men and one woman standing by as their colleague, beer in one hand, typed out the first internet dispatch with the other. A few years ago, I decided to track down the man in the photo. His name is Dave Retz. I asked Retz if anyone there that day had any security concerns about what they were building.

"Absolutely not," he replied. "We were just trying to get the thing working."

Back then, nobody was thinking that this interconnected system, rigged from an old bread truck, would one day become humanity's collective memory, or that it would supply the digital backbone for our modern banking, commerce, transportation, infrastructure, health care, energy, and weapons systems. But, come to think of it, Retz conceded, there had been one ominous blip of what would come.

Two years before they pulled up to Zott's, air-traffic controllers at San Francisco airport started complaining that beams of "unknown origin" were interfering with their radars. As it turned out, SRI's radio frequencies had infiltrated the airport's traffic control. But even then, the idea this invention might one day threaten to bring down airplanes, disrupt water supplies, or rig an election hardly fazed the men and women building its basic blocks. Some four decades later, in 2020, San Francisco International Airport officials had just discovered that the same stealth Russian hackers who were probing our nuclear plants, our grid, and our states had hijacked an internet portal used by airport travelers and employees.

I asked Retz what, if anything, he would take back. His reply was immediate and unequivocal. "Everything can be intercepted," he told me. "Everything can be captured. People have no way of verifying the integrity of these systems. We weren't thinking about this back then. But the fact is," he added ruefully, "everything is vulnerable."

* * *

ONE DECADE AGO, the primary threats to our national security were still, for the most part, in the physical domain: hijackers flying planes into buildings, rogue nations getting ahold of nukes, drug mules tunneling in through the Southern border, the improvised explosive devices tormenting our troops in the Middle East and the homegrown terrorists detonating them in the middle of a marathon. Developing the means to track those threats and stave off the next attack has always been in the NSA's job description. If the next 9/11 struck tomorrow, the first question we would ask ourselves is the same question we asked some two decades ago: How did we miss this?

But in the two decades since 9/11, the threat landscape has been dramatically overhauled. It is now arguably easier for a rogue actor or nation-state to sabotage the software embedded in the Boeing 737 Max than it is for terrorists to hijack planes and send them careening into buildings. Threats that were only hypotheticals a decade ago are now very real. Russia proved it can turn off power in the dead of winter. The same Russian hackers who switched off the safety locks at the Saudi petrochemical plant are now doing "digital drive-bys" of American targets. A rudimentary phishing attack arguably changed the course of an American presidential election. We've seen patients turned away from hospitals because of a North Korean cyberattack. We've caught Iranian hackers rifling through our dams. Our hospitals, towns, cities, and, more recently, our gas pipelines have been held hostage with ransomware. We have caught foreign allies repeatedly using cyber means to spy on and harass innocent civilians, including Americans. And over the course of the coronavirus pandemic, the usual suspects, like China and Iran, and newer players, like Vietnam and South Korea, are targeting the institutions leading our response.

The pandemic is global, but the response has been anything but. Allies and adversaries alike are resorting to cyberespionage to glean whatever they can about each country's containment, treatments, and response. Russian cybercriminals have seized on Americans working from home to break into an untold number of American companies in the Fortune 500.

There is no bottom to these efforts. The week of this writing, the United States was hit by the largest medical cyberattack in its history after cybercriminals held Universal Health Services, a hospital chain with more than four hundred locations, hostage with ransomware. Hundreds of clinical trials were also being held hostage—including the crash effort to develop tests, treatments, and vaccines for the coronavirus—after a ransomware attack hit the company whose software is used to manage those trials. Even countries with no discernible hacking capabilities to speak of are showing new potential. In Nigeria, former scammers have turned to hacking to convince those sheltered-at-home to click on their Covid-themed emails and grant them access to their computers. Hacktivists are having their say too. In retaliation for the murder of George Floyd at the hands of the Minneapolis police, hackers affiliated with Anonymous—a loose hacking collective that had been dormant for the better part of a decade—hacked more than two hundred police departments and FBI stations across the country in support of Black Lives Matter protesters, dumping ten years' worth of law enforcement data online—the largest published hack of American law enforcement agencies. In Israel, officials had just accused Iran of hacking its water facilities in an apparent attempt to cut water to thousands of Israelis stuck at home. As the pandemic peaked in the United States, daily hacking attempts quadrupled. The frequency of attacks and the spectrum of targets was "astronomical, off the charts," a former intelligence operative told me. And those were just the attacks we could detect.

"We're just looking through straws at a much bigger problem," John Hultquist, the threat researcher, told me.

FOR YEARS, INTELLIGENCE agencies rationalized the concealment of digital vulnerabilities as critical to monitoring America's adversaries, to war-planning, to our national security. But those rationalizations are buckling. They ignore the fact that the internet, like so much we are now witnessing in a global pandemic, has left us inextricably connected. Digital vulnerabilities that affect one affect us all. The barrier between the physical and digital worlds is wearing thin. "Everything can be intercepted" is right, and most everything important already has—our personal data, our intellectual property, our chemical factories, our nuclear plants, even our own cyberweapons. Our infrastructure is now virtualized, and only becoming more so as the pandemic thrusts us online with a scope and speed we could never have imagined only weeks ago. As a result our attack surface, and the potential for sabotage, has never been greater.

The United States was correct in concluding that, in terms of offense, its cyber capabilities were far ahead of the pack. NOBUS—the idea that "nobody but us" could find and execute on the vulnerabilities that American agencies discovered—held true for some time. We have the Goslers to thank for that. Stuxnet was a masterpiece. It kept Israeli jets on the ground. Fewer people are dead because of it. It set Iran's nuclear programs back years and arguably helped push Tehran to the negotiating table. But it also showed the world—and nobody more so than its target—what they were missing out on.

One decade later, a global cyberarms race is in full swing. Nations are now investing far more time and money in finding vulnerabilities than the commercial world, and the open-source community, is spending to fix them. Russia, China, North Korea, and Iran are stockpiling their own zero-days and laying their own logic bombs. They know our digital topography well; in too many cases, they are already inside. With the Shadow Brokers leaks, the spread of off-the-shelf hacking tools, and the growing market for digital mercenaries, one could argue the gap between what the United States is capable of and what our enemies can do has sufficiently closed.

The world is on the precipice of a cyber catastrophe. A few years ago, I dismissed these words as alarmist, irresponsible even. Too many used "FUD" to pitch snake oil. The cybersecurity industry pushed so many world-ending scenarios on us, with such frequency, that we became jaded. But after a decade immersed in digital threats, I fear these words have never been truer. We are in a shortsighted race to the bottom and it is now in our urgent national interest to pause and begin to dig our way out.

THEY SAY THE first step in solving a problem is recognizing there is one. This book is my own "left of boom" effort. It is the story of our vast digital vulnerability, of how and why it exists, of the governments that have exploited and enabled it and the rising stakes for us all. While this story may be familiar to some, I suspect it is one few are aware of, and even fewer truly understand. But it is our ignorance of these issues that has become our greatest vulnerability of all. Governments count on it. They've relied on classification requirements and front companies and the technical nature of the issues involved to conceal and confuse one stubborn fact: The very institutions charged with keeping us safe have opted, time and time again, to leave us more vulnerable. My hope is that this book may serve as a wake-up call, to encourage the awareness necessary to solve what may be the most complex puzzle of our digital era.

There is a reason I wrote this book for the lay audience, why I chose to focus primarily on people, not machinery, why I hope it will be "user-friendly." And that is because there are no cyber silver bullets; it is going to take people to hack our way out of this mess. The technical community will argue I have overgeneralized and oversimplified, and indeed, some of the issues and solutions are highly technical and better left to them. But I would also argue that many are not technical at all, that we each have a role to play, and that the longer we keep everyday people in the dark, the more we relinquish control of the problem to those with the least incentive to actually solve it.

Addressing our digital predicament will involve difficult compromises to our national security, to our economy, to the daily conveniences we now take for granted. But the alternative—doing nothing—is leading us down a dangerous path. I'd be lying if I told you I had all the answers. I do not. But I do know we need to start somewhere and I would suggest that we adapt the hacker mindset: start with the ones and the zeroes and work our way up the stack from there.

WE MUST LOCK down the code. Nobody will bother to invest in making the higher-up layers more secure if our basic foundations are still weak. We can't redo the internet or swap out the world's code, nor should we try. But we can significantly raise the bar for the cybercriminals and nation-states looking to profit and wreak havoc on our infrastructure. To do this, we must stop introducing glaring bugs into our code. Part of the problem is the economy still rewards the first to market. Whoever gets their widget to market with the most features before the competition wins. But speed has always been the natural enemy of good security design. Our current model penalizes products with the most secure, fully vetted software.

And yet, the "move fast and break things" mantra Mark Zuckerberg pushed in Facebook's earliest days has failed us time and time again. The annual cost from cyber losses now eclipses those from terrorism. In 2018, terrorist attacks cost the global economy $33 billion, a decrease of thirty-eight percent from the previous year. That same year, a study by RAND Corporation from more than 550 sources—the most comprehensive data analysis of its kind—concluded global losses from cyberattacks were likely on the order of hundreds of billions of dollars. And that was the conservative estimate. Individual data sets predicted annual cyber losses of more than two trillion dollars.

Those costs will only continue to go up as nation-states like North Korea continue to find they can extract far more money and exact far more harm on the web than they can in the physical domain. What are we doing about this? We are still squeezing every last bit of resiliency and security from our digital systems, in the name of profit, speed, and national security. But if there is any good to have come out of the past few years of headline-grabbing attacks, it may be the new phrase I saw graffitied on the wall on a recent visit to Facebook. Someone had crossed out "Move fast and break things" and replaced it with "Move slowly and fix your shit."

Security starts from inception. For too long, we only addressed problems after vulnerable code was already in millions of people's hands, cars, airplanes, medical devices, and the grid. The cybersecurity industry tried to protect vulnerable systems by establishing a digital moat around them with firewalls and antivirus software. It didn't work. It is nearly impossible to think of a company or government agency that has not been hacked. We now need to take what the NSA itself calls a "defense in-depth" approach, a layered approach to security that begins with the code. And the only way to build secure code is to understand why vulnerabilities exist, where they exist, and how attackers exploit them, then use that knowledge to vet code and mitigate attacks, ideally before it hits the market. Today, most software developers and companies still do the bare minimum, testing code only to make sure it works. Security engineers need to be brought in from the start to conduct sanity checks, to vet original code and any code borrowed from third parties.

This is hardly a new idea. Security experts have been arguing for secure design long before the internet. Microsoft's 2002 Trustworthy Computing directive was a turning point. It wasn't perfect. There were missteps and setbacks along the way. Windows vulnerabilities still formed the raw matter for Stuxnet, WannaCry, and NotPetya. But in other ways, it worked. Microsoft used to be a punchline; now it is widely seen as a security leader. The cost of a Microsoft Windows zero-day has gone from next to nothing to one million dollars—a reflection, some argue, of the time and energy required to bypass Microsoft's security. Concerns about Windows have slowly abated, while Adobe and Java have become our collective problem children.

Which brings us to open-source code, the free software code that forms the invisible backbone to much of everything we do online. Companies like Apple and Microsoft maintain proprietary systems but baked inside are the building blocks, constructed from open-source code that is maintained by volunteers who, in theory at least, check one another's work in a peer-review system similar to that found in science or on Wikipedia. Open-source software makes up 80 to 90 percent of any given piece of modern software. Today, the average high-end car contains more than 100 million lines of code—more than a Boeing 787, F-35 fighter jet, and space shuttle. That code powers streaming music, allows hands-free calls, monitors gas levels and speed, and roughly a quarter of it is open-source. As "software eats the world," open-source code has found its way into nearly every device you can think of. Most companies and government agencies that rely on it don't even know what code is in their systems or who maintains it.

We learned this the hard way in 2014 when researchers discovered Heartbleed, the bug in the open-source OpenSSL encryption protocol. Two years went by before anyone noticed a gaping hole had left over a million systems vulnerable. Despite the fact OpenSSL was used by hospital chains, Amazon, Android, the FBI, and the Pentagon, Heartbleed revealed its code had been left to a guy named Steve in England who barely had enough money to eat.

In our brave new world, these unglamorous open-source protocols have become critical infrastructure and we barely bothered to notice. After Heartbleed, the non-profit Linux Foundation and tech companies that relied on OpenSSL stepped up to find and fund critical open-source projects. The Linux Foundation, together with Harvard's Laboratory for Innovation Science, is now midway through a census effort to identify the most critical and widely deployed open-source software in use, with the goal of giving developers the funds, training, and tools to protect it. Separately, Microsoft and Facebook sponsor an internet-wide bug bounty program to pay hackers cash for bugs they turn over in widely used technology. GitHub, the platform for programmers—that is now part of Microsoft—also offers bounties for open-source bugs, and has given the hackers who turn over these bugs legal safe harbor. These efforts are laudable and we need more of them, but they are only a small piece of the puzzle.

Governments have a role to play. After Heartbleed, the European Commission started sponsoring open-source code audits and a bug bounty.

Some U.S. government agencies are now taking baby steps in this direction. The Food and Drug Administration, for example, has been pushing medical device manufacturers to submit a "cybersecurity bill of materials," a list of every commercial, open-source, and off-the-shelf software and hardware component in medical devices that could be susceptible to vulnerabilities. The House Energy and Commerce Committee is also pushing for a Bill of Materials after hackers exploited a piece of unpatched open-source code to breach Equifax, the credit monitor, and hijack data on more than half of all Americans. And most recently, the Cyber Solarium Commission—made up of American lawmakers, administration officials, and cybersecurity experts—recommended the creation of a new National Cybersecurity Certification and Labeling Authority that would give consumers the information needed to assess the security of the tech products and services they buy.

These are the first steps in identifying, prioritizing, supporting, and vetting critical code and the thousands of programmers who maintain it. It will allow end users to know what's in their systems and make their own risk-based decisions about which code they trust and which mandates further review. The same commission also recommended measures that would hold companies liable for damages from hacking incidents that exploit known vulnerabilities—a recommendation that would go a long way towards improving patching.

We also need to start vetting developers themselves. The Linux Foundation recently started awarding digital badges to programmers who take training courses in secure programming and pass certification exams. Jim Zemlin, the foundation's executive director, recently told me he thinks governments should consider mandating the cybersecurity equivalent of a driver's license for programmers who maintain critical code. When you stop to consider that this code ends up in our phones, cars, and weapons systems, his proposal sounds sane.

We also have to address the fact that open-source code developers have themselves become frequent targets for cybercriminals and nation-states in recent years. Attackers have hijacked their accounts to insert backdoors into code baked into millions of systems. Such attacks highlight the need to provide developers with multifactor authentication and other verification tools.

* * *

WE NEED TO rethink the fundamental architecture of our machines. A secure architecture entails identifying the most critical systems—the crown jewels—whether that is customer data, medical records, trade secrets, production systems, or the brakes and steering systems in our cars, and compartmentalizing them from noncritical systems, allowing interoperability only where critical.

"Ideally, you build it like it's broken," is how Casey Ellis, a cybersecurity entrepreneur, put it to me one day. "Companies have to assume they've already been compromised, then figure out how to limit the blast radius."

This model is perhaps most familiar to readers in Apple's "sandboxing" of apps on the iPhone. Apple designed its system so that each app does not have access to other applications or data without an iPhone user's express permission. While attackers can still find critical bugs and "sandbox escapes," Apple has significantly raised the ante, driving up hackers' time and costs. Apple's mitigations are one reason why governments and their brokers are willing to pay hackers two million dollars for a remote iPhone jailbreak. It reflects the labor involved.

On the hardware side, security researchers are currently rethinking the architecture of the microchip, the most fundamental part of our machines. Among the most promising is a joint collaboration between the Pentagon's Defense Advanced Projects Agency (DARPA), SRI, and the University of Cambridge in England. The last big collaboration between SRI and the Pentagon more or less gave birth to the internet. The latest project promises to be just as ambitious. The idea is to redesign computer chips from the inside out, adding contamination chambers that would keep untrusted or malicious code from running on the chips inside our phones, PCs, and servers.

Already, the world's biggest chipmakers, including Arm—which makes processors for most smartphones—have signaled their willingness to incorporate the new design—called CHERI, short for "Capability Hardware Enhanced RISC Instructions" architecture—into their chips. Microsoft, Google, Hewlett-Packard and others are exploring the concept. Lots of questions remain about the performance tradeoffs. And many will inevitably scream economic ruin if the design causes the slightest delay, but given the current cybersecurity horror show playing out, chip and device manufacturers are beginning to entertain some lag time in the name of security.

* * *

FURTHER UP THE stack are the end users, us. They say security is only as good as the weakest link, and we continue to be the weakest link. We are still clicking on malicious links and email attachments. Even when vulnerabilities get fixed, we are not patching them quickly enough. Cybercriminals and nation-states regularly exploit unpatched software. The day patches become available is the day you see the bugs exploited the most. Why? Because we have a horrible track record of running our software updates.

Also, passwords are gone. They were all stolen from organizations who didn't bother to protect them. One day soon, I hope the password will go away. But until we come up with a new model, the easiest way to protect ourselves is to use different passwords across different sites and turn on multi-factor authentication whenever possible. The vast majority of cyberattacks—98 percent—start with phishing attacks that contain no zero-days, no malware. They just trick us into turning over our passwords. Despite the attraction of zero-days, Rob Joyce, the head of TAO, essentially the nation's top hacker, gave a rare talk four years ago, in which he called zero-days over-rated and said unpatched bugs and credential theft is a far more common vector for nation-state attacks.

So-called "password-spraying attacks" have surged in the past three years, in which hackers try common passwords (e.g. "password") across multiple user accounts. It's not rocket science, but it's insanely effective. Password-spraying is all it took for Iranian hacking group, working at the behest of the IRGC, to break into thirty-six private American companies, multiple U.S. government agencies, and NGOs.

Multifactor authentication is the best defense against these attacks. Turn it on, wherever you can, right now.

OUR ELECTIONS. THEY cannot be conducted online. Period. In 2020, with the pandemic in full swing, Delaware, New Jersey, and Colorado were experimenting with online voting. This is lunacy. As J. Alex Halderman, a computer scientist and election security expert, put it to me recently, "these jurisdictions are taking a major risk of undermining the legitimacy of their election results."

To date, there is not a single online voting platform that security experts like Mr. Halderman have not hacked. If one or two academics can hack a system and manipulate it to elect their chosen candidate, so can Russia, China, and every other nation that wants the power to put their man or woman in the White House.

In 2020, we made great strides in improving the security of our voter registration systems. We cannot make the mistake of assuming that because this data is public, it does not need to be protected. Voter registration databases could be locked up with ransomware or manipulated for digital disenfranchisement. All it would take is a hacker slipping into a key district's list to remove registered voters or modify addresses to falsely indicate voters moved out of state. Even just getting into the lists—without manipulating them—could be enough to cast doubt on a contentious election.

THE UNITED STATES needs to reestablish a national cybersecurity coordinator—the position that the Trump administration eliminated in 2018. It's critical we have someone in the White House coordinating a national cybersecurity strategy and running point on the government's response to cyberattacks and cyber threats.

Regulation is not going to get us out of our predicament but by mandating basic cybersecurity requirements, we can make our critical infrastructure more resilient to a cyberattack. The United States is far behind the pack in this regard. Congress has failed, time and time again, to pass any meaningful legislation requiring the companies that manage our most critical functions meet basic standards. In the void, Obama and Trump have each issued executive orders that identify critical infrastructure, set voluntary "best practices" for operators, and encourage the sharing of threat intelligence. These are well intentioned, but so long as ransomware continues to pummel our hospitals and local governments, we must do more.

We could start by passing laws with real teeth that mandate, for instance, that critical infrastructure operators refrain from using old, unsupported software; that they conduct regular penetration tests; that they don't reuse manufacturers' passwords; that they turn on multifactor authentication; and that they airgap the most critical systems. For years, lobbyists at the U.S. Chamber of Commerce have argued that even voluntary standards are too onerous on the private companies that oversee the nation's critical infrastructure. I would argue the cost of doing nothing now outweighs the burden of doing something.

Studies have shown that—digitally speaking—the safest countries in the world, those with the lowest number of successful cyberattacks per machine, are actually the most digitized. The safest are in Scandinavia—Norway, Denmark, Finland, Sweden—and more recently, Japan. Norway, the safest of them all, is the fifth most-digitized country in the world. But Norwegians implemented a national cybersecurity strategy in 2003 and they revisit and update it every year to meet current threats. Norwegian companies that provide "basic national functions"—financial services, electricity, health services, food supply, transportation, heating, media platforms, and communications—are required to have a "reasonable" level of security. The government penalizes companies that do not perform penetration testing, threat monitoring, or adhere to other best security practices. Government employees are required to use electronic IDs, multifactor authentication, and encryption. And Norwegian corporations have made cybersecurity core to their training and corporate culture.

Japan may even be more instructive. In Japan, the number of successful cyberattacks dropped dramatically—by more than 50 percent—over the course of a single year, according to an empirical study of data provided by Symantec. Researchers attributed Japan's progress to a culture of cyber hygiene but also to a cybersecurity master plan that the Japanese implemented in 2005. Japan's policy is remarkably detailed. It mandates clear security requirements for government agencies, critical infrastructure providers, private companies, universities, and individuals. It was the only national cybersecurity plan, researchers discovered in their study, to address "airgapping" critical systems. In the years following its implementation, researchers found that Japanese devices were better protected than other countries with similar GDPs.

We will never build resilience to cyberattacks—or foreign disinformation campaigns, for that matter—without good policy and nationwide awareness of cyber threats. We should make cybersecurity and media literacy a core part of American curriculum. Too many cyberattacks rely on vulnerable American systems, running on software that is not up-to-date or which has not been patched. This is, in large part, an education problem. The same goes for information warfare. Americans are being coopted by disinformation campaigns and conspiracy theories because Americans lack the tools to spot influence operations, foreign and domestic, in real time. As the political scientist Joseph S. Nye put it in the wake of Russia's 2016 interference: "The defense of democracy in an age of cyber information warfare cannot rely on technology alone."

I WOULD ARGUE that the United States—having spawned and sponsored the cyberarms market for vulnerabilities for decades—now needs to use its immense spending power to spawn an arms race for the common good. Gary McGraw, the author of *Software Security*, argues that the government should consider tax credits for companies that develop secure software. So long as governments keep shelling out far more money to hackers to leave vulnerabilities wide open than companies do to close them shut, defense will be handicapped. The government could start by expanding the scope of the Pentagon's bug bounty programs and private initiatives by Synack, HackerOne, and Bugcrowd that privately invite top hackers to hack government networks. It could also expand these programs beyond federal networks to open-source code and critical national infrastructure. It could consider a Google's Project Zero initiative of its own, whereby the government recruits the best hackers from intelligence agencies and the private sector—from the banks, Silicon Valley, and cybersecurity firms—for one- or two-year defense tours of duty. In theory, this would involve deploying the country's best hackers to find and patch vulnerabilities in the nation's most critical code for one year and another year on the ground, helping IT administrators at our nation's hospitals, cities, power plants, pipeline operators, biomedical research labs, and state and local election officials mitigate them.

This is no easy ask. The federal government is hamstrung by a massive trust deficit. Distrust of federal cybersecurity assistance, especially among state and county election officials, is worthy of its own book. Some states, particularly red states, have long been suspicious of federal election assistance and have viewed it as government overreach. It took North Carolina officials three years to greenlight a forensic analysis by the Department of Homeland Security of the computers used to check in voters in Durham County in the 2016 election. This, after widespread computer breakdowns and irregularities left untold numbers of voters disenfranchised in Durham County, a blue county in a swing state, and a leaked NSA report confirmed that the vendor used to check in voters had been compromised by Russian hackers. It was only in late 2019, after three years of headlines and lingering questions, that state officials agreed to a DHS evaluation. (The analysis concluded that technical issues, not hackers, were likely the cause of irregularities).

THIS TRUST DEFICIT is even worse in the private sector. After Snowden private companies, particularly the tech companies that were drawn into the fray, grew incredibly weary of giving the federal government any more information or access than they are compelled to by law. Most American companies and leaders in Washington agree, in theory, that threat-intelligence sharing is critical to defending public and private networks. But companies are still reluctant to do what it would take to set up a channel for the real-time, reliable transmission of threat data to government. Much of this is optics-driven. After Snowden, companies fear that threat-sharing mechanisms—even if they are only used to share data about vulnerabilities, active attacks, and techniques—could be misinterpreted as a government backdoor by their foreign customers in China, Germany, and Brazil.

"What's the obstacle to all of this? It's this trust deficit that we've had lingering since post-Snowden from six years ago," Uber's chief security officer, Matt Olsen, told a cybersecurity audience recently. "I think the government has made some strong steps forward in regaining the trust of the American people on intelligence collection. I think it's done a good job rebuilding relationships with our allies," Olsen said, "but it has not done enough."

This trust deficit has only been furthered by the government's offensive exploitation programs. Heartbleed forced the government to address its Vulnerabilities Equities Process, which we first learned from J. Michael Daniel's public statements and later from a Freedom of Information Act request filed by the Electronic Frontier Foundation, which forced the government to turn over a redacted copy of its VEP policy. More recently, the government has made good faith efforts to disclose more. Rob Joyce, who served as the last White House Cybersecurity Coordinator before that position was eliminated, released a high-level map of the government's VEP in November 2017 because he said it was "the right thing to do." Joyce's document serves as the most comprehensive map we have into the process by which the government keeps or reveals its zero-days. It disclosed the names of the government agencies involved, information that was previously classified. And it reaffirmed that "the primary focus of this policy is to prioritize the public's interest in cybersecurity and to protect core internet infrastructure, information systems, critical infrastructure systems, and the U.S. economy through the disclosure of vulnerabilities discovered by the [U.S. government] absent a demonstrable overriding interest in the use of the vulnerability for lawful intelligence law enforcement or national security purposes." In the annex, the document laid out the key criteria VEP stakeholders use to decide whether a zero-day requires disclosure: its "prevalence, reliance, and severity."

The disclosure was commendable, particularly considering no other country on the planet has done anything close. But when held up against the NSA's EternalBlue exploit, which relied on a bug in one of the most widely used software protocols in the world, it rings hollow. According to every criterion listed—*How widely is it used? Are threat actors likely to exploit it? How severe is it? If knowledge of the vulnerability were to be revealed, what risks could that post for the government's relationships with industry?*—the Microsoft bug underlying EternalBlue should have been turned over years before it was dumped online by the Shadow Brokers. All one has to do is look at the wreckage from its subsequent use by North Korea and Russia to know how pervasive the bug was and at the resulting paralysis at hospitals and shipping hubs and vaccine shortages to see its severity. One former TAO hacker likened the EternalBlue exploit to "fishing with dynamite." And despite the VEP document's claims that zero-days are held only "for a limited time," the NSA held onto EternalBlue for more than five years. Likewise, the Shadow Brokers leaks included a four-year-old Oracle implant that affects some of the most widely used database systems in the world.

It would be naïve, especially given the NSA's breaking-and-entering mission, to require intelligence agencies to turn over every single zero-day they find. Some argue that so long as the government has ways to pierce systems and devices with zero-days, it has less incentive to force companies like Facebook, Apple, and others to weaken the encryption in their products. This was on clear display in the 2016 FBI-Apple case, when the FBI abandoned its efforts to push Apple to weaken the security of systems, after a hacker gave the feds a zero-day exploit to get into the San Bernardino gunman's iPhone.

But it is clear—despite official claims to the contrary—that the VEP is still inherently tilted toward offense rather than defense. I would argue that there are a few commonsense changes that would restore balance. For one, the list of agencies represented in the VEP skews notably towards offense. The Director of National Intelligence, the Department of Justice, including the FBI, the CIA, Cyber Command, and the NSA's offensive elements are all well-represented. And while agencies like the Treasury, State Department, Commerce, Homeland Security, and the Office of Management and Budget—which saw millions of its records siphoned off by the Chinese—may veer more toward disclosure, I would argue more civilian agencies like the Department of Health and Human Services and the Department of Transportation need seats at the table given the onslaught of cyberattacks hitting our hospitals, medical institutions, and transportation systems.

As it stands, the Executive Secretary that oversees the VEP debates is the NSA's head of Information Assurance, the division of the agency responsible for cybersecurity. Researchers at Harvard's Belfer Center for Science and International Affairs have questioned whether NSA can serve as a truly neutral party, even if the official responsible is on the defensive side of the house. They argue that responsibility for the process should be transferred to the Department of Homeland Security and that its implementation be audited by the Inspectors General and the Privacy and Civil Liberties Oversight Board. I agree this would be a good start in restoring faith in the process and ensuring it does become a rubber stamp for the nation's offensive programs.

On a more practical level, VEP should require something akin to zero-day expiration dates. We now have a clear case study in what happens when the NSA holds onto a zero-day in a widely used system for five years. Considering RAND found that the average life span of a zero-day is a little more than a year, we should consider setting expiration dates for less than that. Holding onto zero-days indefinitely, or just until there is clear proof that another adversary is exploiting them against our own interests, is a losing game. (And the losers are us.)

In 2017, a bipartisan group tried to enshrine the VEP into law with the PATCH Act—Protecting our Ability to Counter Hacking Act. The bill would mandate that any zero-days retained be periodically reevaluated and require annual reports to Congress and the public. PATCH stalled in the Senate but its sponsors say they plan to reintroduce it.

We still have yet to see any aggregate numbers about the government's zero-day stockpile. NSA officials have said the notion that they are holding a huge stockpile is an exaggeration. They could back this up by releasing the number of zero-days they disclose and withhold every year, and the average length of time zero-days were withheld. Of course, vulnerabilities are not equal and one vulnerability—take Heartbleed—has the ability to affect a million-plus systems, but more granular data would help reassure the public that the government is not withholding thousands of zero-days in its stockpiles indefinitely.

When the government has disclosed zero-days to vendors, it has not taken credit. Companies regularly credit those who turn over bugs in their products, but when the underlying Microsoft bug in EternalBlue was patched, for example, the attribution was left blank. It may help restore confidence to know when the government turns bugs over for patching. And it would also go a long way to underscoring the severity of a bug, if technology companies and system administrators knew the bug had been discovered by the world's elite hackers. There is recent precedent for this. In 2019, when the GCHQ turned over a major Microsoft vulnerability dubbed BlueKeep, NSA issued an advisory urging users to patch their systems as soon as possible. GCHQ has recently started publicizing the number of zero-days it turns over each year. The United States has taken baby steps in this direction. In 2019, for example, Cyber Command started uploading malware samples it discovered to VirusTotal, a sort of Google search engine for malicious code found in the wild.

The VEP still contains huge loopholes. The most glaring is for the zero-days that the government purchases from third parties. According to the latest VEP disclosures, the government's decision to disclose a vulnerability "could be subject to restrictions by foreign or private sector partners of the [government], such as non-disclosure agreements." In cases where NDAs apply, a zero-day would not even be subject to consideration for disclosure. Giving the government's reliance on zero-days from contractors and hackers, and how commonplace NDA's are in the market, this fine-print exception to the VEP reads like a giant escape clause.

As one of the longest-running and biggest players in the zero-day market, the U.S. has massive spending power. If, starting tomorrow, U.S. government agencies required that any zero-day broker or hacker the government does business with gave them exclusive rights to their tools, as well as the ability to turn them over for patching, we would very likely see this become standard practice. It would also have the added effect of keeping those hackers from selling the same zero-days to foreign governments who might use them to harm U.S. interests. And while I'm dreaming, I think the United States should require that the companies they purchase surveillance tools from—the NSOs and Hacking Teams of the world—not sell to countries with proven track records of using these tools on Americans, or in clear human rights violations—as the Saudis did with Jamal Khashoggi and the UAE did when it used surveillance tools procured by Gamma Group (then Hacking Team), Dark Matter, and NSO to monitor Ahmed Mansoor.

Furthermore, call me crazy, but former NSA hackers should not be hacking the First Lady's emails on behalf of a foreign nation. They should not be tutoring Turkish generals in their tradecraft. We need laws that govern what hackers, brokers, and defense contractors can share with foreign governments, with the critical caveat that we cannot write these rules in a way that would keep defenders from sharing cyber threats across borders. The concern among hackers and cybersecurity researchers is that by prohibiting the trade of exploits across borders, we will be handcuffing defense. On this, I believe we have the ability to write rules that are not overly broad and believe those who argue otherwise grossly exaggerate the hardships that await if we dare change our thinking.

The United States may never sign onto a digital Geneva Convention so long as Russia, China, and Iran continue to outsource much of their dirty work to cybercriminals and contractors. And it will likely never sign onto any agreement that puts its strategic war-planning at a disadvantage. But we need red lines. I believe we can agree on a set of targets that are off-limits for cyberattack, starting with hospitals, election infrastructure, airplanes, nuclear facilities, and so on.

These are the critical assignments of our time. Many will say they are impossible, but we have summoned the best of our scientific community, government, industry, and everyday people to overcome existential challenges before. Why can't we do it again?

As I write these final words, I am sheltering-in-place from a global pandemic. I am watching the world ask the same questions—*Why weren't we better prepared? Why didn't we have enough testing? Enough protective gear? Better warning systems? A recovery plan?*—knowing full well these questions apply to the cyber domain too.

I'm crossing my fingers that the next big cyberattack waits until this pandemic has passed. But finger crossing has never gotten us very far. We don't have to wait until the Big One to get going.

On this, I just keep coming back to the Kiwi hacker, McManus, and his T-shirt that read: SOMEONE SHOULD DO SOMETHING.

ACKNOWLEDGMENTS

The very day I was ushered into the *Times* closet of classified secrets, I received an email from my now husband asking me out on our very first date. I was across the country, working on a project I could not discuss, with no visible end in sight. And so, he flew to meet me in Las Vegas for the one day I'd been granted reprieve to attend the annual Black Hat hacking conference. He arrived just in time to take me to dinner on my birthday. That was our first date and he has been dedicated to me, to my career, and to seeing this story through ever since. There are not many like him, and I am deeply grateful for his love, encouragement, and for playing single dad while I hunkered down and wrapped this up. I also owe a debt of gratitude to my son, Holmes, for not kicking up too much of a fuss while I waddled around the Beltway, stalked cyberweapons factories, and traveled the world with him in my belly. I cannot wait for him to read this book one day, knowing he wrote it with me. And to our most valuable player, Sally Adams, my nanny and now my dearest friend, for caring for him in my absence. Without Sally, there would be no book.

There is no greater professional honor than being able to say the words "Nicole Perlroth from the *New York Times*." Working for the Gray Lady has been the greatest honor of my life. My colleagues truly are my heroes. I have learned more from fellow reporters, editors, copy editors, and photographers in my decade at the *Times* than I ever learned in school. I wake up each day amazed at the Herculean effort it takes to get our paper out. I am keenly aware that the *Times* would never have hired me without the recommendation of my dear mentors and friends, Phil Taubman and Felicity Barringer, who took me under their wing when I was a fledging graduate student, taught me the ropes, and threw my name in the hat when the *Times* went looking for a young, hungry journalist to add to its roster. Glenn Kramon, the assistant managing editor at the *Times*, has been a source of constant support and inspiration. John Geddes, the former *Times* managing editor, took a gamble not only in hiring me but in green-lighting the story of the *Times*' own cyberattack by the Chinese military, a story that arguably helped change the way companies address and acknowledge their own cyberattacks, and shifted the United States' willingness to call out the perpetrators of attacks on American networks. I owe a huge debt to Larry Ingrassia, the former *Times* business editor, for offering me a job and never once underestimating what I was capable of. I want to also thank Damon Darlin and David Gallagher, two of the greatest editors I ever worked with. I am eternally grateful to the Deans: Dean Baquet, the *Times* executive editor, for inviting me into the room of classified secrets, and Dean Murphy, who generously gave me the time to write this book. Pui-Wing Tam and James Kerstetter have built the best technology team around. Both afforded me the time to pursue this side project and the encouragement I needed to stay the course, even when I was on the brink of mental and physical exhaustion. A special thanks to John Markoff, for patiently listening to my (many) questions, offering sources and advice when I took over his beat, and being such an inspiring role-model for so many frightened reporters joining the San Francisco bureau. I have been lucky to work with the very best national security reporters and editors in the business: Scott Shane, David Sanger, Rebecca Corbett, Mark Mazzetti, Matthew Rosenberg, Bill Hamilton, and Thom Shanker. I owe a special thank you to Jeff Cane for editing my final chapters and reassuring me these pages weren't half bad, and to Ewa Beaujon, who fact-checked everything with precision and saved me from many embarrassing errors. In Silicon Valley and beyond, I am grateful for the camaraderie of Brian X. Chen, Nick Bilton, Claire Cain Miller, Jenna Wortham, Mike Isaac, Quentin Hardy, the new moms, and everyone in the baddest bureau around. I also want to thank the Sulzberger family for continuing to support courageous, quality journalism. A special thanks to Arthur Sulzberger Jr. for sharing his closet. And to A. G. Sulzberger for speaking out and standing by David Sanger and myself when the president of the United States singled us out for "treason."

A special thanks also to my competitors, who helped flesh out these pages, and push me every day to be a better writer and reporter. It will never be fun matching one another's stories at 10 P.M. on a Sunday, but ultimately, we are all on the same side. A special shout-out to Joe Menn, Andy Greenberg, Kevin Poulsen, Brian Krebs, Kim Zetter, Ellen Nakashima, and Chris Bing.

The idea for this book started when Danielle Svetcov invited me to dinner. Various agents had solicited my interest in writing a book before, but nobody like Danielle. When I first Googled her, I saw she had represented the authors of several cookbooks, including for some of my favorite San Francisco chefs. Perhaps she could help me score a reservation at State Bird Provisions? Or help me ghostwrite my husband's cookbook? (Remember, Danielle, you promised). Instead, she showed up to dinner with an inch-thick dossier of every story I had ever written for the *Times* that might make for good reading. She had prepared chapter headings and book titles and a list of characters who appeared in my articles who she thought would make for compelling protagonists. She left me no choice but to commit. She stood by my side through brutal publisher dealings, too many medical issues to recall, a marriage, a baby, and was as hands-on an editor as any I have ever worked with. Danielle is not your average agent. She is a legend. A special thanks to my co-agent, Jim Levine, a true mensch of the publishing world, and a voice of calm assurance when I needed it the most.

Anton Mueller and the entire team at Bloomsbury believed in this project when I'm not sure I even believed in it anymore. Anton hustled to make this book work under enormous time pressure, and his deft insights improved every single page. He saw the through line, and he carried this book all the way through to the bitter end. I can never thank you and Bloomsbury enough.

Writing a book is a lonely endeavor. You spend way too much time in your head. A handful of friends from various parts of my life provided much-needed encouragement and laughter along the way: Megan Clancy, Julia Vinyard, Lauren Glaubach, Lauren Rosenthal, Justin Francese, Marina Jenkins, Frederic Vial, Abby and Michael Gregory, Rachel and Matt Snyder, Sarah and Ben Seelaus, Patty Oikawa, Liz Armistead, Bill Broome, Coco and Ethan Meers, Sean Leow, Carolyn Seib, Nate Sellyn, Jen Krasner, Paul Gaetani, Jean Poster, Melissa Jensen, the Dann Family, Paul Thomson, John and Jenna Robinson, Alex Dacosta and Tyson White, and our extended family at Chugach Powder Guides.

To my parents, Karen and Mark Perlroth, and my siblings, Victor and Nina. Whatever success I have had in this life is the direct result of their love and support. When I was little, they would all take turns helping me with home-work assignments and hovering over my term papers. The notion that I would go on to become a journalist and write a book one day, let alone one about a secret, classified market for cyberweapons, is still a mystery. Thank you espe-cially to my brother, for helping me organize my thoughts and plugging in footnotes en route to deliver an address to his shareholders, and to mom, for everything. I am only now realizing all that being "mom" entails.

This book would not have been possible without the hundreds of sources who generously walked me through the ins and outs of this strange beat and the most secretive market on earth. Thank you for your patience in sharing stories, many of which were never meant to be told, and thank you for trusting me to tell them. There are many sources whose names do not appear in these pages but whose help proved invaluable. You know who you are. From the bottom of my heart, thank you.

And, finally, to my big brother, Tristan, who passed years ago and yet I swear I still hear your voice whispering in my ear. More than any other, you taught me life is short and we've got to make it count. This book is for you.

NOTES

Most of the material in this book came from interviews and my reporting for the *New York Times*. As I indicated in the Author's Note, many sources were reluctant to discuss the inner workings of the cyberarms market. Many only agreed to speak on a not-for-attribution basis, that is I could use the information without revealing its source. Whenever possible, I asked these sources to present documentation, in the form of emails, text messages, calendars, contracts, notes and other digital crumbs to corroborate their recollection of events. Any material that is not cited in the bibliography or the notes below came from confidential sourcing and documentary evidence, provided with the understanding that I not reveal the source.

I also relied on a treasure trove of reporting by my peers in the cybersecurity press who have done a remarkable job chronicling nonstop cyber mayhem over the past decade, and I have sought to credit them in the notes to follow. Even in instances where I was able to independently confirm information, I tried to identify the book or article that reported it first, though I am sure I have overlooked some article or publication that beat me to it, and for that, I sincerely apologize.

Some of the best cybersecurity reporting over the past decade—I am proud to say—belongs to my colleagues at the *New York Times*. John Markoff, my predecessor at the *Times*, has been generous with time and source material and collaborated with me on a number of articles that are mentioned in these pages. David Sanger first revealed the real codename "Olympic Games"—for the computer worm the world knew only as Stuxnet. And it was David that pulled me into the reporting on the escalating digital Cold War between the United States and Russia that the President likened to "treason." Scott Shane wrote one of the most sweeping accounts of the NSA's digital capabilities while seated beside me in Sulzberger's closet. David, Scott and I partnered on a number of stories during and after the Shadow Brokers' leaks. Without Azam Ahmed, I would never have uncovered the depths to which Mexico abused NSO's surveillance technology. Mark Mazzetti, Adam Goldman, Ronen Bergman and I wrote a comprehensive account of Dark Matter and NSO Group for the *Times*. And later Mark, Ronen and I reported that a widely downloaded mobile app, called ToTok—a play on the popular Chinese app TikTok—was actually a cleverly disguised Emirati surveillance tool. Matt Rosenberg and I partnered on the more recent Russian attack on Burisma, the Ukrainian company at the heart of President Trump's impeachment. And David, Matt and I continue to cover cybersecurity threats to the 2020 election together. Some of the best reporting on the ethical debates currently playing out in Silicon Valley regarding security and disinformation belongs to my colleagues Sheera Frenkel, Cecilia Kang, Mike Isaac, Daisuke Wakabayashi, Kevin Roose, and Kate Conger. These collaborations have been the highlight of my career and this book would not have been possible without them.

I also want to acknowledge the excellent reporting by my peers at *Wired*, Reuters, the *Washington Post*, and Vice's *Motherboard* site as well as top-notch analysis from cryptographers like Paul Kocher and Peter Neumann and the security researchers at Area 1, Citizen Lab, CrowdStrike, FireEye, Google, Lookout, Microsoft, Recorded Future, Symantec, McAfee, Trend Micro, and others. Among others, I'd like to single out Andy Greenberg, whose chronicle of the NotPetya attack is among the most comprehensive to date, and who provided one of the first glimpses into the zero-day market at *Forbes*. Kim Zetter's book, *Countdown to Zero Day: Stuxnet and the Launch of the World's First Digital Weapon* (Broadway Books), was invaluable in fleshing out the details of the hunt to decipher Stuxnet. Eric Chiu and Liam O'Murchu at Symantec patiently took every call and read over me as I described the zero-days that formed the entry points in Stuxnet. Fred Kaplan's book, *Dark Territory: The Secret History of Cyber War* (Simon & Schuster, 2016), also provided useful context. I found myself frequently citing Joe Menn at Reuters, whose cybersecurity coverage is first rate. Chris Bing and Joel Schectman, also at Reuters, scooped me with their definitive 2019 report on "Project Raven." The dialogue between Walter Deeley and the NSA's analysts quoted in the Project Gunman chapter was taken from Eric Haseltine's book, *The Spy in Moscow Station: A Counterspy's Hunt for a Deadly Cold War Threat* (Thomas Dunne Books, 2019).

A comprehensive list of sources is provided in the Bibliography, available at www .thisishowtheytellmetheworldends.com. What follows in the notes below is not meant to be an all-encompassing list, but a guide to the reporting, security analysis, scholarship, statistics, and sources I referenced and found most useful, should you desire to dig for more.

PROLOGUE

Details of the 2017 NotPetya attack came from my joint reporting at the *Times* with Mark Scott and Sheera Frenkel. Two years later, my colleague Adam Satariano and I were able to provide more detail on the cost of NotPetya as companies like Merck and Mondelez filed lawsuits against insurers who applied a common, but rarely invoked, war exemption in their insurance contracts. Andy Greenberg's work *Sandworm: A New Era of Cyberwar and the Hunt for the Kremlin's Most Dangerous Hackers* (Doubleday, 2019), which was excerpted in *Wired*, cited Tom Bossert, Trump's former Homeland Security Advisor, as a source for the $10-billion price tag of damages caused by the attack. Some sources believe the cost could be much greater, given that many smaller, private companies did not report damages. Much of the detail about these attacks came from interviews conducted in Ukraine. My *Times* colleagues Andrew Kramer and Andrew Higgins helped chronicle the 2017 cyberattacks as they unfolded on the ground. I borrowed from Michael Gordon's *Times* coverage of Russia's annexation of Crimea. I relied on Andrey Slivka's August 20, 2006, *New York Times* travel story, "Joining Tycoons at a Black Sea Playground in Crimea" for a traveler's description of Crimea before it was annexed.

Putin's 2017 remarks about Russia's "patriotic" hackers were chronicled by many news outlets, but I relied on Calamur Krishnadev's June 1, 2017, account in the *Atlantic*, "Putin Says 'Patriotic Hackers' May Have Targeted U.S. Election." I relied on the *Christian Science Monitor*'s Mark Clayton for a description of Russia's interference in the 2014 elections. Andrew Kramer and Andrew Higgins were first to report that the FBI independently attributed Ukraine's 2014 electoral attack to Russia, and also documented the tie-in to Russian state television. Andrew Kramer also chronicled the downing of Malaysian Airlines Flight 17. David Sanger and I covered the North Korean attacks on Sony Pictures, much to the irritation of our families that Christmas in 2014.

Accounts of Russia's hack on Ukraine's power stations came from my interviews, as well as Kim Zetter's detailed reporting in *Wired*: "Inside the Cunning Unprecedented Hack of Ukraine's Power Grid."

Keith Alexander's remark regarding Chinese intellectual property theft, "the greatest transfer of wealth in history," was quoted by Andrea Shalal-Esa in Reuters.

I reported on Iran's bank attacks with my *Times* colleague Quentin Hardy. Michael Corkery and I later reported on North Korea's attack on the Central Bank of Bangladesh. Iran's ransomware attacks on American hospitals, companies, and towns were detailed in a November 2018 Department of Justice indictment, though nobody has been extradited or arrested. Sheldon Adelson's comments prompting Iran's attack on the Las Vegas Sands Casinos were quoted by Rachel Delia Benaim and Lazar Berman in the *Times of Israel*. Bloomberg's Michael Riley and Jordan Robertson wrote the most definitive account of the Sands attacks.

My colleague David Sanger and I first broke news of Russia's attack on the State Department in 2015. Sanger, Shane, and Lipton detailed Russia's other targets in their 2016 tick tock of Russia's attacks on the Democratic National Committee. My colleague Steven Lee Myers wrote a comprehensive account of the Russian cyberattacks on Estonia, though I was able to later find a more direct attribution to Russia's Nashi youth group in the Snowden documents. Russia's attack on France's TV5Monde television station was detailed in Reuters. My colleague Clifford Krauss and I detailed Russia's attack on Petro Rabigh, the Saudi petrochemical plant, in 2018. My colleague David Kirkpatrick detailed Russia's attempts to manipulate the "Brexit" vote in 2017. And David Sanger and I wrote several stories between 2013 and 2019 on Russia's attacks on the American energy grid, and Cyber Command's attacks on Russia's grid. Together with colleagues Michael Wines and Matt Rosenberg, we documented Russia's 2016 attacks on our back-end election apparatus and the questions left unanswered by the local firm North Carolina hired to do its subsequent forensics investigation. Adam Nossiter, David Sanger, and I detailed Russia's attacks on the French elections in 2017, which offered a fascinating window into French cyber preparedness. (The French planted false documents on their network in an effort to lead Russia's hackers astray.) Russia's attacks on the World Anti-Doping Agency were chronicled in a Microsoft Report in late 2019. My colleague Rebecca Ruiz broke the story of Russia's interference in Sochi and deserves full credit for much of the reporting in the 2017 film *Icarus*.

I chronicled the Shadow Brokers leaks in the *Times* with Scott Shane and David Sanger in a series of articles, which included a comprehensive look at how the leaks have rocked the NSA. Andy Greenberg wrote a stupendous account of the impact on Maersk, the Danish shipping conglomerate in *Wired*, "The Untold Story of NotPetya, the Most Devastating Cyberattack in History," in August 2018. Ellen Nakashima at the *Washington Post* was among the first to report that the CIA attributed NotPetya to the Russian Military in a January 12, 2018, story, "Russian Military Was behind 'NotPetya' Cyberattack in Ukraine, CIA Concludes." An excellent analysis of Russia's weaponization of the vaccination "debates" appeared in the October 2018 edition of the *American Journal of Public Health*, "Weaponized Health Communication: Twitter Bots and Russian Trolls Amplify the Vaccine Debate," authored by David A. Broniatowski, Amelia M. Jamison, SiHua Qi, Lulwah AlKulaib, Tao Chen, Adrian Benton, Sandra C. Quinn, and Mark Dredze.

Statistics regarding the rapid adoption of the "Internet of Things" were sourced from a 2017 McKinsey report, "What's New with the Internet of Things?"

The ratio of one hundred cyberwarriors working on offense to only one working on defense was taken from a remark made by Ed Giorgio, who spent thirty years at the NSA, on a 2015 panel at the RSA Conference. Giorgio said that when he was NSA's chief codemaker, he led a group of seventeen cryptographers, and when he led the NSA's codebreakers, he led a group of seventeen hundred cryptanalysts. Giorgio also spent three years at the NSA's British counterpart, GCHA, and noted in the same panel that the same 100:1 ratio was true there.

I relied on *Times* colleagues Alicia Parlapiano and Jasmine C. Lee's chronicle of Russia's 2016 disinformation campaigns in their February 16, 2018 article, "The Propaganda Tools Used by Russians to Influence the 2016 Election," which detailed how Russian trolls posed as Texan secessionists and Black Lives Matter activists. Russia's targeting of African Americans was also explored by David Shane and Sheera Frenkel in their December 17, 2018, story for the *Times*, "Russian 2016 Influence Operation Targeted African-Americans on Social Media." Sanger and Catie Edmonson reported that Russia's interference extended to all fifty states in July 2019. Another colleague, Jeremy Ashkenas, sorted through fringe theories to document evidence that it was indeed Russia—not a "400-pound hacker on his bed" as Trump famously declared—that interfered in 2016. Trump's attacks on the 2016 intelligence are well-documented in Julie Hirschfeld Davis's July 16, 2018, article for the *Times*, "Trump, at Putin's Side, Questions U.S. Intelligence on 2016 Election," and Peter Baker and Michael Crowley were there to document Trump's odd exchange with Putin about meddling in a June 2019 story for the *Times*.

The number of attacks on U.S. networks differ depending on the source, but, for starters, the Pentagon said in 2017 that daily scans, probes and attacks on the Department of Defense's computer networks had risen to eight hundred million "cyber incidents," up from forty-one million daily scans in 2015. Those number were provided by the Pentagon Spokesman, Lt. Col. James Brindle, and are available at the Office of the Deputy Chief of Naval Operations for Information Dominance.

CHAPTER 1: CLOSET OF SECRETS

Luke Harding at the *Guardian* chronicled the paper's back and forth with the GCHQ over its Snowden hard drives in January and February 2014, and even provided video footage of *Guardian* editors destroying Snowden's hard drives. Nicolai Ouroussoff wrote an excellent account of Renzo Piano's architectural design for the *Times* headquarters in November 2007. I worked closely with Jeff Larson and Scott Shane, sorting through the Snowden trove to assess the lengths to which the

NSA had gone to foil digital encryption. Our account, "NSA Able to Foil Basic Safeguards of Privacy on Web," published simultaneously with the *Guardian* and Pro Publica on September 6, 2013. I covered the fallout for the National Institute of Standards and Technology, the federal agency charged with setting cybersecurity standards, in the days that followed. Joe Menn at Reuters took that reporting a step further with an investigation, published that December, which revealed that NSA had paid RSA, a leading cybersecurity company, to use a weak algorithm in its commercial encryption products that the NSA could break.

The vast majority of reporting on how the NSA and GCHQ are capturing data in transit and from the end points came from the Snowden documents and was first reported by Glenn Greenwald, then at the *Guardian*, Barton Gellman, and Laura Poitras, for the *Washington Post*. Subsequent reporting by Poitras also appeared in *Der Spiegel*, with Marcel Rosenbach and Holger Stark. My colleagues James Glanz, Andrew Lehren, and Jeff Larson, then of ProPublica, wrote an account of how the NSA and its Five Eyes partners tap data from mobile apps in January 2014.

David Sanger wrote the most definitive account of Olympic Games in his 2012 book, *Confront and Conceal: Obama's Secret Wars and Surprising Use of American Power* (Broadway Books), which was adapted in article form for the *Times* in June 2012. Kim Zetter's 2014 book, *Countdown to Zero Day*, approached Olympic Games from the perspective of the technical researchers who first discovered it in bits and bytes as it traversed the globe. The most comprehensive technical analysis of the Microsoft zero-day that China used to pierce Google and nearly three dozen other companies was conducted by George Kurtz and Dmitri Alperovich, then at McAfee, who would later go on to cofound CrowdStrike.

A history of the most costly zero-day in American history—the single missing hyphen in the *Mariner 1*'s navigation software—is available in NASA's archives and can be accessed easily by visiting nssdc.gsfc.nasa.gov/nmc/spacecraft/display.action ?id=MARIN1. The *Mariner 1*'s $150 million price tag is the equivalent in today's monies.

Joe Menn wrote the definitive account of the NSA's interception of Yahoo data for Reuters in his article "Exclusive: Yahoo Secretly Scanned Customer Emails for U.S. Intelligence," published in October 2016.

I relied on Paul Fehri's December 17, 2005, *Washington Post* account of the *Times* decision to delay publication of its NSA wire-tapping story. Snowden told the *Advocate*'s Natasha Vargas-Cooper in November 2013 that the *Times* decision to delay that story is why he chose not to take his trove of NSA documents to the *Times*.

The first leak of our work in the closet was published by Ben Smith, then at Buzzfeed, in August 23, 2013. The most comprehensive description of the hotel key card hack was written years later, in August 2017, by Andy Greenberg for *Wired*, "The Hotel Room Hacker."

CHAPTER 2: THE FUCKING SALMON

The industrial security conference I refer to here is the S4 Conference, organized in Miami every year by Dale Peterson. I would later describe some of my conversations with the Italians for an article I wrote with David Sanger for the *Times*: "Nations Buying as Hackers Sell Flaws in Computer Code," in July 2013.

David Sanger's book *Confront and Conceal* is the most comprehensive account of Olympic Games/Stuxnet. Ralph Langner's 2011 TED Talk remains one of the most accessible descriptions of Olympic Games/Stuxnet by a technical expert and can be viewed here: www.ted.com/talks/ralph_langner_cracking_stuxnet_a_21st_century _cyber_weapon#t-615276. I should note that some Israeli publications claim the name "Olympic Games" was a nod to the intelligence agencies of five countries—the U.S., Israel, the Netherlands, Germany, and the UK—but my sources dispute this and say it was a nod to five American and Israeli agencies who collaborated on the development and execution of the computer worm.

CHAPTER 3: THE COWBOY

John Watters, Sunil James, and David Endler were invaluable in providing their time and recollections of iDefense's early days. I was able to confirm their account of the company's bankruptcy from court documents and press releases. I am also indebted to BugTraq's founder, Scott Chasin, hacking alias "Doc Cowboy"; Elias Levy, hacking alias "Aleph One," who took over as BugTraq moderator before it sold to Symantec; and Jeff Forristal, alias "Rainforest Puppy," Saumil Shah, and others who offered countless hours of their time describing the dynamics between hackers and technology vendors during this period.

Scott Culp's impassioned arguments against vulnerability disclosure can still be found on Microsoft's website: "It's Time to End Information Anarchy," originally published October 2001.

Among the more comprehensive and readable accounts of the web's earliest days can be found in Katie Hafner and Matthew Lyon's 1998 book, *Where Wizards Stay Up Late: The Origins of The Internet* (Simon & Schuster). The W3C Organization also maintains a timeline of critical web developments at www.w3.org/History .html. You can still access *Time* magazine's 1993 story, "First Nation in Cyberspace," at content.time.com/time/magazine/article/0,9171,979768,00.html. Gary Trudeau's 1993 Doonesbury cartoon is available at www.gocomics.com/doonesbury/1993 /10/18. I highly recommend reading Glenn Fleishman's December 2000 *Times* story about the famous *New Yorker* cartoon, "Cartoon Captures Spirit of the Internet."

The most comprehensive reporting on Microsoft's court saga was written by my now-deceased mentor Joel Brinkley. I pulled the Bill Gates quote—"How much do we have to pay you to screw Netscape?"—from Brinkley's 1998 article, "As Microsoft Trial Gets Started, Gates's Credibility is Questioned." Mark Zuckerberg relayed his early "Move Fast and Break Things" mantra to Henry Blodget in an October 2009 interview. The full quote was: "Move fast and break things. Unless you are breaking stuff, you are not moving fast enough."

I relied on Jane Perrone's *Guardian* article "Code Red Worm" for the reference to the computer worm. *Scientific American* provided a more comprehensive account in its October 28, 2002 article: "Code Red: Worm Assault on the Web." A detailed history of the Melissa Virus is available on the FBI's website here: www.fbi.gov/news /stories/melissa-virus-20th-anniversary-032519. For a recounting of Nimda, I relied on Charles Piller and Greg Miller's May 2000 account in the *Los Angeles Times*. And for a look back at the impact of Microsoft's Trustworthy Computing initiative, I relied on interviews as well as Tony Bradley's *Forbes* 2014 piece, "The Business World Owes a Lot to Microsoft Trustworthy Computing," and a Microsoft 2012 look back, "At 10-Year Milestone, Microsoft's Trustworthy Computing Initiative More Important Than Ever," for an internal history. Gates's 2002 memo is available on *Wired*'s website: www.wired.com/2002/01/bill-gates-trustworthy-computing.

iDefense's Dave Endler actually started Tipping Point's competitive offering called the Zero-Day Initiative. Endler tweaked the iDefense formula a bit. Instead of one-off bounties, hackers who turned in high-quality bugs were enrolled in frequent flier–like programs that rewarded top-ranked researchers with bonuses up to $20,000. (Watters had invested in Tipping Point in its early days, so in some sense he and Endler were still playing for the same team when he left.)

CHAPTER 4: THE FIRST BROKER

Andy Greenberg was the first to chronicle The Grugq's zero-day business in his 2012 *Forbes* magazine story, "Meet The Hackers Who Sell Spies the Tools to Crack Your

PC (and Get Paid Six-Figure Fees)," which featured the South African exploit broker sitting next to a duffel bag of cash ostensibly from his zero-day sales. The fallout from Greenberg's *Forbes* story was relayed to me in subsequent interviews.

Jimmy Sabien is not the broker's real name. It is also not an alias he chose. I picked it out of thin air. Any likeness to anyone in the zero-day market is pure coincidence.

Regarding the claim that most bugs are the result of human error, a National Research Council study found that the overwhelming majority of security vulnerabilities are caused by "buggy" code. For example, at least a third of the Computer Emergency Response Team (CERT) security advisories since 1997 concern improperly checked software code.

Sabien did not name the specific Hewlett-Packard zero-day exploit he was referring to, but in 2002, two researchers at the Black Hat hacking conference demonstrated an exploit in HP printers that sounded almost identical to the exploit he described. Printers continue to be ripe targets for hackers. Among the exploits the Stuxnet worm relied on was a zero-day in a printer spooler, the software that tells printers what to print. And in 2017, a graduate student identified more than 125 printer vulnerabilities in the National Vulnerability Database dating back 20 years. As recently as 2019, a pair of researchers uncovered 49 vulnerabilities in 6 of the most widely used commercial printers, some of which could be used to remotely access the machines and their contents. See "NCC Group Uncovers Dozens of Vulnerabilities in Six Leading Enterprise Printers," NCC Group press release, August 2019.

The Council on Foreign Relations analyzes trends in military spending, drawing on inflation-adjusted data from the Stockholm International Peace Research Institute and from the U.S. Bureau of Economic Analysis. See www.cfr.org/report /trends-us-military-spending.

The "Rio Grande" quote from James Ellis, the former commander of U.S. Strategic Command, was cited by Patrick Cirenza in his February 2016 piece, "The Flawed Analogy between Nuclear and Cyber Deterrence," for *The Bulletin of the Atomic Scientists*.

CHAPTER 5: ZERO-DAY CHARLIE

I fact-checked Miller's accounts with Andy Greenberg's 2011 *Forbes* story "iPhone Security Bug Lets Innocent-Looking Apps Go Bad." Greenberg also chronicled Apple's subsequent blacklisting of Miller: "Apple Exiles a Security Researcher from Its Development Program for Proof-of-Concept Exploit App."

I chronicled Miller's Jeep hack for the *Times*, but Greenberg's account included a must-watch video with Miller and his co-researcher, Chris Valasek, for *Wired*. The video is available at www.wired.com/2015/07/hackers-remotely-kill-jeep-highway.

Charlie Miller's 2007 White Paper on the zero-day market, "The Legitimate Vulnerability Market: Inside the Secretive World of 0-day Exploit Sales," is still available, complete with the Wite-Out–redacted check for $50,000, at www .econinfosec.org/archive/weis2007/papers/29.pdf.

Miller's later forays into Apple's iOS and Google's Android software are well-documented. I relied on *Times* colleague John Schwartz's 2007 account, "IPhone Flaw Lets Hackers Take Over, Security Firm Says," and *Ars Technica*'s 2008 story on Miller's MacBook Air hack. A YouTube video of a young Miller taking over the MacBook Air in two minutes is available at www.youtube.com/watch?v=n011eIxox6w. Miller's 2007 Black Hat briefing on how he hacked Apple's Mac OS X software is available via the Black Hat podcast here: podcasts.apple.com/gb/podcast/charlie -miller-hacking-leopard-tools-techniques-for/id271135268?i=1000021627342. My colleague John Markoff covered Miller's Android exploit for the *Times* in October 2008, "Security Flaw Is Revealed in T-Mobile's Google Phone." I relied on contemporary accounts from technical publications, including *Computer World* and ZDNet, to fact-check Miller's escapades into the MacBook Pro. Michael Mimoso covered the "No More Free Bugs" movement for the trade publication, Search Security, in 2009. Dino Dai Zovi also provided helpful color and context in interviews.

CHAPTER 6: PROJECT GUNMAN

The most comprehensive account of Project Gunman is the declassified 2007 NSA history, *Learning from the Enemy: The GUNMAN Project* that was based on interviews by Sharon Maneki for the Center for Cryptologic History. It is available at www.nsa.gov/Portals/70/documents/news-features/declassified-documents /cryptologic-histories/Learning_from_the_Enemy.pdf.

While the NSA's official history omitted the name of the allies who tipped the United States off to Russian tradecraft, France and Italy were both mentioned in a May 2017 *Politico* piece, "The Time the Soviets Bugged Congress and Other Spy Tales." I relied on several accounts of Russian spycraft at the U.S. embassies in Moscow. A *Times* May 1964 front-page story headlined "In Moscow, Walls Have Ears (40)," documented the forty bugs American technicians discovered in the structural walls of the American embassy. For a truly fascinating, albeit tragic, account of Russian ingenuity, read Nathaniel Scharping's October 2019 profile of Leon Theremin's life in *Discover* magazine, "Creepy Music and Soviet Spycraft: The Amazing Life of Leon Theremin." It was Theremin's listening device that was implanted in the hand-carved Great Seal of the United States gifted to the American ambassador in 1945, and which outlived four U.S. ambassadors, until its discovery in 1952.

The debacle that was the Americans' bugged embassy in Moscow was detailed in a 1988 *Times* piece by Elaine Sciolino, "The Bugged Embassy Case: What Went Wrong."

The dialogue quoted between Walter G. Deeley and his staff, and several details regarding Deeley's stress at the time, was taken from Eric Haseltine's 2019 book, *The Spy in Moscow Station*.

CHAPTER 7: THE GODFATHER

Gosler's favorite Price Pritchett quote can be found in his 1994 book, *The Employee Handbook of New Work Habits for a Radically Changing World: 13 Ground Rules for Job Success in the Information Age.*

I relied on Matthew Carle's 2013 retrospective on Operation Ivy Bells to describe the undersea cable tap: "40 Years Ago, The Navy's 'Operation Ivy Bells' Ended With a 70s Version of Edward Snowden," published by *Business Insider.*

A history of Los Alamos is available via the U.S. Department of Energy's Office of History and Heritage Resources in "The Manhattan Project: An Interactive History." The reference to Sandia's role in developing 97 percent of America's non-nuclear weapons components is available on Sandia's website: "Evaluating Nuclear Weapons: A Key Sandia Mission." Eric Schlosser provided an entertaining, and disturbing, account of America's nuclear weapons accidents in his 2013 book, *Command and Control: Nuclear Weapons, the Damascus Accident and the Illusion of Safety* (Penguin Press).

Ken Thompson's 1984 Turing Award speech, "Reflections on Trusting Trust," is available here: www.cs.cmu.edu/~rdriley/487/papers/Thompson_1984_Reflection sonTrustingTrust.pdf.

Gosler's Chaperon Experiments were also detailed in a 2016 dissertation by Craig J. Weiner, at George Mason University, titled: "Penetrate, Exploit, Disrupt, Destroy: The Rise of Computer Network Operations as a Major Military Innovation."

The damages estimate from the Morris Worm are taken from Adam Levy's 2016 book, *Avoiding the Ransom: Cybersecurity for Business Owners and Managers* (lulu .com).

The references to the number of lines of code in Linux, the Pentagon's Joint Strike Fighter Aircraft, and Microsoft Vista were sourced from Richard Danzig's 2014 article, "Surviving on a Diet of Poisoned Fruit: Reducing the National Security Risks of America's Cyber Dependencies," published by the Center for a New American Security.

Willis H. Ware's prescient 1967 RAND report was officially titled, "Security and Privacy in Compute Systems," but came to be called the "Ware Report." It is available in George Washington University's National Security Archive: nsar-chive.gwu.edu/dc.html?doc=2828418-Document-01-Willis-H-Ware-RAND-C orporation-P. The 1970 Anderson Report for the Defense Science Board Task Force is available here: csrc.nist.gov/csrc/media/publications/conference-paper/1998 /10/08/proceedings-of-the-21st-nissc-1998/documents/early-cs-papers/ware70 .pdf.

Two comprehensive accounts of NSA's budget and management challenges in the pre-9/11 era are George Cahlink's September 1, 2001 piece, "Breaking the Code," for *Government Executive* magazine, and Roger Z. George and Robert D. Kline's 2006 anthology, *Intelligence and the National Security Strategist: Enduring Issues and Challenges* (National Defense University Press).

The back and forth between William Payne and his former employer Sandia is available in a series of 1997 court documents filed in the U.S. District Court for the District of New Mexico. According to the lawsuit, in 1992, Gosler briefed Sandia employees on Sandia's work for NSA on a "covert channel" which involved "virusing computer software and hardware" and "spiking" the machinery and encryption algorithms. At one point Payne claims he was instructed to break into electronic software on behalf of the FBI, and that Gosler tried to recruit Payne to an NSA Project, which he refused. Payne claims he was fired by Sandia for violating its classified working agreement with NSA. The same lawsuit repeatedly references the *Baltimore Sun* investigation into Crypto AG.

Scott Shane and Tom Bowman provided the first, most in-depth account of the NSA's Crypto AG operation in their December 10, 1995 article for the *Baltimore Sun*, headlined "Rigging the Game." Fifteen years later, Greg Miller at the *Washington Post* added additional reporting on the role played by the CIA and West German intelligence in the Crypto AG operation, including its codename: "Thesaurus" and later "Rubicon." See: www.washingtonpost.com/graphics/2020/world/national -security/cia-crypto-encryption-machines-espionage.

A breakdown of Gosler's threat hierarchy is available in the Department of Defense's Defense Science Board's January 2013 Task Force Report, which Gosler co-chaired: "Resilient Military Systems and the Advanced Cyber Threat," available at nsarchive2.gwu.edu/NSAEBB/NSAEBB424/docs/Cyber-081.pdf.

The NSA's 2013 Black Budget was published by the *Washington Post* in August 2013. The NSA documents that detailed its crypto projects were leaked by Snowden and published in our September 5, 2013, article, "NSA Able to Foil Basic Safeguards of Privacy on the Web."

Additional context on the CIA's role in computer network operations was detailed by Robert Wallace, H. Keith Melton, and Henry R. Schlesinger in their 2008 book, *Spycraft: The Secret History of CIA'S Spytechs, from Communism to Al-Qaeda* (Dutton).

James Woolsey's testimony was quoted by Douglas Jehl in his February 1993 *New York Times* article, "CIA Nominee Wary of Budget Cuts."

The most comprehensive account of the failed Somalia mission was written by Jon Lee Anderson for the *New Yorker* in 2009, "The Most Failed State."

Keith Alexander's "whole haystack" commentary was detailed by Ellen Nakashima and Joby Warrick in their July 2013 *Washington Post* article, "For NSA Chief, Terrorist Threat Drives Passion to 'Collect It All.'"

Michael Hayden's "golden age of signals intelligence" quote was taken from Hayden's 2017 book, *Playing to the Edge: American Intelligence in the Age of Terror* (Penguin Press).

The history of the NSA-CIA intelligence battles is detailed in a CIA memo, dated August 20, 1976, and is available at www.cia.gov/library/readingroom/docs /CIA-RDP79M00467A002400030009-4.pdf. It is also chronicled by Harvey Nelson in his 2008 article, "The U.S. Intelligence Budget in the 1990s," published in the *International Journal of Intelligence and Counterintelligence*.

A history of CIA-developed spy gadgets is available in the CIA archives: "Directorate of Science and Technology: Technology So Advanced, It's Classified." See www.cia.gov/news-information/featured-story-archive/directorate-of-science -and-technology.html.

Henry A. Crumpton, a Gosler trainee at the CIA, detailed Gosler's work at the agency in his 2013 book, *The Art of Intelligence: Lessons from a Life in the CIA's Clandestine Service* (Penguin Press). Gosler also detailed the CIA's role in computer network exploitation in his essay "The Digital Dimension," which was published in the 2005 anthology, *Transforming U.S. Intelligence*, edited by Jennifer E. Sims and Burton Gerber (Georgetown University Press). Gosler summed up the opportunities this way: "The design, fabrication, testing, logistics, maintenance and operation of these systems provide intimate access opportunities to an adversary intent on subtle modifications that will compromise confidentiality, integrity or availability."

Gosler's many intelligence awards are referenced in Alec Ross's 2016 book, *The Industries of the Future* (Simon & Schuster).

CHAPTER 8: THE OMNIVORE

The intelligence failures leading up to 9/11 are best documented in the 9/11 Commission Report, available at 9-11commission.gov/report. Also helpful were the Office of the Inspector General's November 2004 Special Report on Khalid Al-Mihdhar and Nawaf Al-Hazmi, available at oig.justice.gov/special/s0606 /chapter5.htm.

For details on the NSA's wiretapping and surveillance programs, I relied on the following articles: James Bamford's "The NSA Is Building the Country's Biggest Spy Center (Watch What you Say)," *Wired*, 2012, Charlie Savage's coverage in the *New York Times*, including his 2015 piece "Declassified Report Shows Doubts about Value of NSA's Warrantless Spying"; and Peter Baker and David Sanger's 2015 piece, "Why the NSA Isn't Howling Over Restrictions." The reference to "Pizza Hut cases" was taken from Ryan Singel's 2003 *Wired* piece "Funding for TIA All but Dead." Also, Glenn Greenwald and Spencer Ackerman's 2013 coverage off the Snowden leaks in the *Guardian*: "How the NSA Is Still Harvesting Your Online Data" and "NSA Collected U.S. Email Records in Bulk for More Than Two Years under Obama."

For a comprehensive look at NSA's mission creep, I relied on Henrik Moltke's 2019 account in the *Intercept*, "Mission Creep: How the NSA's Game-Changing Targeting System Built for Iraq and Afghanistan Ended Up on the Mexico Border;" Charlie Savage and Jonathan Weisman's 2015 coverage in the *New York Times*, "NSA Collection of Bulk Call Data Is Ruled Illegal," and Scott Shane's 2013 piece in *Counterpunch*, "No Morsel Too Miniscule for All-Consuming NSA." The *Intercept*'s Ryan Gallagher and Peter Maass detailed the NSA's move to hack IT System administrators in 2014, "Inside the NSA's Secret Efforts to Hunt and Hack System Administrators."

The most comprehensive account of AT&T's work with NSA was written by Julia Angwin, Charlie Savage, Jeff Larson, Henrik Moltke, Laura Poitras, and James Risen in their 2015 collaboration for the *New York Times*, "AT&T Helped U.S. Spy on Internet on a Vast Scale."

Barton Gellman and Ellen Nakashima provided the most granular look at the NSA's offensive cyber operations in their 2013 *Washington Post* piece "U.S. Spy Agencies Mounted 231 Offensive Cyber-Operations in 2011, Documents Show." The *Intercept* had a similarly granular take sourced from the same documents on

March 12, 2014, "Thousands of Implants," firstlook.org/theintercept/document/2014
/03/12/thousands-implants.

David Sanger and I wrote an account of the NSA's targeting of Huawei in 2014,
"U.S. Penetrated Chinese Servers It Saw as a Spy Risk."

Der Spiegel published an internal NSA TAO PowerPoint detailing how its TAO
hackers operate. The slideshow is available at www.spiegel.de/fotostrecke/photo
-gallery-nsa-s-tao-unit-introduces-itself-fotostrecke-105372.html.

CHAPTER 9: THE RUBICON

The most comprehensive window into the White House's internal deliberations
leading up to Stuxnet is detailed by David Sanger in his book, *Confront and Conceal*,
and I owe much of the reporting in this chapter to David. I also owe much credit
for this chapter to Kim Zetter's 2014 account, *Countdown to Zero Day*, which does
an excellent job chronicling the urgency by researchers to uncover and dissect the
code behind Stuxnet. Taken together, these books provide the most compelling and
comprehensive looks at the world's first cyberweapon and are more than worthy of
your time. Fred Kaplan's book *Dark Territory* also was helpful in providing a broader
perspective.

One of the best articles describing Israel's sustained pressure campaign on the
Bush White House was written by my *Times* colleagues Ronen Bergman and Mark
Mazzetti in September 2019 and is well worth a re-read: "The Secret History of the
Push to Strike Iran." One unreported aspect to Israel's pressure campaign is that,
starting around 2006, Israel began dumping hundreds of documents containing
Mossad findings at Fort Meade. In 2018, Benjamin Netanyahu made several of those
documents public in an effort to convince President Trump to pull out of the JCPOA
Iran nuclear deal. Several of those very same documents appeared in Netanyahu's
presentation, www.youtube.com/watch?v=_qBt4tSCALA. The Israeli campaign
escalated after a 2007 U.S. National Intelligence Estimate concluded, based on TAO
intelligence, that Iran paused its nuclear weapons program in the lead-up to the 2003
U.S. invasion of Iraq. The debate over the NIE is best chronicled by Gregory F.
Treverton at the RAND Corporation, "The 2007 National Intelligence Estimate on
Iran's Nuclear Intentions and Capabilities," Center for the Study of Intelligence,
May 2013, www.cia.gov/library/center-for-the-study-of-intelligence/csi-publications
/books-and-monographs/csi-intelligence-and-policy-monographs/pdfs/support-to
-policymakers-2007-nie.pdf. An Israeli take on the N.I.E. is detailed by Maj. Gen.
Yaakov Amidror and Brig. Gen. Yossi Kupperwasser, "The US National Intelligence

Estimate on Iran and Its Aftermath: A Roundtable of Israeli Experts," Jerusalem Center for Public Affairs, 2008, jcpa.org/article/the-u-s-national-intelligence -estimate-on-iran-and-its-aftermath-a-roundtable-of-israeli-experts-3.

For contemporary accounts of Israel's pressure campaign on the United States, see also Steven Erlanger and Isabel Kershner's December 2007 *New York Times* article, "Israel Insists That Iran Still Seeks a Bomb." For statistics on U.S. soldier deaths in Iraq in 2007, I relied on numbers available here: www.statista.com/statistics /263798/american-soldiers-killed-in-iraq. And for contemporary polling on Bush's nosediving approval ratings that year, I relied on *USA Today*/Gallup Polls from 2001– 2008, which showed Bush's approval rating in 2007 had nosedived from 90 percent in 2001 to under 40 percent, and continued to drop as low as 27 percent, in 2008, news .gallup.com/poll/110806/bushs-approval-rating-drops-new-low-27.aspx.

I relied heavily on Shane Harris's 2013 *Foreign Policy* profile of Keith Alexander— including Harris's description of Fort Belvoir's *Star Trek*–inspired design—"The Cowboy of the NSA." Additional sources on Alexander's tenure at NSA were Glenn Greenwald's 2013 account in the *Guardian*, "Inside the Mind of Gen. Keith Alexander," as well as the most comprehensive NSA chronicler of all—James Bamford—and his June 12, 2013, *Wired* article, "NSA Snooping Was Only the Beginning: Meet the Spy Chief Leading Us into Cyberwar."

Details of TAO's earlier attempt to sabotage al-Qaeda's communication networks appear in David Sanger's *Confront and Conceal* and Fred Kaplan's *Dark Territory*.

At the time the United States and Israel were plotting to sabotage Natanz' nuclear enrichment program, Iran was still years away from the enrichment levels necessary for the bomb. By 2020, Iran had achieved only 3.7 percent enrichment of the U-235 isotope required for a bomb. In order to reach the level of enrichment necessary for atomic energy, experts say 4 percent U-235 enrichment is required; 90 percent U-235 for an atomic weapon.

For more on the state of Iran's nuclear enrichment program at the time, and the 10 percent of centrifuges Iran was already replacing every year due to natural accidents, see Kim Zetter's excellent 2014 tick tock of Stuxnet, *Countdown to Zero Day*. For a layman's summary of nuclear enrichment, I relied on Charles D. Ferguson's 2011 work, *Nuclear Energy: What Everyone Needs to Know* (Oxford University Press).

My colleagues Michael R. Gordon and Eric Schmitt chronicled Israel's fighter jet exercises to Greece in "U.S. Says Israeli Exercise Seemed Directed at Iran," *New York Times*, June 20, 2008. The most detailed chronology of Israel's previous strike on Syria's nuclear reactor was by Seymour M. Hersh in the *New Yorker*, "A Strike in the Dark," in February 2008. One year later, Dan Murphy, writing for the *Christian Science Monitor* in October 2009, asked, "Could an Israeli Air Strike Stop Iran's Nuclear Program?" His article quotes Dan Halutz, the former head of Israel's air force, who was asked specifically what lengths Israel would be willing to go to stop Iran's nuclear program. Halutz's reply: "2,000 kilometers"—roughly the distance between Tel Aviv and Natanz. For a contemporary account of Israel's attack on Iraq's Osirak nuclear reactor, I also relied on David K. Shipler's 1981 account in the *Times*, "Israeli Jets Destroy Iraqi Atomic Reactor; Attack Condemned by U.S. and Arab Nations," still available here: www.nytimes.com /1981/06/09/world/israeli-jets-destroy-iraqi-atomic-reactor-attack-condemned-us -arab-nations.html.

I relied on David Sanger's reporting on the Israelis' own replica of Natanz at Dimona, which is available in his 2018 book, *The Perfect Weapon*. A contemporary 2008 account of Mahmoud Ahmadinejad's famous photo tour of Natanz was chronicled by William Broad in the *Times*: "A Tantalizing Look at Iran's Nuclear Program," For photos of Ahmadinejad's tour, I relied on 2008 photos published on Iran Watch, which is run by the Wisconsin Project on Nuclear Arms Control, available here: www.iranwatch.org/our-publications/worlds-response/ahmadinejad -tours-natanz-announces-enrichment-progress.

For details on the U.S.-Israel cyberweapon, I relied heavily on interviews with unnamed intelligence officials and analysts. But many only corroborated David Sanger's 2012 account in *Confront and Conceal* and his 2010 work, *The Inheritance: The World Obama Confronts and the Challenges to American Power* (Crown).

For a cost comparison of Stuxnet versus American bunker-buster bombers, I relied on numbers provided by the U.S. Government Accountability Office and price lists of exploits published by Zerodium, the exploit broker. According to the GAO, the U.S. procured twenty-one B-2 bombers at a cost of $44.75 billion, which puts the cost of each aircraft at $2.1 billion (www.gao.gov/archive/1997/ns97181.pdf). According to published zero-day price lists, the most ever advertised for a single zero-day in 2019 was $2.5 million, though the prices continue to rise (zerodium.com /program.html).

As for how Stuxnet got into Natanz, there are mixed reports. In 2019, Kim Zetter and Huib Modderkolk reported in Yahoo News that an Iranian engineer, recruited by the Dutch intelligence agency AIVD, provided "critical data" and "much-needed inside access when it came time to slip Stuxnet onto those systems using a USB flash drive." Other sources provided conflicting accounts, so I leave this as an open question.

For details of Stuxnet's zero-days, I owe a huge debt of gratitude to Eric Chien and Liam O'Murchu at Symantec, who were among the first to publish a detailed analysis of the Stuxnet code. Their analysis was later incorporated into a forensic dissection of Stuxnet by Carey Nachenberg at Stanford University Law School in 2012. I also must thank Ralph Langner, "The German," for being so patient with me as I returned to this subject nearly a decade after he first started dissecting Stuxnet's code. Langner's 2011 Ted Talk on Stuxnet is still one of the most easily digestible analyses there is. It is available here: www.ted.com/talks/ralph_langner _cracking_stuxnet_a_21st_century_cyber_weapon?language=en.

To this day, Iranian officials still maintain that they were able to uncover Stuxnet before it could wreak havoc. The official numbers show otherwise: After steadily building up capacity between 2007 and 2009, the International Atomic Energy Agency (IAEA) records show a gradual drop-off starting in June 2009 that continued over the next year. See David Albright, Paul Brannan, and Christina Walrond, "Did Stuxnet Take Out 1,000 Centrifuges at the Natanz Enrichment Plant?" Preliminary Assessment, Institute for Science and International Security, December 22, 2010, and David Albright, Andrea Stricker, and Christina Walrond's "IAEA Iran Safeguards Report: Shutdown of Enrichment at Natanz Result of Stuxnet Virus?" Institute for Science and International Security Report, November 2010. That November, Ali Akbar Salehi, the head of Iran's Atomic Energy Organization, confirmed to IRNA that the virus had indeed reached Iran: "One year and several months ago, Westerners sent a virus to (our) country's nuclear sites." Yet, he continued, "we discovered the virus exactly at the same spot it wanted to penetrate because of our vigilance and prevented the virus from harming (equipment)." Gen. Michael Hayden's Rubicon comments were taken from his February 2013 speech at George Washington University: www.c-span.org/video/?c4367800/gwu-michael -hayden-china-hacking.

For further reading on the history of Russian and North Korean attacks on U.S. systems, see Craig Whitlock and Missy Ryan, "U.S. Suspects Russia in Hack of Pentagon Computer Network," *Washington Post*, August 6, 2015, and Choe Sang-Hun and John Markoff, "Cyberattacks Jam Government and Commercial Web Sites in U.S. and South Korea," *New York Times*, July 8, 2009. The text of Obama's remarks on cybersecurity and the attacks on his 2008 campaign is available at "Text: Obama's Remarks on Cyber-Security," *New York Times*, May 29, 2009, www.nytimes.com /2009/05/29/us/politics/29obama.text.html. I also relied on the contemporary accounts of Stuxnet written by my *Times* colleagues, including that of John Markoff, my predecessor at the *Times*: "A Silent Attack, but Not a Subtle One," September 27, 2010, as well as Broad, Markoff, and Sanger's 2011 account, "Israeli Test on Worm Called Crucial in Iran Nuclear Delay," January 16, 2011.

Langner generously recalled the moments and days leading up to his TED Talk, but the official program that year was helpful for fact-checking. TED 2011 Program Schedule available at conferences.ted.com/TED2011/program/schedule .php.html.

CHAPTER 10: THE FACTORY

For the quote from Chevron's then information executive on Stuxnet hitting its systems, see Rachael King, "Stuxnet Infected Chevron's IT Network," *Wall Street Journal*, November 8, 2010. The reference to cows being outfitted with digital pedometers is from Nic Fildes, "Meet the 'Connected Cow,'" *Financial Times*, October 25, 2007. The reference to IBM's Watson computer was taken from Markoff's contemporary 2011 article: "Computer Wins on 'Jeopardy!': Trivial, It's Not," *New York Times*, February 16, 2011. Apple's Siri announcement was sourced from Apple: www.apple.com/newsroom/2011/10/04Apple-Launches -iPhone-4S-iOS-5-iCloud. Reporting on the Pentagon's Black Budget and Cyber Command's offensive spy operations was based on Barton Gellman and Ellen Nakashima's article sourced from Snowden leaks: "U.S. Spy Agencies Mounted 231 Offensive Cyber-Operations in 2011, Documents Show," *Washington Post*, August 30, 2013. See also Ryan Gallagher and Glenn Greenwald, "How the NSA Plans to Infect 'Millions' of Computers with Malware," *Intercept*, March 12, 2014. The best public window we have into the NSA's exploits and spy contractions was published by Jacob Appelbaum, Judith Horchert, and Christian Stöcker in *Der Spiegel*, "Catalog Advertises NSA Toolbox," December 29, 2013. Eric Rosenbach, then the Pentagon's Assistant Secretary of Defense for Global Security and Homeland Defense, who would later become the Pentagon's "Cyber Czar," addressed his concerns about the growing market for zero-days getting into foreign adversaries and non-state actors' hands to attack industrial systems in a March 2013 keynote address at the AFCA cybersecurity conference, which is available here (at minute 3:24): www.c-span.org/video/?c4390789/keynote -address-eric-rosenbach. For further reading, see the *Economist*'s 2013 take on the growing digital arms trade: "The Digital Arms Trade," March 2013. Gen. Michael Hayden's commentary regarding the government's "NOBUS" calculation can be found in Andrea Peterson's October 4, 2013, article for the *Washington Post*, "Why Everyone Is Left Less Secure When The NSA Doesn't Help Fix Security Flaws." The comments from an NSA analyst regarding hacking routers was taken from the Snowden leaks and were also detailed in the *Intercept*'s

March 12, 2014, article, "Five Eyes Hacking Large Routers." The NSA's $25.1 million new line item for zero-days was sourced from Brian Fung's August 2013 *Washington Post* article, sourced from a Snowden leak of the Pentagon's Black Budget. The estimate concerning how many zero-days the agency could purchase with those funds was sourced from Stefan Frei's 2013 analysis for NSS Labs, "The Known Unknowns: Empirical Analysis of Publicly Known Security Vulnerabilities." My colleague Scott Shane and I detailed the tax the Snowden leaks had on NSA morale for the *Times*, but they were also covered in a January 2, 2018, article by Ellen Nakashima and Aaron Gregg for the *Washington Post*, "NSA's Top Talent Is Leaving Because of Low Pay, Slumping Morale and Unpopular Reorganization." For a unique empirical analysis of zero-day lifespans, see Lillian Ablon and Andy Bogart's 2017 RAND study, *Thousands of Nights: The Life and Times of Zero-Day Vulnerabilities and Their Exploits*, www.rand.org/pubs/research_reports/RR1751.html. An earlier 2012 study determined the average lifespan of a zero-day was ten months: Leyla Bilge & Tudor Dumitras, "Before We Knew It: An Empirical Study of Zero-day Attacks in the Real World," which was featured at ACM Conference on Computer and Communications Security that year.

The reporting on VRL was my own and was sourced from current and former VRL employees and VRL's website. But shortly after VRL was tipped off to my inquiries, the language regarding its business practices disappeared from VRL's website. I sourced VRL job descriptions off of current and former employees' LinkedIn pages—a good source for understanding which companies participate in the offensive cyberarms trade. It should be noted that some of the other firms mentioned in this chapter—specifically Endgame, Netragard, and Exodus Intelligence—have all claimed to have ceased selling zero-days to government agencies over the past few years. I sourced VRL's contracts with the Pentagon, the air force, and the navy from government contract databases. VRL's marketing claim that it had "unparalleled capabilities" was taken from VRL's CEO himself in a press release timed to the company's acquisition by Computer Sciences Corp.: "CSC Acquires Vulnerability Research Labs, Press Release," *Business Wire*, 2010. For further reading on Trump's abandonment of foreign policy regarding the Kurdish people, see Robin Wright's October 2019 *New Yorker* piece: "Turkey, Syria, the Kurds and Trump's Abandonment of Foreign Policy." A contemporary account of Trump's failure to denounce the Saudis for the brutal murder of Jamal Khashoggi was sourced from Greg Myre's November 20, 2018, account for NPR, "'Maybe He Did, Maybe He Didn't': Trump Defends Saudis, Downplays U.S. Intel."

Finally, Alexander's remarks concerning his fear about the growing likelihood of zero-days ending up in the wrong hands was sourced from James Bamford's June 2013 piece for *Wired*, "NSA Snooping Was Only the Beginning. Meet the Spy Chief Leading Us into Cyberwar."

CHAPTER II: THE KURD

This chapter owes a huge debt of gratitude to Sinan Eren, now with the mobile security firm Fyde, for sharing his story and his experiences working with Immunity. I also could not have written these pages without David Evenden, who shared the story of his experiences in the United Arab Emirates—it should be noted—at great risk to his personal safety.

I first chronicled Revuln's business for the *Times* in March 2013. Two months later, Luigi Auriemma and Donato Ferrante told Reuters' Joseph Menn: "We don't sell weapons, we sell information" in "Special Report: U.S. Cyberwar Strategy Stokes Fear of Blowback."

It should be noted that the debate regarding zero-day disclosure is hardly unique to cybersecurity. Scientists have long sparred over the publication of biological research that could help stop the spread of a virus but could also be weaponized by rogue scientists to create a biological superweapon. When Dutch scientists tried to publish their research into how the lethal H5N1 "Bird Flu" spreads between mammals, a scientific advisory board tried to censor publication. Proponents, including Dr. Anthony Fauci—the same Dr. Fauci that would become America's most recognizable infectious disease specialist—pushed back, arguing that open discussion "makes it easier to get a lot of the good guys involved than the risk of getting the rare bad guy involved." Ultimately, the Bird Flu paper was published without restrictions. See Donald G. McNeil, "Bird Flu Paper Is Published after Debate," *New York Times*, June 21, 2012.

For history on the Wassenaar Arrangement, see Raymond Bonner, "Russia Seeks to Limit an Arms Control Accord," *New York Times*, April 5, 1996.

The reporting on FinSpy's spyware included in the book was taken from my own reporting in the *Times*, often written in conjunction with researchers at the Citizen Lab at the University of Toronto's Monk School of Affairs. In particular, I relied heavily on research conducted by Bill Marczak, a fellow at the Citizen Lab who made these articles possible. For further reading see my 2012 accounts in the *New York Times*: "Researchers Find 25 Countries Using Surveillance Software," March 13, 2013; "Elusive FinSpy Spyware Pops Up in 10 Countries," August 13, 2012; "Software

Meant to Fight Crime Is Used to Spy on Dissidents," August 31, 2012; and "How Two Amateur Sleuths Looked for FinSpy Software," August 31, 2012. For further reading on why security researchers and companies were up in arms about the broad wording included in Wassenaar, see Kim Zetter, "Why an Arms Control Pact Has Security Experts Up in Arms," *New York Times*, June 24, 2015.

The best summary of encryption export rules can be found in a chart on the U.S. Department of Commerce's Bureau of Industry and Security website. The site lists sales to exempt countries, and notes which countries require sellers to obtain a license. It also charts the biannual sales report mandates: www.bis.doc.gov/index.php/documents/new -encryption/1651-740-17-enc-table/file. The reference to the *Little Black Book of Electronic Surveillance* and the renaming of the catalog to *The Big Black Book* in 2016 was sourced from Sharon Weinberger's July 19, 2019, op-ed in the *New York Times*: "Private Surveillance Is a Lethal Weapon Anybody Can Buy," July 19, 2019.

The reference to Kevin Mitnick's emergence in the zero-day exploit trade was sourced from Andy Greenberg's September 24, 2014, *Wired* article, "Kevin Mitnick, Once the World's Most Wanted Hacker, Is Now Selling Zero-Day Exploits." A description of Immunity's Canvas tool was adapted from Immunity's own website, www.immunityinc.com/products/canvas.

For further reading on the mass disappearance of Kurds in Turkey in the 1990s, see Human Rights Watch, "Time for Justice: Ending Impunity for Killings and Disappearances in 1990s Turkey," September 3, 2012, www.hrw.org/report/2012/09 /03/time-justice/ending-impunity-killings-and-disappearances-1990s-turkey.

Some of the reporting in this chapter was also included in a comprehensive account of Dark Matter and NSO that I wrote with my colleagues Mark Mazzetti, Adam Goldman, and Ronen Bergman at the *New York Times*, "A New Age of Warfare: How Internet Mercenaries Do Battle for Authoritarian Governments," March 21, 2019. Three years earlier, without knowing exactly who it was that was leading the UAE's campaign to spy on dissidents, I reported on research conducted by Bill Marczak and John Scott-Railton at the Citizen Lab about the forensic evidence tying these campaigns to the UAE.

In 2019, two reporters at Reuters, Joel Schectman and Christopher Bing, reported for the first time on the attribution tying these campaigns not only to U.S. cyber mercenaries, but to a decade-long effort by senior former White House counterintelligence officials to break into the UAE surveillance market. Their accounts are available at Christopher Bing and Joel Schectman, "Inside the UAE's Secret Hacking Team of American Mercenaries," Reuters, January 30, 2019, and "White House Veterans Helped Gulf Monarchy Build Secret Surveillance Unit," Reuters, December 10, 2019.

I learned about Dark Matter's hacking of FIFA, Qatari officials, and the dragnet that included former First Lady Michelle Obama from my interviews with Evenden and other CyberPoint and Dark Matter employees. This book was the first time any of the details of the hack of Mrs. Obama has been reported. I was able to flesh out the details of Mrs. Obama's visit via contemporary news articles, including the Associated Press, "First Lady Michelle Obama Arrives in Qatar for Speech," November 2, 2015; Nick Anderson, "First Lady Urges Fathers Worldwide to Join 'Struggle' for Girls' Education," *Washington Post*, November 4, 2015; and Paul Bedard, "Michelle Obama's 24 Minute Speech in Qatar Cost $700,000," *Washington Examiner*, December 9, 2015. As it turned out, the Emiratis had good reason to suspect Qatar had bribed FIFA officials to host the 2022 World Cup. As this book was being readied for print, the U.S. Department of Justice revealed in April 2020 that, indeed, Qatar (and Russia) had successfully bribed five members of FIFA's board to lock in Russia as host for the 2018 FIFA men's tournament, and Qatar for the 2022 World Cup. The indictment accused three South American FIFA officials of taking bribes to vote for Qatar. See Tariq Panja and Kevin Draper, "U.S. Says FIFA Officials Were Bribed to Award World Cups to Russia and Qatar," *New York Times*, April 6, 2020.

If you're interested in learning more about Saudi-UAE-Qatari relations—one of the least-understood conflicts in the Middle East—the most comprehensive account, by far, was written by my colleague Declan Walsh for the *New York Times*, "Tiny, Wealthy Qatar Goes Its Own Way, and Pays for It," January 22, 2018.

The details about UAE's "debtors' prison" were fact-checked with Jason DeParle's account, "Migrants in United Arab Emirates Get Stuck in Web of Debt," *New York Times*, August 20, 2011.

In late 2019, my colleague Mark Mazzetti and I revealed just how aggressive and innovative the Emiratis' surveillance had become. In December 2019, we broke the story of ToTok, a seemingly innocuous texting app in the Apple and Google app stores that was secretly an Emirati surveillance tool developed by a Dark Matter subsidiary. Apple and Google subsequently removed the apps from their stores, but as our reporting revealed, by then the Emiratis had successfully collected contacts, faces, voice prints, photos, calls, and text messages from the millions of people around the world who had already downloaded the app.

CHAPTER 12: DIRTY BUSINESS

I owe much to Adriel Desautels for his time and patience walking me through the ins and outs of his exploit brokerage. I corroborated his account of his early 2002 conflict with Hewlett-Packard with legal notices. A contemporary account is available at Declan McCullagh, "HP Backs Down on Copyright Warning," Cnet, August 2, 2002, and Joseph Menn, "Hackers Live by Own Code," *Los Angeles Times*, November 19, 2003.

Among the first public profiles of a spyware seller was a Bloomberg profile of Martin J. Muench—aka MJM—the German spyware entrepreneur behind FinFisher spyware. See Vernon Silver, "MJM as Personified Evil Says Spyware Saves Lives Not Kills Them," Bloomberg, November 8, 2012. MJM has since reappeared with a new company, MuShun Group, and conspicuously opened up new offices in the UAE and Malaysia.

I chronicled FinFisher and Gamma Group's spyware as it was used on dissidents in disparate countries with a big helping hand from the researchers at Citizen Lab in "Ahead of Spyware Conference, More Evidence of Abuse," *New York Times*, October 10, 2012, and "Intimidating Dissidents with Spyware," *New York Times*, May 30, 2016, and later chronicled NSO Group in a number of articles for the *Times*.

After recordings of George Hotz's conversation with and exploit broker leaked online, Hotz denied the deal was ever consummated. The same Apple iOS exploit was later reported to have sold for $1 million to a Chinese company. Asked about the reports, Hotz told the *Washington Post* in 2014, "I'm not big on ethics." See Ellen Nakashima and Ashkan Soltani, "The Ethics of Hacking 101," October 7, 2014. The infamous image of "the Grugq" with a duffel bag of cash he'd make from selling exploits was featured in Andy Greenberg's *Forbes* story, which also included price lists for zero-day exploits: "Shopping for Zero-Days: A Price List For Hackers' Secret Software Exploits," March 23, 2012. According to Andy, these price lists were sourced from Desautels' firm Netragard and other sources in the market. In 2013, my colleague David Sanger and I reported that the top price fetched for an iOS zero-day was $250,000 in "Nations Buying as Hackers Sell Flaws in Computer Code," *New York Times*, July 13, 2013. Later, Zerodium erased any mystery to these lists, publishing a full price list on its website in November 2015, which showed that the price for an iOS exploit had already doubled. For Zerodium's latest prices, see zerodium.com/program.html. Reporters have done yeoman's work chronicling the rise in Zerodium's prices. See Lily Hay Newman, "A Top-Shelf iPhone Hack Now Goes for $1.5 Million," *Wired*, September 29, 2016; Andy Greenberg, "Why 'Zero-day' Android Hacking Now Costs More Than iOS Attacks," *Wired*, September 3, 2019; Lorenzo Franceschi-Bicchierai, "Startup Offers $3 million to Anyone Who Can Hack the iPhone," April 25, 2018. Thanks to a Freedom of Information Act request by the Electronic Frontier Foundation, we learned in 2015 that the NSA was among Vupen's customers. See Kim Zetter, "US Used Zero-Day Exploits Before It Had Policies for Them," *Wired*, March 30, 2015. An entirely new window into the zero-day market arrived when Hacking Team was hacked by the enigmatic hacker "Phineas Fisher" in 2015. Hacking Team's leaked documents are searchable on WikiLeaks here: wikileaks.org/hackingteam/emails. Contemporary accounts of the hack can be read on the *Vice* Motherboard blog here: Lorenzo Franceschi-Bicchierai, "Spy Tech Company 'Hacking Team' Gets Hacked," July 5, 2015; Lorenzo Franceschi-Bicchierai, "The Vigilante Who Hacked Hacking Team Explains How He Did It," April 15, 2016; Franceschi-Bicchierai, "Hacking Team Hacker Phineas Fisher Has Gotten Away With It," November 12, 2018; and at David Kushner, "Fear This Man," *Foreign Policy*, April 26, 2016; Ryan Gallagher, "Hacking Team Emails Expose Proposed Death Squad Deal, Secret U.K. Sales Push and Much More," *Intercept*, July 8, 2015; Cora Currier and Morgan Marquis-Boire, "Leaked Documents Show FBI, DEA and U.S. Army Buying Italian Spyware," *Intercept*, July 6, 2015; Franceschi-Bicchierai, "Hacking Team's 'Illegal' Latin American Empire," *Vice*, April 18, 2016; Joseph Cox, "The FBI Spent $775K on Hacking Team's Spy Tools Since 2011," *Wired*, July 6, 2015; and Mattathias Schwartz, "Cyberwar for Sale," *New York Times*, January 4, 2017.

An excellent analysis of the zero-day market based on Hacking Team's leaks was conducted by researcher Vlad Tsyrklevich, "Hacking Team: A Zero-day Market Case Study," July 26, 22, 2015, tsyrklevich.net/2015/07/22/hacking-team-0day-market. Tsyrklevich picked up a Hacking Team purchase of a Microsoft Office email exploit from an Indian company in Jaipur called Leo Impact Security, the first time in memory that an Indian company selling zero-days appeared on the radar.

The prescient email from Vincenzetti to his team, in which he wrote, "Imagine this: a leak on WikiLeaks showing YOU explaining the evilest technology on earth! ☺" is still available on WikiLeaks at wikileaks.org/hackingteam/emails/emailid/1029632.

Desautels' announcement that he would cease Netragard's zero-day side business was covered by Dan Goodin for *Ars Technica*, "Firm Stops Selling Exploits after Delivering Flash 0-Day to Hacking Team," July 20, 2015.

CHAPTER 13: GUNS FOR HIRE

For a contemporary account of Francisco Partner's private equity investment in NSO Group, see Orr Hirschauge, "Overseas Buyers Snap Up Two More Israeli Cyber Security Firms," *Haaretz*, March 19, 2014.

For the FBI's first public statement on its "going dark" problem, see Valerie Caproni, General Counsel, FBI, Testimony Before the House Judiciary Committee, Subcommittee on Crime, Terrorism, and Homeland Security, February 17, 2011. This later became NSO Group's marketing pitch. For details on NSO Group's pricing, see my 2016 account in the *New York Times*, "Phone Spying Is Made Easy. Choose a Plan," September 3, 2016. In 2015, Hacking Team charged its clients a €200,000 installation fee, plus more for extra features that cost anywhere from €50,000 to €70,000. The leaks showed NSO charged clients $500,000 to hack just five BlackBerry phone users, or $300,000 for five Symbian users on top of that set-up fee. The company also charged an annual 17-percent maintenance fee.

Hacking Team's anxieties over NSO Group's "over the air stealth installation" feature can be read in Hacking Team's leaked emails available on WikiLeaks: wikileaks.org/hackingteam/emails/emailid/6619.

I first learned of NSO Group's relationship with Mexican government agencies from the leaks my source provided in 2016. Partnering with Azam Ahmed in Mexico, we were able to detail the Mexican consumer rights activists, doctors, journalists, international lawyers, and those targets' families in our June 2017 exposés: Nicole Perlroth, "Invasive Spyware's Odd Targets: Mexican Advocates of Soda Tax," *New York Times*, February 12, 2017; Ahmed and Perlroth, "Spyware Meant to Foil Crime Is Trained on Mexico's Critics," *New York Times*, June 19, 2017; and Ahmed and Perlroth, "Using Texts as Lures, Government Spyware Targets Mexican Journalists and Their Families," *New York Times*, Jun 19, 2017. Our reporting ignited street protests and calls for an independent inquiry. Mexico's President conceded his government had acquired NSO's spyware but denied it had been misused. He also issued a vague threat against us, but later recanted. See Ahmed, "Mexican President Says Government Acquired Spyware but Denies Misuse," *New York Times*, June 22, 2017. On the calls for an inquiry, Kirk Semple, "Government Spying Allegations in Mexico Spur Calls for Inquiry," *New York Times*, June 21, 2017. To date, these inquiries have gone nowhere.

NSO Group's relationship with Finland has never been documented. To better understand why Finland would be interested in its spy tools, see Simon Tidsall, "Finland Warns of New Cold War over Failure to Grasp Situation in Russia," *Guardian*, November 5, 2014, and Eli Lake's interview with Finnish president Sauli Niinistö: "Finland's Plan to Prevent Russian Aggression," Bloomberg, June 12, 2019.

Bill Marczak and John Scott-Railton, together with researchers at Lookout, were the first to publish research on NSO Group's spyware in the UAE, and on Ahmed Mansoor specifically. See "The Million Dollar Dissident: NSO Group's iPhone Zero-Days Used against a UAE Human Rights Defender," Citizen Lab,

August 24, 2016, which I covered for the *Times*, Nicole Perlroth, "iPhone Users Urged to Update Software After Security Flaws Are Found," August 25, 2016. See also Perlroth, "Apple Updates iOS to Patch a Security Hole Used to Spy on Dissidents," *New York Times*, August 26, 2016, and Richard Silverstein, "Israel's Cyber Security Firm 'NSO Group' Permits Foreign Intelligence Agencies to Spy on Human Rights Activists," Global Research, June 20, 2017. Later Marczak, Railton, and Citizen Lab researchers Sarah McKune, Bahr Abdul Razzak, and Ron Diebert were able to tie NSO Group's Pegasus software to operations in forty-five countries. See "Hide and Seek: Tracking NSO Group's Pegasus Spyware to Operations in 45 Countries," Citizen Lab, September 18, 2018. Because NSO customers can route their spyware through servers in other countries, or through VPNs—virtual private networks—that reroute traffic through servers around the globe, it is possible some of those countries were only decoys, hosting cloud servers for other countries' operations.

The quotes and personal accounts from Ahmed Mansoor were taken from my interviews with Mansoor before he was imprisoned. Amnesty International and the Gulf Centre for Human Rights have been following the depressing state of Mansoor's case and health. See Amnesty International, "UAE: Activist Ahmed Mansoor Sentenced to 10 Years in Prison for Social Media Posts," May 31, 2018, and the Gulf Centre for Human Rights, "United Arab Emirates: Call for Independent Experts to Visit Ahmed Mansoor, on Liquids Only Hunger Strike since September," February 2, 2020.

For an account of Turkey's abysmal track record on press rights, see Committee to Protect Journalists, "Turkey: A Journalist Prison," December 13, 2016.

CHAPTER 14: AURORA

This chapter relied heavily on interviews with Google researchers, and I owe them a debt of gratitude for granting me the time and license to document the Chinese cyberattack on Google's systems that would later be dubbed Aurora.

For a historical account of the early indications of the Japanese attack on Pearl Harbor, see "Officer Mistook Radar Warning of Pearl Harbor Raid," *Columbus Dispatch*, February 25, 2010.

Where I mention "the Australian outback" I am referring to the Joint American-Australian Defense Facility Pine Gap spy station. See Jackie Dent, "An American Spy Base Hidden in Australia's Outback," *New York Times*, November 23, 2017.

For Sergey Brin's affinity for various modes of cycling transportation, see Richard Masoner, "Sergey Brin Rides an Elliptigo," Cyclelicious, October 21, 2011, also made famous in the film *The Internship*: Megan Rose Dickey, "'The Internship' Movie Is a Two-Hour Commercial for Google," *Business Insider*, May 24, 2013. I also relied on a 2007 profile of Sergey Brin featured in *Moment m*agazine for his biography: Mark Malseed, "The Story of Sergey Brin," May 6, 2007.

Shortly after I reported on China's cyberattack on the *New York Times*, my colleagues David Sanger, David Barboza, and I first outed China's Unit 61398 hacking unit in a February 19, 2013 article, "China's Army Seen as Tied to Hacking against U.S." The reporting, which was based in part by research from Mandiant, the security firm, later led to the indictment of several PLA hackers, who have yet to be extradited. Previously my colleagues and I had reported on the students at Chinese universities that were digitally linked to China's hacks on foreign targets: John Markoff and David Barboza, "2 China Schools Said To Be Tied To Online Attacks," *New York Times*, February 18, 2010; Barboza, "Inquiry Puts China's Elite In New Light," *New York Times*, February 22, 2010; James Glanz and John Markoff, "State's Secrets: Day 7; Vast Hacking by a China Fearful of the Web," *New York Times*, December 5, 2010; and Perlroth, "Case Based in China Puts a Face on Persistent Hacking," *New York Times*, March 29, 2012. For reporting on Chinese cyberespionage on Western targets see also David E. Sanger and Nicole Perlroth, "Chinese Hackers Resume Attacks on U.S. Targets," *New York Times*, May 20, 2013; Perlroth, "China Is Tied To Spying On European Diplomats," *New York Times*, Dec 10, 2013; Perlroth, "China Is Said to Use Powerful New Weapon to Censor Internet," *New York Times*, April 10, 2015; Helene Cooper, "Chinese Hackers Steal Naval Warfare Information," *New York Times*, June 9, 2018; David E. Sanger, Nicole Perlroth, Glenn Thrush, and Alan Rappeport, "Marriott Data Breach Traced to Chinese Hackers," *New York Times*, December 12, 2018; and Nicole Perlroth, Kate Conger, and Paul Mozur, "China Sharpens Hacking to Hound Its Minorities," *New York Times*, October 25, 2019.

I found the most comprehensive contemporary account of Google's struggles with Chinese censorship to be Clive Thompson's 2006 account for the *New York Times Magazine*, "Google's China Problem (And China's Google Problem)," April 23, 2006. The reference to Chinese officials calling Google "an illegal site" can be found here: James Glanz and John Markoff, "Vast Hacking by a China Fearful of the Web," *New York Times*, December 4, 2010, and American lawmakers' criticism of Google, including a Republican congressman's remark that Google had "become evil's accomplice," can be viewed in Congressional testimony available on C-SPAN: "Internet in China: A Tool for Freedom or Suppression," www.c-span.org/video /?191220-1/internet-china. Yahoo's role in outing helping Chinese officials imprison a journalist is described in Joseph Kahn, "Yahoo Helped Chinese to Prosecute Journalist," *New York Times*, September 8, 2005.

For a contemporary account of Google's subsequent pullout from China after the Aurora attack, see Google's then general counsel David Drummond's blog post "A New Approach to China," Google Blog, January 2010. CNN was among the first news outlets to pick up the blog: Jeanne Meserve and Mike M. Ahlers, "Google Reports China-Based Attack, Says Pullout Possible," CNN.com, January 13, 2010. Sergey Brin granted my *Times* colleague Steve Lohr a rare interview in the aftermath of the attack: Steve Lohr, "Interview: Sergey Brin on Google's China Move," *New York Times*, March 22, 2010, in which Brin wrongly predicted that long term, China would have to stop censoring the internet. See also Andrew Jacobs and Miguel Helft, "Google, Citing Cyber Attack, Threatens to Exit China," *New York Times*, January 12, 2010. The articles note that thirty-four companies were tied to the same attack that hit Google, many of them still unknown. Also see Andrew Jacobs and Miguel Helft, "Google May End Venture in China over Censorship," *New York Times*. For a more comprehensive timeline of Google's pullout from China, see Bobbie Johnson, "Google Stops Censoring Chinese Search Engine: How It Happened," *Guardian*, March 22, 2010.

Hillary Clinton's March 2010 remarks on Chinese hacking were covered by Paul Eckert and Ben Blanchard, "Clinton Urges Internet Freedom, Condemns Cyberattacks," Reuters, January 21, 2010.

James Comey's remarks about the breadth of Chinese hacking—"There are two kinds of big companies in the United States. There are those who've been hacked by the Chinese, and those who don't know they've been hacked by the Chinese"—are available in Comey's 2014 interview with CBS News's Scott Pelley, October 5, 2014.

For China's denials and response, see Tania Brannigan, "China Responds to Google Hacking Claims," *Guardian*, January 14, 2010; Brannigan, "China Denies Links to Google Cyberattacks," *Guardian*, February 23, 2010; and Eckert and Blanchard, "Clinton Urges Internet Freedom, Condemns Cyberattacks," China still publicly maintains that it is a victim, not a perpetrator, of cyberattacks to this day.

According to a subsequent forensic analysis of the Aurora attack by McAfee, China's primary goal was to gain access to and potentially modify source code repositories at these high tech, security, and defense contractor companies. "[The SCMs] were wide open," said Dmitri Alperovitch, then a McAfee researcher. "No one ever thought about securing them, yet these were the crown jewels of most of these companies in many ways—much more valuable than any financial or personally identifiable data that they may have and spend so much time and effort protecting."

For the reference to Google's two billion lines of code, see Cade Metz, "Google Is 2 Billion Lines of Code—And It's All in One Place," *Wired*, September 16, 2015.

A source for China's banning of Winnie the Pooh is Javier C. Hernández, "To Erase Dissent, China Bans Pooh Bear and 'N,'" *New York Times*, March 1, 2018.

For Google's change in leadership, priorities, and China strategy, I relied on the following contemporary accounts: Conor Dougherty, "Google Mixes a New Name and Big Ideas," *New York Times*, August 11, 2015; James B. Stewart, "A Google C.F.O. Who Can Call Time-Outs," *New York Times*, July 24, 2015; and Luke Stangel, "Chinese Retail Giant Alibaba Opening New R&D Lab in San Mateo," *Silicon Valley Business Journal*, October 11, 2017. Google's secret plans to re-enter China were first chronicled by the *Intercept*: Ryan Gallagher, "Google Plans to Launch Censored Search Engine in China, Leaked Documents Reveal," August 1, 2018. The ensuing protests by Google employees were covered by my colleagues Kate Conger and Daisuke Wakabayashi in "Google Workers Protest Secrecy In China Project," *New York Times*, August 17, 2018. Ben Hubbard chronicled the Google and Apple apps that allowed Saudi men to track the movements of their female family members: "Apple and Google Urged to Dump Saudi App That Lets Men Track Women," *New York Times*, February 14, 2019. Google's project for the Pentagon and the ensuing backlash were also covered by my colleagues Scott Shane and Daisuke Wakabayashi for the *Times*: "A Google Military Project Fuels Internal Dissent," April 5, 2018. For Google's YouTube troubles, see my colleague Kevin Roose's reporting in "The Making of a YouTube Radical," *New York Times*, June 8, 2019. For an account of the problems with YouTube Kids, see Sapna Maheshwari, "On YouTube Kids, Startling Videos Slip Past Filters," *New York Times*, November 4, 2017, which showed that videos encouraging suicide were slipping past YouTube's filters.

Morgan Marquis-Boire spent several hours, days even, sitting for interviews for this book in late 2016 and 2017. Months after our last interview he was accused of drugging and raping women in a disturbing article published by the technology news site *The Verge*: Chloe Ann-King, "'We Never Thought We'd Be Believed': Inside the Decade-Long Fight to Expose Morgan Marquis-Boire," November 29, 2017, and Sarah Jeong, "In Chatlogs, Celebrated Hacker and Activist Confesses Countless Sexual Assaults," November 19, 2017. I tried multiple times to reach Marquis-Boire for comment, but he did not respond to my inquiries.

CHAPTER 15: BOUNTY HUNTERS

I covered Silicon Valley's ruthless talent wars in my previous job for *Forbes*, back when Google was bumping salaries up by 10 percent to quash defections and Facebook spent tens of millions of dollars to keep just two of its top engineers from defecting to Twitter. Back then, the companies were offering "cubicle decoration budgets" and a year's supply of beer. See Perlroth, "Winners and Losers in Silicon Valley's War for Talent," June 7, 2011.

For further reading on Google's "fuzz farms" see Google Blog, "Fuzzing at Scale," August 2011, security.googleblog.com/2011/08/fuzzing-at-scale.html and for further reading on bug bounties, see Perlroth, "Hacking for Security, and Getting Paid for It," *New York Times*, October 14, 2015; Steven Melendez, "The Weird, Hyper-Incentivized World Of "Bug Bounties," *Fast Company*, January 24, 2014; Andy Greenberg, "Google Offers $3.14159 Million in Total Rewards for Chrome OS Hacking Contest," *Forbes*, January 28, 2013. The account of Missoum Said, the Algerian who made enough money to remodel his family's house and send his parents to Mecca, was based on an interview. I also interviewed Mazen Gamal and Mustafa Hassan, the two hackers who used their winnings to buy an apartment; one used his bounties to purchase an engagement ring. You can read about how Nils Juenemann spent his bounties on schools in Tanzania, Togo, and Ethiopia on his blog: www.nilsjuenemann.de/2012/04/26/ethiopia-gets-new-school-thanks-to-xss

For the other side, I relied on Andy Greenberg's profile of Chaouki Bekrar in *Forbes*, "Meet the Hackers Who Sell Spies the Tools to Crack Your PC and Get Paid Six Figure Fees," *Forbes*, March 21, 2012, and Greenberg, "The Zero-Day Salesman," *Forbes*, March 28, 2012. A FOIA request submitted via public records service MuckRock later revealed that NSA subscribed to Bekrar's services: www.muckrock.com/foi/united-states-of-america-10/vupen-contracts-with-nsa-6593/#file-10505. The contract was dated September 14, 2012, and signed by Bekrar himself.

On the fallout from Hacking Team's 2015 hack, see Kim Zetter, "Hacking Team Leak Shows How Secretive Zero-Day Exploit Sales Work," *Wired*, July 24, 2015, and Lorenzo Franceschi-Bicchierai, "Hacking Team Has Lost Its License to Export Spyware," *Vice*, April 6, 2016. It was after Hacking Team's hack put the zero-day market under regulators' microscope that Vupen rebranded as Zerodium. See Dennis Fisher, "VUPEN Founder Launches New Zero-Day Acquisition Firm Zerodium," Threatpost, July 24, 2015. Zerodium subsequently became the first exploit broker to advertise a $1 million payout for a remote Apple iOS jailbreak:

twitter.com/Zerodium/status/645955632374288384, the highest that had been offered publicly.

For Google's bounty increases, see Aaron Holmes, "Google Is Offering a $1.5 Million Reward to Anyone Who Can Pull Off a Complex Android Hack," *Business Insider*, November 22, 2019.

After a whistleblower accused Pentagon contractors of outsourcing the Pentagon's programming to coders in Russia, both contractors denied the accusations. CSC claimed it knew nothing of the NetCracker, the subcontractor's conduct, even though the whistleblower claimed all parties knew. Nevertheless, both CSC and NetCracker settled the case for a combined $12.75 million in penalties—including a $2.3 million payout to NetCracker's whistleblower.

The account of Facebook and Microsoft's early bug bounty programs were compiled from interviews with Michiel Prins, Jobert Abma, Alex Rice of HackerOne, Merijn Terheggen, Katie Moussouris of Luta Security, Bill Gurley of Benchmark, and others who patiently sat for several hours', in some cases days', worth of interviews.

The discovery of Flame and its aftermath for Microsoft was captured from interviews with Microsoft employees. I chronicled Flame's discovery for the *New York Times*: "Microsoft Tries to Make Windows Updates Flame Retardant," June 4, 2012. The theory that Microsoft had somehow been infiltrated by CIA and NSA spies was first put forth by Mikko Hypponen, a Finnish security researcher, in Kevin Fogarty, "Researcher: CIA, NSA May Have Infiltrated Microsoft to Write Malware," IT World, June 18, 2012.

The first Snowden leaks of the NSA's Prism project appeared in June 2013 in the *Guardian*, which described Prism as a "team sport" between tech companies and the FBI, CIA, and NSA: Glenn Greenwald and Ewen MacAskill, "NSA Prism Program Taps into User Data of Apple, Google and Others," June 7, 2013. One month later, Greenwald, MacAskill, Laura Poitras, Spencer Ackerman, and Dominic Rushe reported that Microsoft was giving the NSA access to encrypted messages: "Microsoft Handed the NSA Access to Encrypted Messages," *Guardian*, July 12, 2013. My colleague Claire Cain Miller and I reported on the aftermath for the technology companies, as officials overseas threatened to "Balkanize" the internet: "N.S.A. Spying Imposing Cost on Tech Firms," *New York Times*, March 22, 2014.

The tech press covered the debut of Microsoft's bug bounty program in 2013. See Andy Greenberg, "Microsoft Finally Offers to Pay Hackers for Security Bugs with $100,000 Bounty," *Forbes*, June 19, 2013. For further reading on the economics of bug bounty programs and a useful summary of their origins, I recommend Andreas Kuehn and Milton Mueller's 2014 paper, "Analyzing Bug Bounty Programs: An Institutional Perspective on the Economics of Software Vulnerabilities," available here: papers.ssrn.com/sol3/papers.cfm?abstract_id=2418812, as well as Mingyi Zhao, Aron Laszka, and Jens Grossklag's 2017 article in the *Journal of Information Policy*, "Devising Effective Policies for Bug-Bounty Platforms and Security Vulnerability Discovery," which discusses the signal-to-noise ratio of Facebook, Google, Bugcrowd, and HackerOne's bug bounty programs. Bill Gurley provided background on Benchmark's early investment in HackerOne.

My colleagues David Sanger, Michael Shear, and I covered the Chinese attack on the Office of Personnel Management for the *Times*: "Attack Gave Chinese Hackers Privileged Access to U.S. Systems," June 21, 2015.

For contemporary accounts of the Pentagon's first bug bounty program, I relied on interviews with Pentagon officials, private security firms that participated in the bounty programs, and public reports including Lisa Ferdinando, "Carter Announces, 'Hack the Pentagon' Program Results," DOD News, June 17, 2016; U.S. Department of Defense, "D.O.D. Expands 'Hack the Pentagon' Crowdsourced Digital Defense Program," press release, October 24, 2018; Jason Murdock, "Ethical Hackers Sabotage F-15 Fighter Jet, Expose Serious Vulnerabilities," *Newsweek*, August 15, 2019; Aaron Boyd, "DOD Invests $34 Million in Hack the Pentagon Expansion," Nextgov.com, October 24, 2018.

The remarks attributed to Chris Inglis, the retired NSA deputy director, were taken from Dan Geer's 2014 Black Hat Talk, the text of which is available here: "Cybersecurity as Realpolitik," 2014 Black Hat Talk, August 6, 2014, geer.tinho.net /geer.blackhat.6viii14.txt.

CHAPTER 16: GOING DARK

Among the most damning of the Snowden leaks was the October 2013 *Washington Post* article showing the Post-it note on which an NSA analyst had illustrated how the agency was getting into Google and Yahoo's data centers. The drawing included a smiley face at the exact point where Google's data lay unencrypted. See Barton Gellman and Ashkan Soltani, "NSA Infiltrates Links to Yahoo, Google Data Centers Worldwide, Snowden Documents Say," October 30, 2013. See also Gellman and Soltani's addendum with Todd Lindeman: "How the NSA Is Infiltrating Private Networks," *Washington Post*, October 30, 2013. The reference to the NSA's theft of emails and contacts was taken from Gellman and Soltani's October 14, 2013, story:

"NSA Collects Millions of Email Address Books Globally." On the subsequent fallout for tech companies and the NSA, see "Google's Schmidt: NSA Spying on Data Centers Is 'Outrageous,'" *Wall Street Journal*, November 4, 2013; Mike Masnick, "Pissed Off Google Security Guys Issue FU to NSA, Announce Data Center Traffic Now Encrypted," Techdirt, November 6, 2013; Alexei Oreskovic, "Google Employees Lash Out at NSA over Reports of Cable Tapping," Reuters, November 6, 2013; as well as my own reporting with Vindu Goel and David Sanger for the *New York Times*: "Internet Firms Step Up Efforts to Stop Spying," December 5, 2013; and David E. Sanger and Nicole Perlroth, "Internet Giants Erect Barriers to Spy Agencies," *New York Times*, June 6, 2014. I highly recommend Steven Levy's comprehensive account, "How the NSA Almost Killed the Internet," *Wired*, January 7, 2014.

The first public account of Google's Project Zero debut appeared in *Wired*. See Andy Greenberg, "Meet 'Project Zero,' Google's Secret Team of Bug-Hunting Hackers," July 15, 2014. The first major vulnerability Project Zero uncovered in Microsoft touched off a war of words between the two companies. For more on that, see Steve Dent, "Google Posts Windows 8.1 Vulnerability before Microsoft Can Patch It," Engadget.com, January 2, 2015, and Lorenzo Franceschi-Bicchierai, "How Google Changed the Secretive Market for the Most Dangerous Hacks in the World," *Vice*, September 23, 2019. For coverage of Jung Hoon Lee's—aka Lokihardt—exploits, see Russell Brandom, "A Single Researcher Made $225,000 (Legally!) by Hacking Browsers This Week," *The Verge*, March 20, 2015. Project Zero chronicles its zero-day discoveries on a tracking sheet, available at googleprojectzero.blogspot .com/p/0day.html.

Tim Cook, I would argue, has made himself more accessible to reporters than any other CEO in Silicon Valley. Cook has used these opportunities to speak about his personal views on everything from privacy to immigration, and these sit-downs proved incredibly helpful to this book, and this chapter in particular. Other Silicon Valley CEOs: Take note!

Cook publicly his described an early encounter with KKK Klansmen to Bloomberg: "Tim Cook Speaks Up," October 30, 2014. See also Tim Cook's 2019 Commencement Address to Stanford University: www.youtube.com/watch ?v=2C2VJwGBRRw. My colleagues Matt Richtel and Brian X. Chen chronicled Cook's early days as CEO: "Tim Cook, Making Apple His Own," *New York Times*, June 15, 2014.

In the wake of the Snowden revelations, then President Obama held several closed-door meetings with Cook and other tech executives in 2014. In December, Obama met with Cook as well as Marissa Mayer of Yahoo, Dick Costolo of Twitter,

Eric Schmidt of Google, Sheryl Sandberg of Facebook, Brian Roberts of Comcast, Randall Stephenson of AT&T, Brad Smith of Microsoft, Erika Rottenberg of LinkedIn, and Reed Hastings of Netflix. The official agenda was to discuss improvements to Healthcare.gov, but the conversation was quickly diverted to government surveillance and the public's fading trust. See Jackie Calmes and Nick Wingfield, "Tech Leaders and Obama Find Shared Problem: Fading Public Trust," *New York Times*, December 17, 2013. For an account of their earlier August meeting, see Tony Romm, "Obama, Tech Execs Talk Surveillance," *Politico*, August 8, 2013. The articles do not mention the letters Cook received, which were sourced to other interviews.

Cook's debut of Apple's new iPhone security was covered extensively by the tech press. See John Kennedy, "Apple Reveals 4.7-inch and 5.5-inch iPhone 6 and iPhone 6 Plus Devices," Siliconrepublic.com, September 9, 2014, as well as David Sanger and Brian Chen, "Signaling Post-Snowden Era, New iPhone Locks Out N.S.A.," *New York Times*, September 26, 2014.

For Comey's "Going Dark" publicity tour, see Igor Bobic and Ryan J. Reilly, "FBI Director James Comey 'Very Concerned' about New Apple, Google Privacy Features," *Huffington Post*, September 25, 2014; James Comey, "Going Dark: Are Technology, Privacy, and Public Safety on a Collision Course?" Full Remarks, Brookings Institution, October 16, 2014; and Scott Pelley's *60 Minutes* interview with Comey on June 21, 2015.

For a comprehensive account of the NSA's efforts to crack encryption, see the account I wrote with ProPublica for the *Times*: Nicole Perlroth, Jeff Larson, and Scott Shane, "NSA Able to Foil Basic Safeguards of Privacy on Web," September 5, 2013.

I relied on Adam Nagourney, Ian Lovett, Julie Turkewitz, and Benjamin Mueller's account of the San Bernardino shooters: "Couple Kept Tight Lid on Plans for San Bernardino Shooting," *New York Times*, December 3, 2015. The subsequent back and forth between Apple and FBI was covered extensively by my colleagues and I at the *New York Times*: Mike Isaac, "Why Apple Is Putting Up a Fight Over Privacy with the F.B.I.," February 18, 2016; Cecilia Kang and Eric Lichtblau, "F.B.I. Error Led to Loss of Data in Rampage," March 2, 2016; and Eric Lichtblau and Katie Benner, "Apple Fights Order to Unlock San Bernardino Gunman's iPhone," *New York Times*, February 17, 2016.

Security researchers at the time dubbed the software changes the FBI was asking Apple to make "GovtOS"—a spoof on its iOS mobile software. See Mikey Campbell, "Apple Rails against FBI Demands for 'Govtos' in Motion to Vacate Decryption Request," Appleinsider.com, February 25, 2016.

Tim Cook personally addressed the company's battle with the FBI in interviews and in a message to Apple's customers: "A Message to Our Customers," February 16, 2016. In private, Apple employees argued that the government could not be trusted with the keys to Apple's communications when it could not even keep its own employees' data secure, as it had failed to do in the Chinese hack of the U.S. Office of Personnel Management.

The quote from Carole Adams, the mother of one of the San Bernardino shooting victims, in support of Apple was sourced from Tony Bradley, "Apple vs. FBI: How Far Can the Government Go in the Name of 'National Security'?" *Forbes*, February 26, 2016.

For Comey's unprecedented admission that the FBI had paid hackers some $1.3 million to unlock the gunman's iPhone, see Eric Lichtblau and Katie Benner, "FBI Director Suggests Bill for iPhone Hacking Topped $1.3 Million," *New York Times*, April 21, 2016.

Unfortunately, reporters were quick to latch onto Cellebrite's claims that it had provided the FBI with the iPhone hack. That reporting was inaccurate. For an example of erroneous coverage at the time, see Jonathan Zalman, "The FBI Is Apparently Paying an Israel-Based Tech Company $15,278.02 to Crack San Bernardino Killer's iPhone," *Tablet*, March 24, 2016.

Four months after the FBI disclosed it had purchased the iPhone exploit, Apple announced at the Black Hat conference that August that it would start paying hackers bounties for bugs. For years, as Google, Facebook, Microsoft, and others offered bounties, Apple was a notable holdout. Executives argued that it was already getting tipped off to vulnerabilities by researchers and should not need to create a financial incentive. They also said Apple would never be able to match the government's prices. But after the FBI controversy drew attention to the lucrative market for iOS exploits, Apple caved, offering bounties as high as $200,000 to a closed list of invited hackers who could find holes in Apple's firmware, the software that lies closest to the bare metal of the machine. As the underground market prices for iOS exploits went up, and iOS exploits started popping up in NSO and other spyware through 2019, Apple bumped its top bounty to $1 million for "zero-click" remote exploits that would allow hackers to break into the kernel of its devices undetected. It also opened its bounty program to the public and in 2020, Apple started giving hackers special iPhones in a new program called the iOS Security Research Device Program. The phones contained unique features—such as debugging capabilities—that made it easier for hackers to find their weaknesses, such as probing portions of Apple's iOS operating system and memory, that are not accessible on its locked-down commercial iPhones.

There is a paucity of data on the number of countries with sophisticated offensive cyber exploitation programs. But Richard Ledgett, the former deputy director of NSA, told an audience in 2017 that "well over 100" countries around the world are now capable of launching cyberattacks. See Mike Levine, "Russia Tops List of 100 Countries That Could Launch Cyberattacks on US," ABC News, May 18, 2017. This corroborated what I, along with David Sanger, reported for the *Times* in July 2013, "Nations Buying as Hackers Sell Flaws in Computer Code."

CHAPTER 17: CYBER GAUCHOS

The reference to Argentina's import restrictions causing high-def TVs to cost twice as much and arrive six months late was taken from Ian Mount, "A Moveable Fiesta," *New York*, February 17, 2006.

For results of the International Collegiate Programming Contest (ICPC), the oldest and most prestigious programming contest of its kind, see://icpc.baylor.edu /worldfinals/results, which shows American teams are regularly losing to teams from Russia, Poland, China, South Korea, and other countries.

For the drop in NSA morale, see Ellen Nakashima and Aaron Gregg, "NSA's Top Talent Is Leaving Because of Low Pay, Slumping Morale and Unpopular Reorganization," *Washington Post*, January 2, 2018.

For an account of Cesar Cerrudo's traffic light hack, see my article for the *Times*: "Traffic Hacking: Caution Light Is On," June 10, 2015.

The reference to Brazil's issues with cybercrime was sourced from Janes Intelligence Review: "Brazil Struggles with Effective Cybercrime Response," 2017. A McAfee Report found that Brazil lost between $7 billion and $8 billion from cybercrime, including password theft, credit card fraud, and other cyberattacks. Fifty-four percent of Brazilian cyberattacks originate within the country: McAfee and Center for Strategic International Studies, "Economic Impact of Cybercrime—No Slowing Down," February 2018.

For more on Dan Geer's 2014 Black Hat Keynote, see Tim Greene, "Black Hat Keynote: U.S. Should Buy Up Zero-day Attacks for 10 Times Going Rate," Network World, August 7, 2014.

For a contemporary account of the 2001 Argentina protests, see Uki Goni, "Argentina Collapses into Chaos," *Guardian*, December 20, 2001.

Alfredo Ortega, the "Cyber Gaucho," was familiar with the challenges nuclear scientists faced. Most trekkers do not realize it, but for years Patagonia was home to a secret nuclear lab. In the late 1940s, an Austrian-born German scientist, Ronald Richter, convinced Argentina's president to build a nuclear "Thermotron." Three years and more than $400 million later, the project failed and Richter was jailed for fraud. But the project helped lay the foundation for Argentina's first nuclear reactor. Fifty years later, Argentina's nuclear reactors are world-renowned and used primarily for medical testing and treatment. Ortega told me a childhood friend of his in Patagonia was a nuclear physicist, who barely traveled these days, fearing kidnapping. "If he goes to Qatar or Saudi Arabia, the risk of kidnapping is too great," Ortega told me. "He'd be forced to develop nuclear weapons programs." The same, he feared, would be true for him if word ever got out that he sold exploits—particularly those that could cross nuclear air gaps, hack satellites, or subvert the global supply chain. More about Argentina's nuclear reactors are documented in Charles Newbery, "Argentina Nuclear Industry Sees Big Promise in Its Small Reactors," *Financial Times*, September 23, 2018.

For a relatively recent account of U.S.-Argentine relations, I highly recommend Graciela Mochkofsky, "Obama's Bittersweet Visit to Argentina," *New Yorker*, March 23, 2016. And for further reading into Christina Kirchner's troubles with hedge funds and conspiracy theory that the United States wants her dead, see Agustino Fontevecchia, "The Real Story of How a Hedge Fund Detained a Vessel in Ghana and Even Went for Argentina's Air Force One," *Forbes*, October 5, 2012, and Linette Lopez, "The President of Argentina Thinks the US Wants Her Dead," *Business Insider*, October 2, 2014.

In 2002, Bill Clinton declassified documents showing that amid vast human rights violations by Argentina's military in June 1976, then Secretary of State Henry Kissinger told Argentine Foreign Minister Cesar Augusto Guzzetti to get on with it. "If there are things that have to be done, you should do them quickly," Kissinger told Guzzetti at a meeting in Santiago that year. "We have followed events in Argentina closely. We wish the new government well. We wish it will succeed. We will do what we can to help it succeed . . . If you can finish before Congress gets back, the better." National Security Archive Electronic Briefing Book No. 133, posted Dec 4, 2003, nsarchive2.gwu.edu/NSAEBB/NSAEBB104/index.html.

CHAPTER 18: PERFECT STORM

This chapter relied heavily on interviews with American officials and executives. It also borrowed from the reporting I conducted for the *Times* between 2012 and 2019. For the most comprehensive account of the Saudi Aramco cyberattack, see Nicole Perlroth, "In Cyberattack on Saudi Firm, U.S. Sees Iran Firing Back," *New York Times*, October 23, 2012.

For Abdullah al-Saadan's remarks concerning Iran's goal, see "Aramco Says Cyberattack Was Aimed at Production," Reuters, December 9, 2012.

The reference to the surge in cyberattacks on American targets was taken from my article "Hacked Vs Hackers, Game on," *New York Times*, December 3, 2014. For the Obama administration's unsuccessful efforts to pass cybersecurity legislation in what was an era dominated by constant cyberattacks, see Michael S. Schmidt and Nicole Perlroth, "Obama Order Gives Firms Cyberthreat Information," *New York Times*, February 12, 2013; Nicole Perlroth, David E. Sanger, and Michael S. Schmidt, "As Hacking against U.S. Rises, Experts Try to Pin Down Motive," *New York Times*, March 3, 2013; Nicole Perlroth, "Silicon Valley Sounds Off on Failed Cybersecurity Legislation," *New York Times*, August 3, 2012.

For further reading on Iran's cyber force, see Ashley Wheeler, "Iranian Cyber Army, the Offensive Arm of Iran's Cyber Force," September 2013, available at www .phoenixts.com/blog/iranian-cyber-army. As for Tehran's boasts that Iran now has the fourth-biggest cyber army in the world, see "Iran Enjoys 4th Biggest Cyber Army in the World," FARS (Tehran), February 2, 2013.

For the cost of Chinese intellectual property theft, see Intellectual Property Commission Report, "The Theft of American Intellectual Property: Reassessments of The Challenge and United States Policy," published May 22, 2013, updated February 27, 2017. For my reporting on the *Times*'s own hack by China see "Chinese Hackers Infiltrate *New York Times* Computers," *New York Times*, Jan 30, 2013. For the impact those disclosures had on other victims of Chinese cyber-attacks see Perlroth, "Some Victims of Online Hacking Edge into the Light," *New York Times*, February 20, 2013. See also our story outing the PLA's Unit 61398: David E. Sanger, David Barboza, and Nicole Perlroth, "Chinese Army Unit Is Seen as Tied to Hacking against U.S.," *New York Times*, February 18, 2013, and the DOJ's subsequent indictments, see "U.S. Charges Five Chinese Military Hackers for Cyber Espionage Against U.S. Corporations and a Labor Organization for Commercial Advantage," Department of Justice, May 19, 2014, which we covered for the *Times* here: Sanger and Perlroth, "Hackers from China Resume Attacks on U.S. Targets," May 19, 2013. Not one year later, working with CrowdStrike, I identified a second PLA hacking unit responsible for cyberattacks on U.S. targets: "2nd China Army Unit Implicated in Online Spying," *New York Times*, June 9, 2014.

For Iran's attacks on U.S. banks, see Nicole Perlroth and Quentin Hardy, "Bank Hacking Was the Work of Iranians, Officials Say," *New York Times*, January 8, 2013. For Iran's shift from disruptive denial-of-service attacks to more destructive cyberattacks see Perlroth and Sanger, "Cyberattacks Seem Meant to Destroy, Not Just Disrupt," *New York Times*, March 28, 2013.

J. Michael Daniel was also generous in giving me the time to walk me through the government's initial response to the Iranian attack on the Bowman Avenue Dam in Westchester County, New York. Three years after the attack, the U.S. Justice Department unsealed an indictment against the seven Iranians it said were responsible for the attacks on the banks and the breach at Bowman Avenue Dam: Hamid Firoozi, Ahmad Fathi, Amin Shokohi, Sadegh Ahmadzadegan (better known by his alias "Nitrojen26"), Omid Ghaffarinia ("PLuS"), Sina Keissar, and Nader Saedi ("Turk Server"). Collectively they worked at the two IRGC front companies: ITSecTeam and Mersad. It was Firoozi who penetrated the Bowman Dam. According to the indictment, Firoozi repeatedly accessed information about the dam's operation—its water levels, pressure levels, and its locks and gates. The indictment never made clear whether Firoozi had meant to hack the much larger Bowman Dam in Oregon, or whether he was just studying up on the way American dams operated, a trial run for some future attack. Either way, American officials had to assume the motive was hostile. For further reading on the DOJ indictments, see David E. Sanger, "U.S. Indicts 7 Iranians in Cyberattacks on Banks and a Dam," *New York Times*, March 24, 2016, and Joseph Berger, "A Dam, Small and Unsung, Is Caught Up in an Iranian Hacking Case," *New York Times*, March 25, 2016. For Iran's hack of U.S. navy computers, see Julian E. Barnes and Siobhan Gorman, "U.S. Says Iran Hacked Navy Computers," *Wall Street Journal*, September 27, 2013.

Starting in 2009, researchers started publicly attributing Chinese cyberattacks to universities in China that received funding from China's 863 program. See Michael Forsythe and David E. Sanger, "China Calls Hacking of U.S. Workers' Data a Crime, Not a State Act," *New York Times*, December 2, 2015. China has denied this.

For a contemporary account of Panetta's speech on the USS *Intrepid*, see Elisabeth Bumiller and Thom Shanker, "Panetta Warns of Dire Threat of Cyberattack on U.S.," *New York Times*, October 11, 2012. Panetta cited the attacks on American banks and Aramco and said he hoped his speech would be a "clarion call" for American action to combat cyber threats. As he stepped off the lectern, some in the audience of business leaders accused him of hyperbole and overheated rhetoric aimed at Congress. But Panetta rejected that notion, telling a reporter after the speech, "The whole point of this is that we simply don't just sit back and wait for a goddamn crisis to happen. In this country we tend to do that."

For the most complete account of Iran's hack on Sands Casino, see Ben Elgin and Michael Riley, "Now at the Sands Casino: An Iranian Hacker in Every Server," Bloomberg, December 12, 2014.

To see Sheldon Adelson's remarks that touched off the Iranian attack on his casino, see Rachel Delia Benaim and Lazar Berman, "Sheldon Adelson Calls on U.S. to Nuke Iranian Desert," *Times of Israel*, October 24, 2013. In response, Iran's Supreme Leader Ayatollah Ali Khamenei said Adelson "should receive a slap on the mouth." As meek as that sounded, the next time Khamenei mentioned slapping anyone was in 2019, when Iran launched twenty-two missiles on the U.S.-Iraqi military base in retaliation for the U.S. drone strike that killed Iran's General Qassem Suleimani. Khamenei called the strike a "slap in the face" then, too.

North Korea's strangely similar attack on Sony Pictures in December 2014 was chronicled by my colleague David Sanger and me for the *Times*: "U.S. Said to Find North Korea Ordered Cyberattack on Sony," December 17, 2014. For media coverage of the Sony leaks, see Sam Biddle, "Leaked: The Nightmare Email Drama Behind Sony's Steve Jobs Disaster," *Gawker*, December 9, 2014, and Kevin Roose, "Hacked Documents Reveal a Hollywood Studio's Stunning Gender and Race Gap," *Fusion*, December 1, 2014.

For the Obama Administration's response, see David E. Sanger, Michael S. Schmidt, and Nicole Perlroth, "Obama Vows a Response to Cyberattack on Sony," *New York Times*, December 19, 2014. For coverage of North Korea's internet outage one week later, see Nicole Perlroth and David E. Sanger, "North Korea Loses Its Link to the Internet," *New York Times*, December 22, 2014.

Former American officials deny they played any part in North Korea's December 22, 2014, internet outage. They point to the Obama Administration's subsequent January 2015 sanctions as their official response. That month, the White House sanctioned ten senior North Korean officials and the intelligence agency it said was the source of "many of North Korea's major cyberoperations." Interestingly, of the ten, two were North Korean representatives in Iran, a major buyer of North Korean military technology, which offered a glimpse into potential cooperation and knowledge sharing between the two countries.

Sanger and I first broke the story of the Iranian attack on State Department personnel for the *Times*. See "Iranian Hackers Attack State Dept. via Social Media Accounts," November 24, 2015.

After Trump launched his trade war with China and abandoned the Iran nuclear deal, Chinese and Iranian cyberattacks on U.S. companies resumed. See Perlroth, "Chinese and Iranian Hackers Renew Their Attacks on U.S. Companies," *New York Times*, February 18, 2019.

The account of Xi Jinping's first meeting with Putin, and his remark "We are similar in character" were sourced to Jeremy Page, "Why Russia's President Is "Putin the Great" in China," *Wall Street Journal*, October 1, 2014. For arrests made in Xi's early tenure, see Ivan Watson and Steven Jiang, "Scores of Rights Lawyers Arrested after Nationwide Swoop in China," CNN, July 15, 2015. I highly recommend Evan Osnos's *New Yorker* article about Xi Jinping, "Born Red," April 6, 2015.

The Obama Administration's strategy to curtail Chinese cyber theft was taken from dozens of interviews with former officials and also documented in Susan Rice's memoir, *Tough Love: My Story of the Things Worth Fighting For* (Simon & Schuster, 2019). Ellen Nakashima scooped Rice, et al, on impending sanctions with her August 30, 2015, article, "U.S. Developing Sanctions Against China over Cyberthefts," *Washington Post*.

For coverage of the September 2015 Xi-Obama agreement to cease hacking for commercial gain, see Julie Hirschfield Davis and David E. Sanger, "Obama and Xi Jinping of China Agree to Steps on Cybertheft," *New York Times*, September 25, 2015. For descriptions of the festivities that September, I relied on Associated Press, "Obama Hosts Lavish State Dinner for China's President Xi Jinping," September 25, 2015.

After the agreement was reached that September, iSight Intelligence, a division of FireEye, reported an immediate 90 percent drop in the frequency of Chinese cyberespionage. See David E. Sanger, "Chinese Curb Cyberattacks on U.S. Interests, Report Finds," *New York Times*, June 20, 2016, and Ken Dilanian, "Russia May Be Hacking Us More, But China Is Hacking Us Much Less," NBC News, October 12, 2016.

CHAPTER 19: THE GRID

This chapter relied heavily on my interviews with current and former Homeland Security officials who sounded the warnings throughout 2012 and 2013 about attacks on the U.S. grid. For the surge in attacks in 2012, see also David Goldman, "Hacker Hits on U.S. Power and Nuclear Targets Spiked in 2012," CNN, January 9, 2013; Nicole Perlroth, "Tough Times at Homeland Security," *New York Times*, May 13, 2013; and Perlroth, "Luring Young Web Warriors Is Priority. It's Also a Game," *New York Times*, March 25, 2013.

For further reading on cyber threats to the grid, I recommend Ted Koppel's *Lights Out* (Broadway Books, 2015). Koppel's book included the text of the 2010 confidential letter that R. James Woolsey, John Deutsch, James Schlesinger, William Perry, Stephen Hadley, Robert McFarlane, and others sent to Congress.

In his 2013 State of the Union Address, Obama confirmed what DHS employees had been telling me for a year: "Now our enemies are also seeking the ability to sabotage our power grid, our financial institutions, and our air traffic control systems," Obama said. "We cannot look back years from now and wonder why we did nothing in the face of real threats to our security and our economy." Earlier that day, Obama had signed an executive order that encouraged better threat-sharing between the government and the private companies that oversee the nation's infrastructure—offering a weakened alternative to legislation. The order was all carrot and no stick, because requiring utilities and infrastructure providers to beef up their cybersecurity required Congressional approval.

One year later, officials and private security researchers had grown confident in their attribution: Russian hackers were to blame. See Perlroth, "Russian Hackers Targeting Oil and Gas Companies," New York Times, June 30, 2014.

In the article that June, I noted researchers believed the attacks were industrial trade theft: "The motive behind the attacks appears to be industrial espionage—a natural conclusion given the importance of Russia's oil and gas industry," the researchers said. "But the manner in which the Russian hackers are targeting the companies also gives them the opportunity to seize control of industrial control systems from afar, in much the same way the U.S. and Israel were able to use the Stuxnet computer worm in 2009 to take control of an Iranian nuclear facility."

Later researchers changed their tune. The real goal wasn't intellectual property theft; these attacks were the planning stages for cyberwar.

For an account of Russia's calls for an international cyberweapons ban, see Andrew E. Kramer and Nicole Perlroth, "Expert Issues a Cyberwar Warning," New York Times, June 3, 2012. For Russian anxieties about cyber escalation, see Timothy Thomas, "Three Faces of the Cyber Dragon: Cyber Peace Activist, Spook, Attacker," Foreign Military Studies Office, 2012. For details of Russia's Internet of Things—or IoT—market, see MarketWatch, "Russia Internet of Things (IoT) Market Is Expected to Reach $74 Billion By 2023," October 17, 2019.

For statistics on Russia's GDP, purchasing power parity, and population growth, I relied on the CIA World Factbook, the World Bank GDP Ranking, and the Wilson Center studies on Russia's demographic trends.

John Hultquist of FireEye's research on the Russian GRU unit, known to private researchers as Sandworm, proved invaluable to my reporting for the Times. But by far, the most comprehensive account of Hultquist's Sandworm discovery is written by Andy Greenberg: Sandworm: A New Era of Cyberwar and the Hunt for the Kremlin's Most Dangerous Hackers, and articles by Kim Zetter, "Russian Sandworm Hack Has Been Spying on Foreign Governments for Years," Wired, October 14, 2014.

Hultquist's research compelled further dissection of Sandworm's tools by two researchers at Trend Micro, another security firm. See Kyle Wilhoit and Jim

Gogolinski, "Sandworm to Blacken: The SCADA Connection," Trend Micro, October 16, 2014.

Two weeks later, the Industrial Control System Cyber Emergency Team, ICS-CERT, published a security advisory detailing Sandworm's hacks not only on GE's software, but on Siemens and Advantech, which both sell software that enables connections to industrial infrastructure. The advisory is no longer available but is detailed in various press reports. See Michael Mimoso, "BlackEnergy Malware Used in Attacks Against Industrial Control Systems," Threatpost, October 29, 2014.

The U.S. State Department and the UK's National Cyber Security Centre officially outed Russia's Sandworm/BlackEnergy Group as a division of Russia's GRU Unit 74455 on February 20, 2020. In a rare attribution statement, American and British officials tied the group to a series of 2019 cyberattacks in Georgia, which disrupted several thousand Georgian government and privately run websites and interrupted the broadcasts of at least two major TV stations. See Secretary of State Michael R. Pompeo, "The United States Condemns Russian Cyber Attack Against the Country of Georgia," press statement, February 20, 2020.

The U.S. Justice Department had previously named GRU Unit 74455 in an October 4, 2018, indictment against seven Russian military officers. The indictment accused members of the unit of setting up the social media accounts and other hacking infrastructure used by GRU Unit 26165—known as "Fancy Bear"—to hack anti-doping officials and organizations investigating Russia's use of chemical weapons. See Department of Justice, "U.S. Charges Russian GRU Officers with International Hacking and Related Influence and Disinformation Operations," press release, October 4, 2020. GRU Unit 74455 was also mentioned repeatedly in the Mueller Report as one of the two units that hacked into the Democratic National Committee and assisted in the release of documents stolen by Unit 26165. The Mueller Report also claimed that Unit 74455 was responsible for the hacks of computers belonging to U.S. state boards of elections, secretaries of state, and U.S. companies that supplied software and other election technology. The unit maintains several departments, one of which is the group Hultquist's team coined Sandworm; another is the group CrowdStrike refers to as Cozy Bear. See Special Counsel Robert S. Mueller III, "Report on the Investigation into Russian Interference in the 2016 Election," Volume I and II, March 2019.

In Ukraine, Oleksii Yasinsky and Oleh Derevianko spent days walking me through the timeline of the Sandworm attacks on Ukraine's media organizations, power grid, and NotPetya. For further reading on Russia's attacks on Ukraine, I highly recommend Kim Zetter's *Wired* story "Inside the Cunning, Unprecedented

Hack of Ukraine's Power Grid," March 3, 2016, and Greenberg's book *Sandworm*, the most complete account of Russia's attacks on Ukraine to date.

CHAPTER 20: THE RUSSIANS ARE COMING

The most complete press account of the DNC hacks was reported by my colleagues Eric Lipton, David E. Sanger, and Scott Shane, "The Perfect Weapon: How Russian Cyberpower Invaded the U.S.," *New York Times*, December 13, 2016.

The discovery of Heartbleed was followed by a Bloomberg report which claimed the NSA had known about the Heartbleed bug for years—a claim the NSA and White House vehemently denied. See Michael Riley, "NSA Said to Exploit Heartbleed Bug for Intelligence for Years," April 11, 2014, and David E. Sanger and Nicole Perlroth, "U.S. Denies It Knew of Heartbleed Bug on the Web," *New York Times*, April 11, 2014.

The back-and-forth compelled the White House to disclose, for the first time, the Vulnerabilities Equities Process by which the government decides which zero-days to keep and which to turn over for patching. See J. Michael Daniel, "Heartbleed: Understanding When We Disclose Cyber Vulnerabilities." White House Blog, April 28, 2014, and our coverage in the *Times*: David Sanger, "Obama Lets NSA Exploit Some Internet Flaws, Officials Say," April 13, 2014.

I am indebted to J. Michael Daniel for the hours he spent discussing (as much as he could publicly, anyway) the calculus the government must make when deciding to keep or disclose software vulnerabilities. I am also eternally grateful to Howard Schmidt, who preceded Daniel as White House Cybersecurity Coordinator and who sadly passed away in March 2017. Over the course of several interviews, Schmidt discussed the dilemma our own government now faced as the market for zero-days moves abroad. Some of those interviews made their way into the *Times*, others helped inspire and inform this book. See Nicole Perlroth and David E. Sanger, "Nations Buying as Hackers Sell Flaws in Computer Code," *New York Times*, July 13, 2013.

For the Netherlands' role in outing Russia's Cozy Bear hackers, see Huib Modderkolk, "Dutch Agencies Provide Crucial Intel about Russia's Interference in U.S.-Elections," *de Volkskrant*, January 25, 2018.

For our coverage of EternalBlue, the NSA exploit that was leaked and formed a critical role in North Korea's WannaCry attacks and shortly thereafter Russia's NotPetya attacks, see Perlroth and Sanger, "Hackers Hit Dozens of Countries Exploiting Stolen NSA Tool," May 12, 2017; Perlroth, "A Cyberattack 'the World Isn't Ready For,'" *New York Times*, June 22, 2017; and Perlroth and Scott Shane, "In

Baltimore and Beyond, a Stolen NSA Tool Wreaks Havoc," *New York Times*, May 25, 2019.

This chapter relied heavily on the Senate Committee on Intelligence's "Report on Russian Active Measures Campaigns and Interference in the 2016 U.S. Election." I recommend every American read the Senate report and the Mueller Report in their entirety. See www.intelligence.senate.gov/sites/default/files/documents /Report_Volume2.pdf and Special Counsel Robert S. Mueller, III, Volumes I and II, "Report on the Investigation into Russian Interference in the 2016 Presidential Election," March 2019. Available at www.justice.gov/storage/report.pdf.

The first public inkling of Russia's efforts to infiltrate the 2016 election came in June 2016: Ellen Nakashima, "Russian Government Hackers Penetrated DNC, Stole Opposition Research on Trump," *Washington Post*, June 14, 2016.

The most definitive account was later reported by colleagues Eric Lipton, David E. Sanger, and Scott Shane, "The Perfect Weapon: How Russian Cyberpower Invaded the U.S.," December 13, 2016.

Lost in the maelstrom was subsequent reporting that year which showed that the *New York Times* Moscow bureau was also the attempted target of a Russian cyberattack. There was no evidence that Russia's hackers were successful. See Nicole Perlroth and David E. Sanger, "*New York Times*'s Moscow Bureau Was Targeted by Hackers," *New York Times*, August 23, 2016.

For the most definitive press accounts of Russia's social media influence efforts, see Scott Shane and Mark Mazzetti, "Inside a 3-Year Russian Campaign to Influence U.S. Voters," *New York Times*, February 16, 2018.

I also relied on the United States indictment of Russia's Internet Research Agency. See Indictment, *United States v. Internet Research Agency, et al.* Case 1:18-cr-00032-DLF, D.D.C., February 16, 2018. For further reading on Yevgeny Prigozhin, who ran point on Russia's Internet Research Agency's efforts to influence the 2016 election, see Neil MacFarquhar, "Yevgeny Prigozhin, Russian Oligarch Indicted by U.S., Is Known as 'Putin's Cook,'" *New York Times*, February 16, 2018. After Prigozhin was indicted for his role in the 2016 election interference, he told the Russian state news agency Ria Novosti, "The Americans are very impressionable people; they see what they want to see. I have a lot of respect for them. I am not upset at all that I ended up on this list. If they want to see the devil, let them see him."

For a more granular view of how Russia's social media influence campaign played out locally, see Stephen Young, "Russian Trolls Successfully Peddled Texas Pride in 2016, Senate Reports Say," *Dallas Observer*, December 19, 2018.

For the hack of VR Systems, see the NSA's leaked report published by Matthew Cole, Richard Esposito, Sam Biddle, and Ryan Grim, "Top Secret NSA Report Details Russian Hacking Effort Days before 2016," *Intercept*, June 5, 2017.

Note that the *Intercept*, and Americans, only know about Russia's hack of VR Systems because of a leaker named Reality Winner, who shared a classified NSA report with reporters at the *Intercept*. Even DHS was apparently in the dark about the extent to which Russian hackers infiltrated the back-end vendors that supply the critical software and equipment for U.S. elections. The *Intercept* made a fateful reporting error. They sent the NSA a scanned copy of the leaked report. The report appeared to have been folded and creased, which tipped officials off that the document had been printed and hand-carried out by an employee. From there, U.S. government investigators only had to look to see who printed the document—six people had—and discovered one of them, Reality Winner, had been in contact with *Intercept* from her work computer in the NSA's Augusta, Georgia facility, albeit on an unrelated topic. When FBI agents showed up at Winner's home, she confessed to removing the classified report and mailing it to the *Intercept*. The report contained dots, denoting a serial number invisible to the naked eye, that allowed the NSA to match the document back to a machine in its offices. This was a tragic lapse in reporters' "operational security." In August 23, 2018, Ms. Winner was sentenced to sixty-three months in prison. See Amy B. Wang, "Convicted Leaker Reality Winner Thanks Trump after He Calls Her Sentence So Unfair," *Washington Post*, August 30, 2018.

For our early reporting on Guccifer 2.0, see Charlie Savage and Nicole Perlroth, "Is DNC Email Hacker a Person or a Russian Front? Experts Aren't Sure," *New York Times*, July 27, 2016. *Vice* Motherboard's Lorenzo Franceschi-Biccierai played a critical role in outing Guccifer 2.0 when he interviewed him "Why Does DNC Hacker 'Guccifer 2.0' Talk Like This?" Motherboard, June 23, 2016.

For contemporary accounts of the DNC leaks and their impact, see Sam Biddle and Gabrielle Bluestone, "This Looks Like the DNC's Hacked Trump Oppo File," *Gawker*, June 15, 2016; Kristen East, "Top DNC Staffer Apologizes for Email on Sanders' Religion," *Politico*, July 23, 2016; Mark Paustenbach, "Bernie Narrative," via WikiLeaks, May 21, 2016, Wikileaks.org/dnc-emails; Meghan Keneally, "Debbie Wasserman Schultz Booed at Chaotic Florida Delegation Breakfast," ABC News, July 25, 2016; Rosaline S. Helderman and Tom Hamburger, "Hacked Emails Appear to Reveal Excerpts of Speech Transcripts Clinton Refused to Release," *Washington Post*, October 7, 2016; and David E. Sanger and Nicole Perlroth, "As Democrats Gather, a Russian Subplot Raises Intrigue," *New York Times*, July 24, 2016.

My colleague Scott Shane provided the most in-depth look at the fake accounts Russia generated to influence the 2016 election: "The Fake Americans Russia Created to Influence the Election," *New York Times*, September 7, 2017.

For further reading on the Obama Administration's early deliberations on how to respond to Russia's 2016 interference, I recommend David Sanger's *The Perfect Weapon: War, Sabotage, and Fear in the Cyber Age* (Crown, 2018).

Initially the Department of Homeland Security reported that Russia targeted voter registration systems in twenty-one states. That number was later adjusted to include all fifty states. See David E. Sanger and Catie Edmondson, "Russia Targeted Election Systems in All 50 States, Report Finds," *New York Times*, July 25, 2019.

For an in-depth look at how the current administration evaluates threats to 2020, see Matthew Rosenberg, Nicole Perlroth, and David E. Sanger, "'Chaos Is the Point': Russian Hackers and Trolls Grow Stealthier in 2020," *New York Times*, January 10, 2020, and Sanger, Perlroth, and Rosenberg, "Amid Pandemic and Upheaval, New Cyberthreats to the Presidential Election," *New York Times*, June 7, 2020. McConnell's beleaguered response to election security is detailed in Steve Benen, "McConnell's Response to Russian Attack Is Back in the Spotlight," MSNBC, February 19, 2018. For Trump's public remarks concerning his doubts Russia meddled in 2016, see Michael D. Shear, "After Election, Trump's Professed Love for Leaks Quickly Faded," *New York Times*, February 15, 2017; Cristiano Lima, "Trump on RT: Russian Election Interference 'Probably Unlikely,'" *Politico*, September 8, 2016; and First Presidential Debate, CNN, September 26, 2016.

For further reading on John Brennan's warning to Russia to not interfere, see Matt Apuzzo, "Ex-CIA Chief Reveals Mounting Concern over Trump Campaign and Russia," *New York Times*, May 23, 2017.

The reference to disinformation experts reporting Russian *kompromat* did little to influence 2016, see Christopher A. Bail, Brian Guay, Emily Maloney, Aidan Combs, D. Sunshine Hillyguus, Friedolin Merhout, Deen Freelon, and Alexander Volfovsky, "Assessing the Russia Internet Research Agency's Impact on the Political Attitudes and Behaviors of American Twitter Users in Late 2017," *Proceedings of the National Academy of Sciences of the United States* 117, no. 1 (November 25, 2019).

The following articles suggest otherwise: Matthew Yglesias, "What Really Happened in 2016, in 7 Charts," *Vox*, September 18, 2017; Jens Manuel Krogstad and Mark Hugo Lopez, Pew Research Center, "Black Voter Turnout Fell in 2016, Even as a Record Number of Americans Cast Ballots," May 12, 2016; and Brooke Seipel, "Trump's Victory Margin Smaller Than Total Stein Votes in Key Swing States," *The Hill*, December 1, 2016.

In response to Russia's 2016 interference, the Obama administration ultimately sanctioned Russians, kicked thirty-five Russian diplomats—including many spies—out of the U.S. and closed two Russian diplomatic properties. See Mark Mazzetti and Michael S. Schmidt, "Two Russian Compounds, Caught Up in History's Echoes," *New York Times*, December 29, 2016. I relied on a CBS news account of the smoke pouring out of Russia's San Francisco embassy: "Black Smoke Pours from Chimney at Russian Consulate in San Francisco," September 2, 2017.

CHAPTER 21: THE SHADOW BROKERS

The first blip of Shadow Brokers came in the form of barely legible tweets and a post on Pastebin, a site that allows people to post anonymously. See Shadow Brokers, "Equation Group Cyber Weapons Auction—Invitation," August 13, 2016.

For an early account of the questions swirling in law enforcement, intelligence agencies, and cybersecurity circles, see David Sanger, "Shadow Brokers' Leak Raises Alarming Question: Was the NSA Hacked?" *New York Times*, August 16, 2016.

Cisco was forced to immediately warn customers that its firewalls were vulnerable: "Cisco Adaptive Security Appliance SNMP Remote Code Execution Vulnerability," Cisco Security Advisory Alerts, August 17, 2016, and Thomas Brewster, "Cisco and Fortinet Confirm Flaws Exposed by Self-Proclaimed NSA Hackers," *Forbes*, August 17, 2016.

For text of the Shadow Brokers' early manifesto, see Bruce Sterling, "Shadow Brokers Manifesto," *Wired*, Aug 19, 2016.

For the NSA's hacking of Angela Merkel's cell phone, see "NSA Tapped German Chancellery for Decades, WikiLeaks Claims," July 8, 2015. For Snowden's tweet regarding the leaks, see Edward J. Snowden, Twitter, August 16, 2016, twitter.com /Snowden/status/765515619584311296.

The Shadow Brokers' early auction proposal went nowhere. See Andy Greenberg, "No One Wants to Buy Those Stolen NSA-Linked Cyberweapons," *Wired*, August 16, 2016.

For the NSA reaction to the leaks see Scott Shane and Nicole Perlroth, "Security Breach and Spilled Secrets Have Shaken the NSA to Its Core," *New York Times*, November 12, 2017.

For text of Biden's remarks on *Meet the Press*, see Joe Lapointe, "Despite All Other News, Sunday Shows Keep Covering Trump's Sex Scandal," *Observer*, October 17, 2016.

After the Shadow Brokers published the web addresses of the NSA's decoy servers around the globe, hackers immediately started dissecting the list and publishing their findings on a site called MyHackerHouse.com. The site was subsequently taken down. www.myhackerhouse.com/hacker-halloween-inside-shadow-brokers-leak.

For further reading on the case of the CIA leaker, see Adam Goldman, "New Charges in Huge C.I.A. Breach Known as Vault 7," *New York Times*, Jun 18. 2018, and Nicole Hong, "Trial of Programmer Accused in C.I.A. Leak Ends with Hung Jury," *New York Times*, March 9, 2020.

For further reading on Kaspersky's role in accessing an NSA employees' documents from his home computer see Scott Shane, David Sanger, Nicole Perlroth, "New NSA Breach Linked to Popular Russian Antivirus Software," October 5, 2017, and Perlroth and Shane, "How Israel Caught Russian Hackers Scouring the World for U.S. Secrets," *New York Times*, October 10, 2017. Kaspersky denies that it exfiltrated the NSA contractor's classified documents on purpose and said it collected them inadvertently and deleted them. For Kaspersky's full technical response see "Preliminary Results of the Internal Investigation into Alleged Incidents by US Media," www.kaspersky.com/blog/internal-investigation-preliminary-results/19894 and its FAQ: "What Just Hit the Fan: FAQs," www.kaspersky.com/blog/kaspersky-in-the-shitstorm/19794. Kaspersky's explanation was enough for many security researchers and clients, but others—notably the U.S. government—have banned Kaspersky's antivirus products from their networks. It didn't help that the company has long been dogged by allegations it is a Russian front. See Andrew Kramer and Nicole Perlroth, "Expert Issues a Cyberwar Warning," *New York Times*, June 4, 2012.

For Jake Williams's blog post on the Shadow Brokers' leaks see "Corporate Business Impact of Newest Shadow Brokers Dump," Rendition Infosec, April 9, 2017. For the Shadow Brokers' response, see The Shadow Brokers, "Response to Response to DOXing," Steemit Post, 2017.

Microsoft's 2017 patch made no mention of who had turned over the critical vulnerability that was used by the NSA's Eternal exploits. See "Microsoft Security Bulletin MS17-010- Critical Microsoft Security Update for Windows SMB Server," March 14, 2017, docs.microsoft.com/en-us/security-updates/securitybulletins/2017/ms17-010.

CHAPTER 22: THE ATTACKS

For contemporary reporting on the WannaCry attack, see Nicole Perlroth and David Sanger, "Hackers Hit Dozens of Countries Exploiting Stolen NSA Tool," *New York*

Times, May 12, 2017, and Perlroth, "More Evidence Points to North Korea in Ransomware Attacks," *New York Times*, May 22, 2017. Homeland Security Adviser Thomas Bossert first addressed the WannaCry attacks on ABC's *Good Morning America*: "Unprecedented Global Cyberattack Is 'an Urgent Call' to Action, Homeland Security Adviser Says," May 15, 2017. For a more comprehensive look at North Korea's cyber capabilities see David Sanger, David Kirkpatrick, Nicole Perlroth, "The World Once Laughed at North Korean Cyberpower. No More," *New York Times*, October 15, 2017.

Marcus Hutchins, the British hacker, who intervened in WannaCry and stopped the attack in its tracks was later arrested for writing malware. See Palko Karasz, "He Stopped a Global Cyberattack. Now He's Pleading Guilty to Writing Malware," *New York Times*, April 20, 2019. For further reading on Hutchins, see Andy Greenberg, "The Confessions of Marcus Hutchins, the Hacker Who Saved the Internet," *Wired*, May 12, 2020.

For a look at how the attack hit China, see Paul Mozur, "China, Addicted to Bootleg Software, Reels from Ransomware Attack," *New York Times*, May 15, 2017. A study by BSA, a trade association for software vendors, found that 70 percent of software installed in China was not properly licensed in 2015. Russia, at 64 percent, and India, 58 percent, were close behind. By using unlicensed, unpatched software, these computers would not have been able to download Microsoft's patch. It was a reminder that, as vulnerable as the United States may be to cyberattacks, piracy has made countries like China, Russia and India equally vulnerable. The BSA Study is available at www.bsa.org/~/media/Files/StudiesDownload/BSA_GSS_US.pdf.

Bossert publicly attributed the WannaCry attack to Russia in December 2017: Thomas P. Bossert, "It's Official: North Korea Is behind WannaCry," *New York Times*, December 18, 2017.

For Microsoft's earlier efforts to push back on government surveillance see Rory Carroll, "Microsoft and Google to Sue over US Surveillance Requests," *Guardian*, August 31, 2013, and Spencer Ackerman and Dominic Rushe, "Microsoft, Facebook, Google and Yahoo release US Surveillance requests," *Guardian*, February 3, 2014. For Brad Smith's response to WannaCry, see "The Need for Urgent Collective Action to Keep People Safe Online: Lessons from Last Week's Cyberattack," May 14, 2017, blogs .microsoft.com/on-the-issues/2017/05/14/need-urgent-collective-action-keep-people -safe-online-lessons-last-weeks-cyberattack.

The early accounts of Ukraine's response to NotPetya were taken from interviews with Dmitro Shymkiv. For his role in Ukraine's 2014 protests, se: Serhiy Kvit, "What the Ukrainian Protests Mean," University World News, Jan 8, 2014.

I wrote about the NotPetya attacks for the *Times*: Nicole Perlroth, Mark Scott, and Sheera Frenkel, "Cyberattack Hits Ukraine Then Spreads Internationally," *New York Times*, June 27, 2017. Andy Greenberg, citing Thomas Bossert, provided the first public estimate of damages: "The Untold Story of NotPetya, the Most Devastating Cyberattack in History," *Wired*, August 22, 2018. Shymkiv and others say $10 billion is a gross underestimate, noting the private companies that never reported damages. My colleague Adam Satariano and I covered the battles between victims and their insurers, who refused to pay for damages at Merck and Mondelez citing a common but rarely invoked war exemption clause in their contracts. See Adam Satariano and Nicole Perlroth, "Big Companies Thought Insurance Covered a Cyberattack, They May Be Wrong." *New York Times*, April 15, 2019.

For Brad Smith's remarks to the United Nations, see "Remarks on Cybersecurity and a Digital Geneva Convention," United Nations, November 9, 2017: www .youtube.com/watch?v=EMG4ZukkClw. For Richard Clarke's earlier proposal that countries agree not to conduct cyberattacks on civilian infrastructure, see Richard A. Clarke and Robert K. Knake, *Cyber War: The Next Threat to National Security and What to Do about It* (HarperCollins, 2010).

For a chronicle of Russian cyberattacks on America's grid, and vice versa, see David E. Sanger and Nicole Perlroth, "U.S. Escalates Online Attacks on Russia's Power Grid," *New York Times*, June 15, 2019.

CHAPTER 23: THE BACKYARD

Scott Shane and I reported on the long tail of EternalBlue, which included attacks in San Antonio, Texas; Allentown, Pennsylvania; and finally, in the NSA's own backyard, in Baltimore. See "In Baltimore and Beyond, A Stolen NSA Tool Wreaks Havoc," *New York Times*, May 25, 2019. See also Yami Virgin, "Federal Agents Investigate Attempted Hacking at Bexar County Jail," Fox San Antonio, January 31, 2019.

The story provoked a huge backlash from exploit developers like Dave Aitel and the agency itself, which denied EternalBlue had played a role in the Baltimore attack. As it turned out, Baltimore had been hit by multiple attacks, one of which involved EternalBlue and another that involved a ransomware called Robinhood. Investigators at Microsoft, which has the best telemetry into EternalBlue's presence on its systems and contracted with the city of Baltimore, confirmed the presence of EternalBlue and initially concluded the NSA's exploit had helped spread ransomware across Baltimore's systems. The final consensus was that Baltimore's attackers had spread their ransomware manually, while a separate group of attackers had used the NSA's tool for another purpose—which is still unknown. However, it's worth noting that by the time Baltimore was hit, the NSA's tool was popping up everywhere. Amit Serper, a researcher at Cybereason, said his firm had responded to EternalBlue attacks at three different American universities and found vulnerable servers in major cities like Dallas, L.A., and New York City. Jake Williams, the former TAO analyst, had responded to EternalBlue attacks on multiple American municipalities. After the NSA's top hacker, Rob Joyce, publicly claimed EternalBlue was not used in the Baltimore attack, Microsoft told its clients in Baltimore that it planned to put out a statement confirming our reporting. Ultimately, Baltimore refused to give the company permission to disclose its findings for reasons unknown but ostensibly out of a desire to make the publicity go away, or simply not wanting to publicize the fact that it had not patched for the Microsoft bug. Many security researchers said the fault was not the NSA's for losing its tools but Baltimore's for not patching its systems. But it is worth noting that cities and towns often operate on shoestring cybersecurity budgets and IT administrators oversee tangled networks of older, out-of-date systems. For the back and forth see Scott Shane and Nicole Perlroth, "NSA Denies Its Cyberweapon Was Used in Baltimore Attack, Congressman Says," *New York Times*, May 31, 2019, and the *Wall Street Journal* confirming multiple cyberattacks on Baltimore, see Scott Calvert and Jon Kamp, "Hackers Won't Let Up in Their Attack on U.S. Cities," *Wall Street Journal*, June 7, 2019.

For China's detection and reuse of the NSA's exploits see Nicole Perlroth, David E. Sanger and Scott Shane, "How Chinese Spies Got the NSA's Hacking Tools, and Used Them for Attacks," *New York Times*, May 6, 2019. The Chinese group identified by Symantec in their report is alternatively known as Buckeye, Gothic Panda, or APT3 by different security firms. According to leaked NSA reports I had access to, the group, based in Guangzhou and known previously by the classified NSA codename "Legion Amber," is one of several contractors working for China's Ministry of State Security and was previously caught hacking American weapons developers and scientific research labs.

For further reading on U.S.-China relations see Evan Osnos, "The Future of America's Contest with China," *New Yorker*, January 6, 2020. For more on China's alleged nuclear weapons development, see Associated Press, "China Denies U.S. Allegations It's Testing Nuclear Weapons," April 16, 2020. For a comprehensive account of China's surveillance of its ethnic Uighur minorities, see Paul Mozur, "One Month, 500,000 Face Scans: How China Is Using A.I. To Profile a Minority," *New York Times*, April 14, 2019, and Austin Ramzy and Chris Buckley, "Absolutely No Mercy': Leaked Files Expose How China Organized Mass Detentions of Muslims," *New York Times*, November 16, 2019.

Google's discovery of China's iOS exploits is detailed here: Ian Beer, "A Very Deep Dive into iOS Exploit Chains Found in the Wild," Google Project Zero,

August 29, 2019. For further reading on this campaign, see Nicole Perlroth, Kate Conger, and Paul Mozur, "China Sharpens Hacking to Hound Its Minorities, Far and Wide," *New York Times*, October 22, 2019. For the Pentagon's Nitro Zeus operation to take out Iran's grid in the event of a broader conflict, see David E. Sanger and Mark Mazzetti, "U.S. Had Cyberattack Plan if Iran Nuclear Dispute Led to Conflict," *New York Times*, February 16, 2016. For sourcing on Cyber Command's cyberattacks on Iran in retaliation for the attacks on oil tankers, see Julian E. Barnes and Thomas Gibbons-Neff, "U.S. Carried Out Cyberattacks on Iran," *New York Times*, June 22, 2019. And for an account of Iran's cyberattacks on hundreds of Western companies in recent years, see Robert McMillan, "Iranian Hackers Have Hit Hundreds of Companies in Past Two Years," *Wall Street Journal*, March 6, 2019.

In October 2019, Microsoft reported that Iran was probing at least one presidential campaign, which we confirmed was Trump's. See Nicole Perlroth and David E. Sanger, "Iranian Hackers Target Trump Campaign as Threats to 2020 Mount," *New York Times*, October 4, 2019. Despite Trump's claims that "all is well" after Iran's retaliatory attacks on a U.S.-Iraqi army base, later, we would learn that the attacks inflicted serious brain injuries on more than one hundred American soldiers. See Bill Chappell, "109 U.S. Troops Suffered Brain Injuries in Iran Strike, Pentagon Says," NPR, February 11, 2020.

The reference to the *Washington Post* not letting Khashoggi's death go unanswered was sourced to the *Washington Post* Editorial Board, "One Year Later, Our Murdered Friend Jamal Has Been Proved Right," *Washington Post*, September 30, 2019. For a contemporary account of the Saudis' hack of Bezos, see Karen Weise, Matthew Rosenberg, and Sheera Frenkel, "Analysis Ties Hacking of Bezos' Phone to Saudi Lader's Account," *New York Times*, January 21, 2020. Trump's fixation with the Wild West was evident in his 2020 State of the Union speech: Wyatt Earp, Annie Oakley, and Davy Crockett were all mentioned in Trump's 2020 State of the Union. Jessica Machado, "Trump Just Gave Americans a Lesson in White History," *Vox*, Feb 5, 2020.

The Trump White House eliminated the cybersecurity coordinator position in May 2018. See Nicole Perlroth and David E. Sanger, "White House Eliminates Cybersecurity Coordinator Role," *New York Times*, May 15, 2018. On the easing of Russian sanctions, see Donna Borak, "Treasury Plans to Lift Sanctions on a Russian Aluminum Giant Rusal," CNN Business, December 19, 2018. On the warnings to Homeland Security Secretary Kirstjen Nielsen to stop bringing up election security to the president, see Eric Schmitt, David E. Sanger, Maggie Haberman, "In Push for 2020 Election Security, Top Official Was Warned: Don't

Tell Trump," *New York Times*, April 24, 2019. On Trump's continued willingness to accept foreign *kompromat* on a political opponent, see Lucien Bruggeman, "'I Think I'd Take It': An Exclusive Interview, Trump Says He Would Listen if Foreigners Offered Dirt on Opponents," ABC News, June 13, 2019. For Trump joking with Putin about offing journalists, see: Julian Borger, "Trump Jokes to Putin They Should 'Get Rid' of Journalists," *Guardian*, June 28, 2019. And on the very real possibility that Russia's 2016 hack of our election systems was a mere trial run for 2020, and Russia's continued efforts to evade social media controls by renting legitimate accounts and using VPNs, see Matthew Rosenberg, Nicole Perlroth, David E. Sanger, "'Chaos Is the Point': Russian Hackers and Trolls Grow Stealthier in 2020," *New York Times*, January 10, 2020. For more on Russia's continued meddling in American debates on guns, immigration and race, see Kevin Roose, "Facebook Grapples with a Maturing Adversary in Election Meddling," *New York Times*, August 1, 2018. For a more granular look at Russia's social media impersonation tactics, see my June 2020 article in the *Times* about Russia's role in spewing a divisive conspiracy theory during the Iowa caucus: Perlroth, "A Conspiracy Made in America May Have Been Spread by Russia," June 15, 2020.

For Cyber Command's cyberattack on Russian servers during the 2018 midterm elections, see David E. Sanger, "Trump Loosens Secretive Restraints on Ordering Cyberattacks," *New York Times*, September 20, 2018, and Julian E. Barnes, "U.S. Begins First Cyberoperation Against Russia Aimed at Protecting Elections," *New York Times*, October 23, 2018. For Russia's subsequent hacking attempts on the DNC servers in 2018, see Perlroth, "DNC Says It Was Targeted Again by Russian Hackers after '18 Election," *New York Times*, January 18, 2019.

For contemporary accounts of the ransomware attacks in Riviera Beach and Palm Springs, Florida, see Alexander Ivanyuk, "Ransomware Attack Costs $1.5 Million in Riviera Beach, Fl.," Acronis Security Blog, June 24, 2019, and Sam Smink, "Village of Palm Springs confirms cyberattack," West Palm Beach TV, June 20, 2019.

The details linking ransomware attacks to Russian cybercriminal outfits have been documented by various security firms over the years. For an early account, see Kaspersky, "More than 75 Percent of Crypto Ransomware in 2016 Came from Russian-Speaking Cybercriminal Underground," February 14, 2017, usa.kaspersky .com/about/press-releases/2017_more-than-75-of-crypto-ransomware-in-2016-came -from-the-russian-speaking-cybercriminal-underground. Of note, Russian ransomware authors behind the "Sigrun" Ransomware family offered to decrypt data belonging to Russian Victims for free. Alex Svirid, a security researcher, first tweeted this observation in May 31, 2018. A Malwarebytes security researcher replied with emails between a Russian ransomware author and two victims—one in the United

States, the other in Russia—proving Svirid's point. See Lawrence Abrams, "Sigrun Ransomware Author Decrypting Russian Victims for Free," Bleeping Computer, June 1, 2018. For a technical analysis of how ransomware authors search for and avoid computers with Russian keyboards, see SecureWorks, Revil Sodinokibi Ransomware. For a contemporaneous account, I relied on interviews with researchers at CrowdStrike in 2019 and 2020.

For data on ransomware payouts, I found estimates varied widely. An FBI analysis of Bitcoin wallets and ransom notes found that between October 2013 and November 2019, $144,350,000 was paid in Bitcoins to ransomware authors. That was the conservative estimate. In 2020, an Emsisoft analysis of some 450,000 incidents projected that ransomware demands could exceed $1.4 billion in 2020 in the United States alone. As for the total cost to businesses—ransom payout plus downtime—Emsisoft estimated the total cost of ransomware attacks in the U.S. exceeded $9 billion. See Emsisoft, "Report: Cost of Ransomware in 2020. A Country-by-Country Analysis," February 11, 2020. For a fascinating account of the rise in ransomware payouts and attacks, and the role of the cyber insurance industry in encouraging victims to pay up, see Renee Dudley, "The Extortion Economy: How Insurance Companies are Fueling the Rise in Ransomware Attacks," ProPublica, August 27, 2019.

The suspected links between the ransomware hitting American towns and cities and threats to U.S. election infrastructure is based on more than a dozen interviews I conducted with American officials and private researchers throughout 2019 and 2020. But in June 2020, a confidential FBI report warning of the "likely" convergence of ransomware and attacks on election infrastructure appeared in the Anonymous "Blue Leaks." The FBI report described the ransomware attack in Louisiana and a subsequent January 2020 ransomware attack in Tillamook County, Oregon, which rendered voter registration systems inaccessible. The report concluded ransomware attacks on "county and state government networks likely will threaten the availability of data on interconnected election servers, even if that is not the actors' intention." See Federal Bureau of Investigation Executive Analytical Report, "(U//FOUO) Ransomware Infections of US County and State Government Networks Inadvertently Threaten Interconnected Election Servers," May 1, 2020. For a contemporary account of the November Louisiana ransomware attack, see Mark Ballard, "Louisiana: Cyberattack Has No Impact on State's Elections," Government Technology, November 25, 2019.

For further reading on Mitch McConnell's attempts to block election security bills, see Nicholas Fandos, "New Election Security Bills Face a One-Man

Roadblock: Mitch McConnell," *New York Times*, June 7, 2019. The efforts earned McConnell the moniker "Moscow Mitch." In September 2019, Sen. McConnell finally agreed to a bill authorizing $250 million for election security. The bill did not require that states use the funds for secure hand-marked paper ballot machines, the gold standard, and it was a far cry from the $1 billion in election security legislation Democrats had proposed. See Philip Ewing, "McConnell, Decried as 'Moscow Mitch' Approves Election Security Money," NPR, September 20, 2019.

　For further reading on Trump's unfounded CrowdStrike conspiracy theory, see Scott Shane, "How a Fringe Theory About Ukraine Took Root in the White House," *New York Times*, October 3, 2019. I personally addressed the conspiracy theory in a November 25, 2019, interview with CNN's Jim Sciutto: www.youtube.com/watch ?v=TLShgL7iAZE. For CrowdStrike's response, see "CrowdStrike's Work with the Democratic National Committee: Setting the record straight," CrowdStrike Blog, January 22, 2020. The events leading to Trump's impeachment are now well-known, but the declassified transcript of Trump's July 25, 2019 call with Volodymyr Zelensky is still available here: www.whitehouse.gov/wp-content/uploads/2019/09/Unclassified09 .2019.pdf.

For further reading on Russia's 2019 hack of Burisma, see my article with Matthew Rosenberg, "Russians Hacked Ukrainian Gas Company at Center of Impeachment," *New York Times*, January 13, 2020. Ukraine's prosecutors announced in June 2020 they found no evidence implicating Hunter Biden in a case audit. See Ilya Zhegulev, "Ukraine Found No Evidence Against Hunter Biden in Case Audit: Former Top Prosecutor," Reuters, June 4, 2020. The intelligence reports concluding Russia was, once again, meddling to reelect Trump in 2020 and boosting Sen. Sanders are described by my colleagues Adam Goldman, Julian E. Barnes, Maggie Haberman, and Nicholas Fandos, here: "Lawmakers Are Warned That Russia Is Meddling to Re-Elect Trump," *New York Times*, February 20, 2020. For the specific intel reports concerning Russia's boosting of Sen. Sanders, see Julian E. Barnes and Sydney Ember, "Russia Is Said to Be Interfering to Aid Sanders in Democratic Primaries," *New York Times*, February 21, 2020. For the NSA's report concerning Russia's exploitation of an email transfer protocol, see National Security Agency, "Exim Email Transfer Agent Actively Exploited by Russian GRU Cyber Actors," May 2020.

For further reading on Russian attacks on American infrastructure, including nuclear facilities, see Nicole Perlroth, "Hackers Are Targeting Nuclear Facilities, Homeland Security Dept. and FBI Say," *New York Times*, July 6, 2017; Department of Homeland Security, "Alert (TA18-074A): Russian Government Cyber Activity Targeting Energy and Other Critical Infrastructure Sectors," March 15, 2018, www .us-cert.gov/ncas/alerts/TA18-074A; Nicole Perlroth and David E. Sanger, "Cyberattacks Put Russian Fingers on the Switch at Power Plants, U.S. Says," *New York Times*, March 15, 2018; and Sanger and Perlroth, "U.S. Escalates Online Attacks on Russia's Power Grid," *New York Times*, June 15, 2019. For an early account of the Petro Rabigh attack in Saudi Arabia, see Nicole Perlroth and Clifford Krauss, "A Cyberattack in Saudi Arabia Had a Deadly Goal. Experts Fear Another Try," *New York Times*, March 15, 2018. For the later attribution to Russia's Central Scientific Research Institute of Chemistry and Mechanics, see FireEye, "Triton Attribution," October 23, 2018, www.fireeye.com/blog/threat -research/2018/10/triton-attribution-russian-government-owned-lab-most-likely -built-tools.html.

For further reading on Nakasone's role at Cyber Command and on Operation Nitro Zeus, see David Sanger's *The Perfect Weapon*.

For Trump's outing of sensitive classified intelligence to Russian officials, see Matthew Rosenberg and Eric Schmitt, "Trump Revealed Highly Classified Intelligence to Russia, in Break with Ally, Officials Say," *New York Times*, May 15, 2017.

For coverage of Trump's response to Sanger and my article detailing Cyber Command's attacks on the Russian response—"the virtual act of treason"—see Erik Wemple, "'Virtual Act of Treason': The *New York Times* Is Blowing Trump's Mind," *Washington Post*, June 17, 2019, and A.G. Sulzberger's op-ed in the *Wall Street Journal*, "Accusing the *New York Times* of 'Treason,' Trump Crosses a Line," June 19, 2019. For further reading on my colleague David Kirkpatrick's arrest in Egypt, see Declan Walsh, "Egypt Turns Back Veteran *New York Times* Reporter," *New York Times*, February 19, 2019. And for Declan's own account of how the U.S. Embassy failed to act in his own imminent arrest, see Declan Walsh, "The Story behind the *Times* Correspondent Who Faced Arrest in Cairo," *New York Times*, September 24, 2019.

For the statistic on American computers being attacked every thirty-nine seconds, see Michel Cukier, "Study: Hackers Attack Every 39 Seconds," University of Maryland, A. James Clark School of Engineering, February 9, 2017.

EPILOGUE

In February, DHS's Cybersecurity agency warned that ransomware attacks were now hitting pipeline operators. See Homeland Security, Cybersecurity and

Infrastructure Security Agency, "Ransomware Impacting Pipeline Operations," February 18, 2020, www.us-cert.gov/ncas/alerts/aa20-049a.

On Juneteenth 2020, Anonymous—the loose hacking collective that had been dormant for the better part of the decade—reemerged with a hack of more than two hundred police departments and FBI "Fusion centers," intelligence gathering centers, across the country. The leaks, which Anonymous dubbed Blue Leaks, amounted to a decade's worth—269 GB—of sensitive data, the largest public hack of U.S. law enforcement agencies to date. Reporters, law enforcement, and activists were still sorting through the leaks at the time of this writing. Among the FBI documents included in the dump, was a May 1, 2020, report detailing two specific ransomware attacks, the first in Louisiana in November 2019, the other in Tillamook County, Oregon, in January 2020, that affected election infrastructure. The FBI report ominously concluded it was likely ransomware attacks would have an impact on American election infrastructure as the 2020 elections neared.

For numbers on the rise of cyberattacks during the Covid-19 pandemic see Dmitry Galov, Kaspersky Labs, "Remote Spring: The Rise of RDP Brute force Attacks," April 29, 2020, securelist.com/remote-spring-the-rise-of-rdp-bruteforce-attacks /96820.

For further reading on the hacking of vaccine data, see David Sanger and Nicole Perlroth, "U.S. to Accuse China of Hacking Vaccine Data," *New York Times*, May 11, 2020.

For the declining costs of terrorism and the rising costs of cyberattacks, I relied on the Institute for Economics & Peace's 2019 Global Terrorism Index, available at visionofhumanity.org/app/uploads/2019/11/GTI-2019web.pdf, and RAND's 2018 analysis on global cyber costs. See Paul Dreyer, Therese Jones, Kelly Klima, Jenny Oberholtzer, Aaron Strong, Jonathan William Welburn, and Zev Winkelman, "Estimating the Global Cost of Cyber Risk," Rand Corporation, 2018.

For the prevalence of open-source software, see Sonatype, "The 2016 State of the Software Supply Chain," www.sonatype.com/hubfs/SSC/Software_Supply _Chain_Inforgraphic.pdf?t=1468857601884. The statistic regarding the "100 million lines of code" in high-end cars, see Synopsys, "Managing and Securing Open Source Software in the Automotive Industry," www.synopsys.com/content/dam/synopsys /sig-assets/guides/osauto-gd-ul.pdf. For further reading on Heartbleed and the problem of keeping enough eyeballs (and funding) on open-source code, see my 2014 article in the *Times*: "Heartbleed Highlights a Contradiction in the Web," April 18, 2014. For an early report on the Internet Bug Bounty Program see: Jaikumar Vijayan, "Security Researchers Laud Microsoft, Facebook Bug Bounty Programs," Computer

World, November 8, 2013. A technical description of the CHERI initiative is available at www.cl.cam.ac.uk/techreports/UCAM-CL-TR-850.pdf. The effort recently got a $45 million confidence boost from the UK government. For further insight into credential theft and the role it still plays in sophisticated nation-state espionage, see Rob Joyce's 2016 talk: "NSA TAO Chief on Disrupting Nation State Hackers," USENIX Enigma 2016, www.youtube.com/watch?v=bDJb8WOJYdA. See also Rob Joyce's November 15, 2017 description of the government's Vulnerabilities Equities Policy and Process, the most comprehensive account provided by any government to date: www.whitehouse.gov/sites/whitehouse.gov/files/images/External%20-%20 Unclassified%20VEP%20Charter%20FINAL.PDF.

The UK's GCHQ has been increasingly forthcoming about its own VEP process and recently started publicizing the number of zero-days it discloses each year. See Joseph Cox, "GCHQ Has Disclosed Over 20 Vulnerabilities This Year, Including Ones in iOS," *Vice*, Motherboard, April 29, 2016.

The reference to Iran's breach of thirty-six U.S. companies, government agencies, and NGOs and predilection for "password-spraying" was sourced from the United States Southern District of New York's March 2018 Indictment of Iranian hackers. See assets.documentcloud.org/documents/4419747/Read-the-Justice-Dept-indictment -against-Iranian.pdf. For Norway and Japan's cybersecurity rankings and details of their respective national cybersecurity policies see V. S. Subrahmanian, Michael Ovelgonne, Tudor Dumitras, and B. Aditya Prakash, "Global Cyber-Vulnerability Report," Springer, 2015. A personal thanks to Subrahmanian, who walked me through the country-by-country analysis.

For DHS's final assessment on the 2016 election issues in Durham County, North Carolina, see "Digital Media Analysis for Durham County Board of Elections," Department of Homeland Security, October 23, 2019, static.politico.com/c5/02 /66652a364a2989799fd6835adb45/report.pdf.

On Silicon Valley's lingering mistrust of government and Uber security executive Matt Olsen's 2019 remarks addressing this, see Jeff Stone, "Mistrust Lingers between Government, Industry on Cyber Information Sharing," Cyberscoop, October 2019.

Several people proved invaluable to me in the eleventh hour as I put together proposals in the epilogue. Among them were Paul Kocher, among the most thoughtful people working in cybersecurity today, who proved a critical soundboard. I also owe a thank you to Peter Neumann at SRI, the "original white hat," who patiently walked me through the CHERI project and provided some much-needed perspective over noodles. Jim Zemlin at the Linux Foundation proved invaluable in discussing open-source software security. Gary McGraw helped inform my thinking on incentive structures. Casey Ellis took my long-distance calls while hunkered Down Under and kindly shared his thinking on how the pandemic is affecting our cybersecurity. My many conversations with Jim Gosler, who never once did not pick up the phone, informed my thinking on what needs to be done to right this ship. I am also indebted to Mike Assante, who passed away in July 2019. We only know a glimmer of what Mike did behind the scenes to draw awareness to the vulnerability of our critical infrastructure. A few weeks before he passed of cancer, Mike sent me an email that described our national cyber predicament. It read: "Please continue to give infrastructure cyber risk the attention it deserves. We are at an important point in which old engineering assumptions and safety design do not fully account for software and the ability to manipulate reality."

INDEX

A NOTE ON THE AUTHOR

NICOLE PERLROTH is a staff writer at the *New York Times,* where she covers cybersecurity and digital espionage. She previously covered Silicon Valley as a writer and editor at *Forbes* magazine. She graduated from Princeton and Stanford and lives in Northern California with her husband, their son, and their dog.